Public Policy for Academic Quality

HIGHER EDUCATION DYNAMICS

VOLUME 30

SCOPE OF THE SERIES

Higher Education Dynamics is a bookseries intending to study adaptation processes and their outcomes in higher education at all relevant levels. In addition it wants to examine the way interactions between these levels affect adaptation processes. It aims at applying general social science concepts and theories as well as testing theories in the field of higher education research. It wants to do so in a manner that is of relevance to all those professionally involved in higher education, be it as ministers, policy-makers, politicians, institutional leaders or administrators, higher education researchers, members of the academic staff of universities and colleges, or students. It will include both mature and developing systems of higher education, covering public as well as private institutions.

For other titles published in this series, go to
http://www.springer.com/series/6037

David D. Dill · Maarja Beerkens
Editors

Public Policy for Academic Quality

Analyses of Innovative Policy Instruments

Springer

Editors
Dr. David D. Dill
University of North Carolina
Chapel Hill NC 27599-7090
USA

Dr. Maarja Beerkens
University of Twente
CHEPS
7500 AE Enschede
Netherlands

ISSN 1571-0378
ISBN 978-94-007-3222-3 ISBN 978-90-481-3754-1 (eBook)
DOI 10.1007/978-90-481-3754-1
Springer Dordrecht Heidelberg London New York

Softcover reprint of the hardcover 1st edition 2010

Printed on acid-free paper

Springer is part of Springer Science+Business Media (www.springer.com)

Preface

This book is a product of the Public Policy for Academic Quality Research Program (PPAQ), a project based in the Department of Public Policy at the University of North Carolina at Chapel Hill. The major goal of the Research Program was to provide policymakers and other stakeholders in higher education analyses of innovative academic quality assurance policies designed to influence academic standards. These policy analyses attempt to offer assessments of the goals, implementation problems, and impacts of the new quality assurance policies developed around the world in as fair-minded and objective a manner as possible, reflecting the larger public interest. In this volume we have drawn these policy analyses together and provided commentary on them based on insights drawn from related research and background papers prepared as part of the PPAQ Research Program.

We want to express our deep appreciation to the many scholars and doctoral students who contributed to this project over the years. We especially wish to acknowledge the very generous support of the Research Program by the Ford Foundation and to express our grateful thanks to Dr. Jorge Balan formally of the foundation for his wise guidance and support.

As we anticipated when we began this project the regulation of academic quality has now become a major national and international issue in public policy. We have no illusions that the contents of this volume will provide the final answers to the many questions and issues involved in this global debate, but we hope that the evidence we present will help inform the design of public policies that better contribute to the public good.

Chapel Hill, NC, USA — David D. Dill
Enschede, The Netherlands — Maarja Beerkens

Contents

Contributors

Maarja Beerkens Centre for Higher Education Policy Studies (CHEPS), University of Twente, Enschede, The Netherlands

David W. Breneman University of Virginia, Charlottesville, VA, USA

Joaquim Brugué Autonomous University of Barcelona, Barcelona, Spain

David D. Dill University of North Carolina, Chapel Hill, NC, USA

Elaine El-Khawas George Washington University, Washington, DC, USA

Peter T. Ewell National Center for Higher Education Management Systems (NCHEMS), Boulder, CO, USA

Kerri-Lee Harris Centre for the Study of Higher Education, University of Melbourne, VIC, Australia

Lee Harvey, Copenhagen Business School, Frederiksberg, Denmark

Richard James Centre for the Study of Higher Education, University of Melbourne, Parkville, VIC, Australia

Barbara M. Kehm International Centre for Higher Education Research (INCHER), University of Kassel, Kassel, Germany

Richard Lewis Centre for Higher Education Research and Information (CHERI), Open University, London, UK

Craig McInnis Australian education consultancy firm PhillipsKPA, Richmond Victoria, Australia

William F. Massy Business and Finance, Stanford University, USA

Simon Schwartzman Institute for the Study of Labor and Society (IETS), Rio de Janeiro, Brazil

Bjørn Stensaker NIFU STEP – The Norwegian Institute for Studies of Innovation, Research and Higher Education, Oslo, Norway

Josep M. Vilalta Catalan Association of Public Universities (ACUP), Barcelona, Spain

Gareth L. Williams Institute of Education, University of London, London, UK

Notes on Contributors

Maarja Beerkens is a Research Associate at the Centre for Higher Education Policy Studies (CHEPS) at the University of Twente in the Netherlands.

David W. Breneman is University Professor, Newton and Rita Meyers Professor in Economics of Education, and Director of the Public Policy program in the Batten School of Leadership and Public Policy at the University of Virginia, USA.

Joaquim Brugué is Professor Emeritus of Public Policy and Political Science at the Autonomous University of Barcelona in Spain.

David D. Dill is Professor of Public Policy at the University of North Carolina at Chapel Hill in the USA.

Elaine El-Khawas is Professor of Education Policy at George Washington University, Washington, DC, USA.

Peter T. Ewell is Vice President of the National Center for Higher Education Management Systems (NCHEMS) in Boulder, Colorado, USA.

Kerri-Lee Harris is Senior Lecturer in Higher Education Centre for the Study of Higher Education at the University of Melbourne, Australia.

Lee Harvey is Visiting Professor at the Copenhagen Business School, Frederiksberg, Denmark.

Richard James is Professor of Higher Education and Centre Director, CSHE Centre for the Study of Higher Education at the University of Melbourne, Australia.

Barbara M. Kehm is Professor of Higher Education Research at the University of Kassel and Managing Director of the International Centre for Higher Education Research (INCHER) at the University of Kassel, Germany.

Richard Lewis is a consultant and Associate of the Centre for Higher Education Research and Information (CHERI) of the Open University in the UK and immediate Past President of the International Network of Quality Assurance Agencies in Higher Education (INQAAHE).

Dr. Craig McInnis is Principal Associate with the Australian education consultancy firm PhillipsKPA, and former Professor and Director of the Centre for the Study of Higher Education at the University of Melbourne, Australia.

William F. Massy is Professor Emeritus and former Vice president for Business and Finance, Stanford University, USA.

Simon Schwartzman is Senior Researcher at the Institute for the Study of Labor and Society (IETS), Rio de Janeiro, and former Professor of Political Science at the University of São Paulo in Brazil.

Bjørn Stensaker is a research professor at NIFU STEP – The Norwegian Institute for Studies of Innovation, Research and Higher Education in Oslo, Norway.

Josep M. Vilalta is the Executive Secretary of the Catalan Association of Public Universities (ACUP) in Barcelona, Spain.

Gareth L. Williams is Emeritus Professor of Educational Administration at the Institute of Education, University of London, UK, where he was founding head of the Centre for Higher Education Studies.

Chapter 1
Introduction

David D. Dill and Maarja Beerkens

At the opening of the twenty-first century, the structure of higher education in most countries of the world has undergone significant change as a result of new social demands for expanded access, technological developments, and global market forces. In this period of change the traditional concerns with access and cost have been supplemented by a new concern of policy makers with academic quality (Brennan and Shah 2000). As a consequence, new public policies on academic quality and new forms of academic quality assurance have rapidly emerged in many countries and have just as swiftly migrated across continents and around the globe. One indirect measure of the diffusion of these new public policies is the development of an international association of public and independent entities engaged in academic quality assurance – the International Network of Quality Assurance Agencies in Higher Education (INQAAHE). In 1990 when it held its inaugural meeting in Hong Kong, the INQAAHE had 25 members from 17 nations, primarily represented by the "Westminster" countries. By 2009 when the INQAAHE held its Ninth International Conference in Abu Dhabi, it had some 200 organizational members from 79 nations, with extensive representation from every continent save Antarctica.

While there has also been a commensurate increase in the literature on academic quality, a relatively small amount of this scholarship directly addresses the design, implementation, and impacts of these new policies and practices (see, for example, Westerheijden et al. 2007; OECD 2008). The rich and growing public debate about academic quality regulation within and across countries is therefore not well informed by evenhanded examinations of the strengths and weaknesses of these new regulatory instruments. The goal of this volume is to help fill this void with relevant policy analyses. The chapters that follow scrutinize new and innovative instruments of academic quality assurance in teaching and learning activities, utilizing the knowledge of informed scholars around the world, and provide comprehensible, easily accessible evaluations. The analyses will be as fair-minded as

D.D. Dill (✉)
University of North Carolina, Chapel Hill, NC, USA
e-mail: ddill@email.unc.edu

D.D. Dill, M. Beerkens (eds.), *Public Policy for Academic Quality*, Higher Education Dynamics 30, DOI 10.1007/978-90-481-3754-1_1,

possible, assessing the relative costs and benefits of the respective policies from the perspective of the overall "public interest."

We recognize that the framework of rules and regulations affecting academic quality within a state, province, or country as well as in the larger global community is still evolving, and continued experimentation and evaluation of quality assurance policies is needed. Our intent, therefore, is to provide information and analyses that can help inform and enrich the ongoing public debate about the appropriate regulation of academic quality.

In the sections to follow we provide an overview of what we mean by "academic quality," why academic quality regulation may be necessary, the nature of the new forms of academic quality regulation, and the orientation of the policy analyses that follow in this volume.

What We Mean by "Academic Quality"

As policy makers in various countries have debated policies designed to assure academic quality, there has been extensive dispute about the meaning of the term (Green 1994). Many academics have argued that "academic quality" is amorphous, non-measurable, or so ambiguous a concept as to be not appropriate for government regulation. Early writers on academic quality regulation (Ball 1985; Bogue and Saunders 1992) were fond of quoting the novelist Robert Pirsig's classic phrase from *Zen and the Art of Motorcycle Maintenance*: "what the hell is quality?"

> Quality . . . you know what it is, yet you don't know what it is. But that's self-contradictory. But some things *are* better than others, that is they have more quality. But when you try to say what the quality is, apart from the things that have it, it all goes *poof*! There's nothing to talk about. But if you can't say what Quality is, how do you know what it is, or how do you know that it even exists? If no one knows what it is, then for all practical purposes, it doesn't exist at all. But for all practical purposes it really *does* exist. What else are the grades based on? Why else would people pay fortunes for some things and throw other things in the trash pile? Obviously some things are better than others . . . but what's the "betterness?" So round and round you go, spinning mental wheels, and nowhere finding any place to get traction. What the hell is Quality? What is it? (Pirsig 1974, p. 179).

Yet there is an element of academic gamesmanship in this definitional debate. As Pirsig suggests in this quotation, professors routinely identify and differentiate academic quality when they grade student's work. Many of the core processes of academic life – subject examinations, external examiners, as well as review processes for professional meetings, academic journals, and the award of research grants – are predicated based on a professional ability to identify and evaluate academic quality in student learning and academic research. While academics may vigorously debate the meaning of academic quality when confronted with potential government quality regulations, few professors have rejected a Nobel Prize because the process whereby her or his work was selected was too ambiguous!

From a public policy perspective we would argue that academic quality is best defined as equivalent to academic standards – the level of knowledge and skill achieved by graduates as a result of their academic program or degree (Eustace 1991). During their higher education, students develop knowledge, skills, and abilities – "human capital" – that over their lifetimes provide private benefits to them as well as social benefits to the larger society. This human capital perspective (Becker 1964) provides the primary logic for the public subsidies of higher education in all countries as well as for the more recent spread of mass higher education around the world. We use the term human capital here in its broadest meaning to include not only the contributions that educated graduates make to the economy, but also the nonmonetary benefits they contribute to society through improved parenting, healthier lifestyles, greater civic participation, and increased social cohesion (Haveman et al. 2003).

The conception of human capital outlined above provides a means of defining academic quality in the public interest. From this perspective the public interest is best served by an institutional framework of policies, rules, and norms (North 1990) that maximizes in as efficient and equitable a manner as possible the academic standards attained by graduates. Not surprisingly, it is this conception of academic quality as academic standards that most often is articulated in current national policies on academic quality (Brennan et al. 1997). Consistent with human capital theory, these policies increasingly focus on improving academic outcomes, the educational "value-added" of an academic program or degree (Dill 2000).

Within the field of higher education, Astin (1985) has most clearly articulated this perspective on academic quality in his "talent development model." Astin argued that the major purpose of a university is to develop the talents of its students to their maximum potential. This development is achieved by facilitating changes in students' intellectual capacities and skills, values, attitudes, interests, habits, and mental health. Institutions that provide the largest amount of developmental benefits to students in Astin's view, therefore, possess the highest academic quality.

Academic quality, understood as academic standards in student achievement, is also a necessary component of any discussion of cost and access in higher education (Berdahl and Spitzberg 1991). Policy makers must consider whether the rapidly increasing public investment in higher education is purchasing more, less, or comparable levels of academic achievement among students. Without some knowledge of the relationship between the level of public investment in higher education and the level of academic achievement produced, the public debates about higher education cost can be seriously misleading. Even if a government introduces market forces into higher education, which may lead, as in the USA, to institutions with varying levels of academic achievement, there is an important public interest in academic standards. If the market is to function efficiently, individual consumers need to be able to fairly evaluate the relative value-added by colleges and universities of widely varying cost (Dill and Soo 2004). For example, will an education at an expensive, well-established university in every case lead to higher student achievement than an education from a newly established distance learning institution? Similarly, policy makers in most countries who are concerned with access to higher education must confront the often-unasked question, "access to what" (Massy 2003)? Investments in access without a commensurate concern with the level of learning outcomes produced by institutions of higher education inevitably may come to be seen as "a deception and a new form of discrimination" (Moodie 1991, p. 9).

This lack of connection between academic cost, access, and quality is also reflected, as noted earlier, in the substantial disparity in the volume of policy research addressing these respective regulatory issues in higher education. Policy-related research on quality assurance regulation is often national in orientation and, while growing, is still small in comparison to the amount of policy research on higher education cost and access. While we have no illusions that this volume can by itself address this lack of balance, we believe that systematic analyses of academic quality policies utilizing existing research and evidence, conducted by knowledgeable experts, and made available in an accessible form can make a substantial contribution to current policy debates.

Is Regulation Needed?

At the outset, a useful distinction can be drawn between internal and external academic quality assurance. Internal quality assurance refers to those policies and practices whereby academic institutions themselves monitor and improve the quality of their education provision, while external quality assurance refers to supra-institutional policies and practices whereby the quality of higher education institutions and programs is assured. Individual universities have always possessed policies and practices designed to assure the quality of education, but academic institutions have also always operated within a national policy framework designed by the state to assure academic standards.

As suggested above, the combined impacts of globalization and massification have radically altered the traditional relationship between the state and institutions of higher education and motivated policy makers to seek new means for assuring academic quality in higher education (OECD 2008). First, the global demand for skilled human capital has motivated changes in the degree frameworks of many countries as policy makers sought international recognition of the credentials granted by their country's higher education institutions. These new degree frameworks also encouraged a rapid proliferation of new academic programs in many countries, thereby testing established national practices for assuring academic standards. Second, the rapid growth of higher education systems has provided incentives for the development of private institutions, including cross-border franchise and virtual universities, which have posed novel challenges to national systems of external quality assurance, particularly those based on central control of public institutions. Third, the competitive forces unleashed by globalization and massification have required institutions of higher education to become more responsive to rapidly changing labor markets and to student program interests. Consequently, institutions in many countries have sought increased flexibility and autonomy from traditional state quality assurance regulations so that they can react more swiftly to changing social demands by establishing new academic programs, reconfiguring existing programs, and eliminating outdated programs. Fourth, the rapidly expanding social demand for higher education has been caused in large part by students' desire to acquire the increasing private benefits available to individuals with higher degrees. The empirical reality of the growing private benefits of academic degrees has altered the traditional debate about higher education finance, encouraging many countries to require students and their families to pay a larger share of higher education costs. Consequently, as previously noted, the new public policies on academic quality assurance also seek to respond to public concerns that institutions provide educational value for money. In sum, the traditional external processes for assuring academic quality have significant limitations in the new, more competitive and demanding environment of higher education.

There is also emerging evidence that the internal processes by which universities have traditionally monitored and maintained academic standards may be inadequate to the new demands of mass higher education (Dill 1999). For example, a survey of Australian university administrators inquiring into how they evaluated the academic standards of their universities observed:

> ... when we asked how they knew, there was no VC or dean who had any valid or reliable means of knowing about the intellectual standards of their university's degrees, e.g. how they might have changed over time, how they compared between departments or how they compared with other universities (Anderson et al. 2002, p. 36).

Changes in the nature of academic work have also weakened the effectiveness of the existing internal mechanisms for academic quality assurance. The exponential growth of academic knowledge and the increasing specialization of research have made the traditional reliance on disciplinary norms a less reliable means of assuring academic standards in subject fields within colleges and universities (Clark 1996).[1] Studies of academic work at the subject level in the USA confirm the existence of an increasingly fragmented, atomistic, academic culture (Lattuca and Stark 1994; Massy et al. 1994). Not only do professors in many subjects do much of their teaching alone, but also because disciplinary subfields are defined quite narrowly, many academics find it almost impossible to discuss their teaching with other members

[1] Commenting on the contribution that disciplinary fragmentation makes to the complexity of higher education systems, Clark (1996) observed, "in mathematics, 200,00 new theorems are published each year, periodicals exceed 1,000, and review journals have developed classification scheme that includes over 4,500 subtopics arranged under 62 major topic areas. In history, the output of literature in the two decades of 1960–1980 was apparently equal in magnitude to all that was published from the time of the Greek historian Thucydides in the fourth century B.C. to the year 1960. In psychology, 45 major specialties appear in the structure of the American Psychological Association, and one of these specialties, social psychology, reports that it is now comprised of 17 subfields .... In the mid-1990s, those who track the field of chemistry were reporting that 'more articles on chemistry have been published in the past 2 years than throughout history before 1900.' Chemical Abstracts took 31 years to publish its first million abstracts, 18 years for its second million, and less than 2 years for its most recent million. An exponential growth of about 4–8% annually, with a doubling period of 10–15 years, is now seen as characteristic of most branches of science" (pp. 421–422).

of staff. Collective debate about the content of the curriculum, about pedagogic methods, and about means of assuring and improving the academic standards of programs has become increasingly difficult and rare. In many subjects, US academic staff expressed the belief that the field's diversity prevented achieving a consensus on what students should be taught. This lack of agreement is exacerbated by the rapid expansion of multidisciplinary and interdisciplinary subjects, because in these emerging fields academic staff can no longer rely on disciplinary norms to define academic standards.

This growing fragmentation of academic work means that in many disciplines and subjects, shared information on student learning no longer exists or is not easily obtainable. These changes pose a "collective action dilemma" with significant implications for institutional efforts to assure academic standards (Dill 2007). That is, for an individual member of academic staff to decide that participating in a collective effort to assure or improve student learning is more important than an equivalent hour spent on her or his own research or teaching, he or she needs to make a prediction as to the learning benefits generated by this cooperative activity. But if few incentives exist to produce evidence on student learning, then the individual will necessarily conclude that investing time in cooperative efforts to assure or improve academic standards is not rational.

This observed deterioration of the traditional collegial mechanisms for assuring academic standards within US colleges and universities is likely to have broader implications. The increasing specialization of academic work is inherent in the advancement of science and therefore affects all systems of higher education. In addition, as other nations "massify" their systems of higher education, rapidly expanding their academic offerings and providing access to a much more varied group of students than in the past, they are adopting modular forms of instruction, methods of continuous assessment, and credit-based systems similar to those in the USA. As a consequence, the traditional internal mechanisms for assuring academic standards are coming under strain in all countries.

Competing missions among universities is another factor that puts teaching and learning activities under great stress. The personal priorities of academic staff tend to lean toward research rather than teaching activities, because of either intrinsic interests of individual staff or future career perspectives (Fairweather 2000). Also for universities, financial and reputational rewards for research activities have considerably increased in recent years, especially in Australia and Europe, which has placed research management in universities at the center of attention. While the extent to which teaching and research are competing or supplementing activities is still open to debate (Hattie and Marsh 1996), it is clear that the adoption of strong policies regulating research quality without balancing policies regulating teaching quality will negatively affect the teaching mission of universities. In the worst case, when information about teaching quality is inadequate, research quality may become a proxy for institutional quality in the eyes of the public and contribute to the degradation of the teaching mission in the long run. As illustrated by the cases of this volume, a well-designed academic quality policy may help to restore a needed balance and increase awareness about academic standards within the university.

The dramatically altered environment of institutions of higher education has helped to reveal the inadequacy of both the traditional internal university practices for assuring academic standards and the limitations of existing approaches to public regulation (Brennan and Shah 2000). In their search for a national framework that will encourage innovation in academic programs while maintaining and improving academic standards, policy makers are experimenting with many innovative forms of academic quality assurance. Exactly what form such regulation should take and how extensive it should be is an issue that is deserving of increased policy research and public debate.

Forms of Regulation

The concept of regulation is most often associated with a binding set of governmental rules to be applied by a public agency over specific activities – the so-called command and control perspective. But regulation can also be understood more broadly as all state actions designed to influence social behavior valued by the public (Baldwin and Cave 1999). In a similar spirit Clark's (1983) classic "triangle of coordination" emphasized three possible approaches to coordinating or controlling behavior in academic institutions: state authority, the academic oligarchy (i.e., professional control), and the market. From this perspective the state has a number of policy alternatives to command and control approaches for assuring academic standards. Academic quality potentially could be assured by professional self-regulation, which is "enforced" by government structuring or oversight, or by the competitive market, which is in turn steered by appropriate competition and disclosure laws designed to ensure that institutions of higher education provide adequate services to consumers. From this broader perspective the creation through legislation of a public agency such as the Quality Assurance Agency (QAA) in the UK for the purpose of conducting assessments of academic quality in universities would represent one mode of public regulation. But so also would state recognition of professional accrediting agencies as a means of assuring academic quality, or state policies that facilitate consumer sovereignty in a competitive market for higher education by mandating the provision of university information on academic program quality. Each of these mechanisms represents a possible approach to the public regulation of academic quality.

In fact, while the traditional national frameworks for academic quality assurance varied from country to country, they had generally followed three modal forms similar to those outlined by Clark (1983): the European model of central control of quality assurance by state educational ministries, the US model of decentralized quality assurance combining limited state control with market competition, and the British model in which the state essentially ceded responsibility for quality assurance to self-accrediting universities (Dill 1992). In the UK, up until the election of the Thatcher government in the 1980s, the assurance of academic quality in the publicly supported university sector was delegated to the academic profession itself, which monitored and assured the standard of university degrees through collective mechanisms such as the external examiner system. In contrast, ministries of education on the continent were much more active in setting standards for universities. They established and monitored regulations on university admissions, academic appointments, program curricula, and end-point examinations. In the USA, as higher education rapidly expanded following Word War II, the federal Congress explicitly adopted a market-based approach to academic quality assurance as a supplement to the existing tradition of regional and professional accreditation (Leslie and Johnson 1974). In the 1972 re-authorization of the Higher Education Act, Congress rejected the entreaties of the higher education community to enact formula-based, enrollment-driven federal aid to academic institutions. Instead, legislators argued that providing aid directly to students was the most efficient and effective means to equalize opportunities in higher education and to harness market forces for enhancing the quality of higher education.

In accordance with this broader conception of regulation, Table 1 outlines the generic policy approaches (in bold) and new policy instruments for academic quality assurance (in italics) we analyze in the following chapters. As noted, each

Table 1. New public policy instruments for the assurance of academic quality

Professional (self) regulation	Market regulation	State (direct) regulation
External examining *External Examining (UK)* **Professional accreditation and licensure** *Teacher Accreditation (USA)*	**Information provision – university rankings** *CHE-Ranking (Germany)* *National Survey of Student Engagement (USA)* *Course Experience Questionnaire and Graduate Survey (Australia)** **Information provision – system rankings** *State Report Card (USA)*	**Specification of standards** *National Qualifications Framework (Australia)* *Subject Benchmarking (UK)* **Program assessment and accreditation** *Subject Assessments (Denmark)* *Subject Accreditation (Germany)* *Medical Accreditation (UK)* **Institutional accountability** *Academic Audit (Hong Kong)* *Performance-based contracting (Catalonia, Spain)* **Information provision** *Course Experience Questionnaires and Graduate Surveys (Australia)** *National Assessment of Courses (Brazil)*

*The Australian Course Experience Questionnaire and Graduate Survey instrument is an interesting combination of state-mandated information and market-based dissemination. To aid comparative analysis we have grouped this policy with the other market-based instruments in Part II, Market Regulation of Academic Quality.

instrument assumes one of the three loci of authority. Professional or self-regulation clearly assumes producer sovereignty in which academics themselves are principally responsible for defining and enforcing the rules and norms assuring the quality of academic provision. This places greatest emphasis on traditional voluntary practices carried out by professional bodies including accreditation of academic programs and institutions by professional associations as well as collective professional practices such as external examining. For the market to work effectively as a means of assuring academic standards, it is necessary for students and their families to achieve effective consumer sovereignty through informed choice of academic programs. Quality assurance practices associated with this perspective include the provision of information and rankings by commercial, non-profit, or government agencies, which are designed to provide academic quality information to students and policy makers. Finally, state or direct regulation of academic quality assumes the sovereignty of the state in defining and enforcing academic standards. The new instruments emphasized by the state have adopted different approaches to academic quality assurance. Some countries have made efforts to articulate specific standards for all study fields and/or for higher education degrees as a guideline or benchmark for universities. The National Qualifications Framework in Australia and subject benchmarks in the UK are examples of such policies. The most direct means of government monitoring of academic quality in universities is likely assessment and accreditation of individual programs. In contrast, an institutional accountability approach to quality assurance employs performance contracts or an academic audit. In the former case, universities individually negotiate their targets with the state, and in the latter case the university itself maintains responsibility for its quality assurance, while the state assures only that the university takes this responsibility seriously. Finally, the state may attempt to assure academic quality by providing or mandating better information on academic performance.

Several key points can be derived from these simple distinctions. First, in a number of cases in this volume, the locus of authority is an indication of the instrument originator rather than a limitation on who can carry it out. A number of quality assurance practices such as accreditation or academic audit are essentially generic processes that can be conducted voluntarily under the auspices of academic professional organizations such as the Teacher Education Accreditation Council (TEAC) in the USA or the European University Association (EUA), or can be a requirement of national policy carried out by agencies established by or affiliated with the state as is the case with the academic audits conducted by the University Grants Committee in Hong Kong or accreditations conducted by the General Medical Council (GMC) in the UK. Similarly, quality rankings can be produced by the academic profession as in the world university league table published by the Shanghai University, by the private or non-profit sector as in the commercially produced rankings of the *US News and World Report* or the CHE Rankings in Germany, or by the state as in the Graduate Surveys produced in Australia. Second, while it is often argued that professional self-regulation or market forces represent serious alternatives to state regulation of academic standards, professional or market-based quality assurance practices are usually dependent on the state for their effective functioning. That is, if professional self-regulation or market forces are to successfully protect the public interest in the assurance of academic standards, they must be reinforced by law or formally recognized and/or subsidized by the state. For example, the current influence of supposedly voluntary accreditation in the USA derives almost entirely from the fact that the national government utilizes institutional accreditation to determine college and university eligibility for federal student aid. Similarly, more valid commercial rankings such as those of the *Guardian* in the UK, the *Good University Guide* in Australia, or the *Globe and Mail* in Canada are greatly reliant upon government subsidized or produced data on universities (Dill and Soo 2005).

In sum, effective professional self-regulation and/or market regulation is best understood as an alternative state approach for assuring academic quality. For this reason we have purposely included a number of instruments that were initially developed by voluntary or non-profit entities (e.g., UK external examining, TEAC Accreditation, the National Survey of Student Engagement, and the CHE Rankings), since these types of instruments also could become important components of a national policy framework.

A "Public Interest" Perspective to Policy Analysis

Professor Ulrich Teichler once whimsically observed that the main difference between research on higher education policy and on "mad cow" disease is that when the mad cow researchers present their findings, the mad cows are not in the room! Academics may not be "mad" in this sense, but with regard to the topic of academic quality regulation they are often easily incensed. Academic staff's experience with and criticisms of academic quality regulation are of significant importance to policy makers, especially given the complexities of implementing such policies in the necessarily decentralized world of academic work and given society's understandably strong support for academic freedom. But by the same token, academics, who carry out the vast majority of research on academic quality regulations, have a clear self-interest in the design of any such policies. Therefore, there is a real need for research on academic quality policies that is genuinely objective and evenhanded.

Our analyses attempt to address this need by adopting a "public interest" perspective. That is, while all researchers necessarily have value biases, we aspired to produce analyses that are as balanced as possible. This was pursued first by adopting as outlined below a common format for all of the analyses of policies and practices presented in this volume. Second, the analyses attempt to assess both the intended and unintended impacts of new regulatory policies – the relative costs and benefits of these policies to all stakeholders, not just to the members of the academic community. Third, the analyses have been carried out by experienced researchers with specific knowledge of the relevant policy and related research literature.

Each policy analysis examines one quality assurance policy instrument and the experience implementing it in a specific country or context. The analyses address

- the nature of the relevant higher education system in order to help readers understand the national context prior to the introduction of the new quality assurance policy;
- the perceived problems that led to the adoption of the quality assurance instrument;
- the nature of the policy instrument – the specific components of the policy and what it is designed to accomplish;
- the existing evidence regarding the impacts of the instrument, positive or negative, intended or unintended;
- where available, information on the financial costs of the instrument;
- the relationship of the analyzed instrument to comparable policies implemented in other countries.

The main purpose of each policy analysis is to provide an in-depth analysis of the design and implementation of the policy. It will inform readers about the impacts of the policy as well as guide them through associated risks, debates, strengths, and weaknesses. The analysis also will enable a reader to consider the possible effectiveness of the instrument in another political and academic environment.

As noted, the goal of these analyses is to provide information on innovative instruments and practices. By employing a similar framework for each policy analysis, we hope to enable policy makers and other stakeholders to consider different options, to compare their effects, and to see their relative advantages and disadvantages. This collected set of analyses should therefore be a helpful resource for designing or revising existing quality assurance policies in any country.

Organization of the Volume

In the three sections that follow, we will introduce and analyze the policy instruments listed in Table 1. In Part I we will explore the instruments of professional or self-regulation of academic quality. In Part II we will discuss the instruments of market regulation, and in Part III we will assess the new state instruments for regulating academic quality. In our concluding chapter we will summarize what we have learned about the strengths and weaknesses of each regulatory approach and attempt to synthesize our findings into the national framework conditions necessary for the effective assurance of academic standards in the new environment of higher education.

References

Anderson, D., Johnson, R., Saha, L. (2002). *Changes in academic work: Implications for universities of the changing age distribution and work roles of academic staff.* Canberra: Commonwealth of Australia. http://www.dest.gov.au/archive/highered/otherpub/academic_work.pdf Accessed 12 April 2007.

Astin, A. W. (1985). *Achieving educational excellence: A critical assessment of priorities and practices in higher education*. San Francisco, London: Jossey-Bass Publishers.
Baldwin, R., Cave, M. (1999). *Understanding regulation: Theory, strategy, and practice*. Oxford: Oxford University Press.
Ball, C. (1985). What the hell is quality? In D. Urwin (Ed.), *Fitness for purpose: Essays in higher education*. Guildford, UK: SRHE and NFER/Nelson.
Becker, G. S. (1964). *Human capital*. New York: Columbia University Press.
Berdahl, R. O., Spitzberg Jr., I. J. (1991). Quality and access as interrelated policy issues. In R. O. Berdahl, G. C. Moodie, I. J. Spitzberg Jr. (Eds.), *Quality and access in higher education: Comparing Britain and the United States* (pp. 164–171). Buckingham, UK: SRHE and Open University Press.
Bogue, E. G., Saunders, R. L. (1992). *The evidence for quality: Strengthening the tests of academic and administrative effectiveness*. San Francisco, London: Jossey-Bass Publishers.
Brennan, J., de Vries, P., Williams, R. (Eds.) (1997). *Standards and quality in higher education*. London: Jessica Kingsley.
Brennan, J., Shah, T. (2000). *Managing quality in higher education: An international perspective on institutional assessment and change*. Buckingham, UK: OECD, SRHE & Open University Press.
Clark, B. R. (1983). *The higher education system: Academic organization in cross-national perspective*. Berkeley: The University of California Press.
Clark, B. R. (1996). Substantive growth and innovative organization: New categories for higher education research. *Higher Education*, 32(4), 417–430.
Dill, D. D. (1992). Academic administration. In B. R. Clark, G. Neave (Eds.), *The encyclopedia of higher education* (pp. 1318–1329). Oxford: Pergamon Press.
Dill, D. D. (1999). Academic accountability and university adaptation: The architecture of an academic learning organization. *Higher Education*, 38(2), 127–154.
Dill, D. D. (2000). Designing academic audit: Lessons learned in Europe and Asia. *Quality in Higher Education*, 6(3), 187–207.
Dill, D. D. (2007). Are public research universities effective communities of learning?: The collective action dilemma of assuring academic standards. In R. L. Geiger, C. L. Colbeck, R. L. Williams, C. K. Anderson (Eds.), *The future of the American Public Research University* (pp. 187–203). Rotterdam: Sense Publishers.
Dill, D. D., Soo, M. (2004). Transparency and quality in higher education markets. In P. Teixeira, B. Jongbloed, D. Dill, A. Amaral (Eds.), *Markets in higher education: Rhetoric or reality?* (pp. 61–86). Dordrecht, the Netherlands: Kluwer.
Dill, D. D., Soo, M. (2005). Academic quality, league tables, and public policy: A cross-national analysis of university ranking systems. *Higher Education*, 49(4), 495–533.
Eustace, R. B. (1991). Gold, silver, copper: standards of first degrees. In R. O. Berdahl, G. C. Moodie, I. J. Spitzberg Jr. (Eds.), *Quality and access in higher education: Comparing Britain and the United States* (pp. 29–41). Buckingham, UK: SRHE and Open University Press.
Fairweather, J. S. (2000). Diversification or homogenization: How markets and governments combine to shape American Higher Education. *Higher Education Policy*, 13(1), 79–98.
Green, D. (Ed.) (1994). *What is quality in higher education?* Buckingham, UK: SRHE and Open University Press.
Hattie, J., Marsh, H. W. (1996). The relationship between research and teaching: A meta- analysis. *Review of Educational Research*, 66(4), 507–542.
Haveman, R. H., Bershadker, A., Schwabish, J. A. (2003). *Human capital in the United States from 1975 to 2000: Patterns of growth and utilization*. Kalamazoo, MI: W. E. Upjohn Institute for Employment Research.
Lattuca, L. R., Stark, J. S. (1994). Will disciplinary perspectives impede curricular reform? *Journal of Higher Education*, 65(4), 401–426.
Leslie, L. L., Johnson, G. P. (1974). The market model and higher education. *Journal of Higher Education*, 45(1), 1–20.

Massy, W. F. (2003). *Honoring the trust: quality and cost containment in higher education*. Bolton, MA: Anker Publishing.

Massy, W. F., Wilger, A. K., Colbeck, C. (1994). Overcoming hollowed collegiality. *Change*, 26(4), 10–20.

Moodie, G. C. (1991). Setting the scene. In R. O. Berdahl, G. C. Moodie, I. J. Spitzberg Jr. (Eds.) *Quality and access in higher education: Comparing Britain and the United States* (pp. 1–10). Buckingham, UK: SRHE and Open University Press.

North, D. (1990). *Institutions, institutional change, and economic performance*. Cambridge: Cambridge University Press.

OECD (2008). *Tertiary education for the knowledge society, Vol 1*, P. Santiago, K. Tremblay, E. Basri, E. Arnal (Eds). Paris: OECD.

Pirsig, R. (1974). *Zen and the art of motorcycle maintenance: An inquiry into values*. New York: Morrow.

Westerheijden, D. F., Stensaker, B., Rosa, M. J. (2007). *Quality assurance in higher education: Trends in regulation, translation and transformation*. Dordrecht: Springer.

Part I
Professional (Self) Regulation of Academic Quality

David D. Dill and Maarja Beerkens

Historically self-regulation has been the dominant mode for assuring academic quality in the US and the UK higher education. In the USA the academic standards of colleges and universities were supposedly assured by six regional accreditation agencies, such as the North Central Association of Colleges and Schools, which are voluntary organizations. In addition a number of professional associations in fields such as chemistry, law, and medicine have taken on the role of accrediting or licensing-related academic programs in their respective fields. In the UK the long tradition of university subject examinations and of external examiners was believed to assure the reputed "gold standard" of UK higher education.

As a means of achieving the public interest, self-regulation has predictable strengths and weaknesses (Baldwin and Cave 1999). Advantages include access to greater expertise and technical knowledge that can lead to more reasonable rule making and greater voluntary compliance, lower cost to the public, and more rapid adjustment of self-regulatory rules to changing circumstances. Disadvantages include the potential for regulatory "capture" in which self-regulation comes to serve the private interests of the professional members rather than the interest of the public, a charge that has been made in the USA with regard the resistance of traditional academic accrediting agencies to the emergence of distance learning and/or for-profit higher education institutions. Professional self-regulation may also be insensitive to the needs of less influential but affected groups, such as students. Finally, the public accountability or transparency of self-regulation is a further source of concern.

Of the many instruments of academic quality assurance to be examined in this volume arguably external examining, as conducted in the UK and some Scandinavian countries, most clearly addresses academic standards. External examiners traditionally assessed the actual performance of students on subject examinations used to award degrees in the university sector. Richard Lewis' analysis of external examining in the UK provides an invaluable insight into this quality assurance instrument. External examining emerged in the early-19th-century England as a professional practice and was explicitly encouraged by the government in subsequent university charters. However, the practice was not formally regulated nor codified until, under pressure from the Thatcher government to assure academic standards, the Committee of Vice Chancellors and Principals published the first *Code of Practice* in 1986. Subsequent research revealed that only a minority of universities were following these professional standards, challenging academic assertions on the effectiveness of the practice (Warren Piper 2004). The role of external examiners has recently been given fresh impetus by changes in quality assurance arrangements introduced as a result of a UK governmental initiative. In return for the lightening of the burden on institutions through the abandonment of externally organized program review, institutions are now required to make more information publicly available including summaries of external examiners' reports.

Lewis' analysis traces the history of external examining and discusses the relevant section of the Quality Assurance Agency's *Code of Practice*, which is the nearest equivalent to a national system of regulations governing the external examining system. He also discusses the ways in which the system has responded to changes that have occurred, especially the substantial growth in student numbers and the modularization of degree programs.

The other primary mode of professional self-regulation – voluntary academic accreditation – evolved first in the USA due to the unique structure of the national academic system. Because academic accreditation has been actively advocated as a potentially valuable instrument for assuring academic quality outside the USA as well (Schwarz and Westerheijden 2004), an extended discussion of the US experience with accreditation may prove useful before introducing the voluntary professional accreditation instrument to be analyzed in this part. The US constitution does not explicitly cite education as a federal concern; therefore, authority over education has been reserved for the states. Following the founding of the Republic, the states granted charters for new colleges and universities liberally and generally adopted a laissez faire approach to coordinating and controlling higher education. By the end of the 19th century, less than a quarter of the states had adopted regulations governing educational programs or degrees and over half of the states exercised no supervision over their colleges once incorporated (Brubacher and Rudy 1968). The mobility of students across state borders and the growing diversity of admissions standards among the many new public and private institutions eventually led to "educational chaos" (Orlans 1975) and made clear that some new mechanism was needed to develop and enforce the definition of a college. In the 1880s preparatory school administrators and collegiate institutions voluntarily banded together in regional associations to standardize college admissions requirements and accredit secondary schools. Six regional associations were eventually formed, and over time these associations also adopted standards for the accreditation of colleges and universities.

From its outset, academic accreditation was advocated as a means of addressing academic standards (Selden 1960). However, unlike external examining in the UK, the US academic accreditation has never attempted to directly assess the academic performance of students in colleges and universities. Instead the earliest accreditation standards focused on quantitative indictors of resources and inputs such as buildings, volumes in the library, and the number of faculty members with Ph.D. degrees. In response to increasing criticisms of these quantitative standards by the institutions, the accreditation approach was gradually modified in the 1930s to a more subjective, improvement-oriented process assessing the educational mission each institution had set for itself (Selden 1960).

The context for academic accreditation in the USA was dramatically altered by the federal government with the Veterans Readjustment Assistance Act (Korean War GI Bill) of 1952 (Orlans 1975). This Congressional statute and succeeding legislation established that college and university eligibility for federal student aid was to be determined primarily by the six regional accrediting agencies, but also that the Commissioner of Education (now Secretary of Education) would be awarded general oversight of the accreditation agencies. In subsequent years the power of the Secretary was steadily expanded by federal statute to include authority to recommend accrediting agencies for federal recognition and to establish explicit criteria for the evaluations conducted by recognized agencies. Many have argued that this federal action effectively eliminated the "voluntary" nature of institutional accreditation in the USA, since no major US college or university could effectively function without the support of federal student aid funds (Ewell 2008).

In the post–World War II years, as mass higher education came to be seen as a strategic investment crucial to economic development, criticisms of the contribution of institutional accreditation to academic quality continued to mount (Ewell 2008). The agencies were faulted not only for their failure to examine evidence of student achievement, but also for their lack of interest in an institution's means of "internal accountability" – the collective faculty processes for assuring academic standards (Graham et al. 1995). In response to these criticisms and increasing pressure for reform from the states and federal government, all six of the regional accreditation associations revised their standards and criteria in the period between 1984 and 1994, to place greater emphasis on student assessment practices (Nettles et al. 1998).

Nonetheless, a decade later a national review by the Secretary of Education's Committee on the Future of Higher Education (United States Department of Education 2006), noting the continuing decline in US college graduation rates and the significant recent drop in the literacy scores of college graduates, renewed the call for a new national accreditation framework and reaffirmed the need for accreditation standards focused primarily on measurable academic quality outcomes rather than inputs.

In sum, despite its long history, there is little empirical evidence to date that US institutional accreditation as traditionally organized has been an effective instrument for assuring academic standards. Some recent institutional accrediting initiatives, such as the Western Association of Schools and Colleges' Educational Effectiveness Review, may offer potentially valuable new approaches, but they are still in the early stages of implementation.

In addition to institutional accrediting by regional associations, voluntary specialized accrediting associations that evaluate particular programs or schools within colleges and universities have also flourished in the USA. The influence of specialized accreditation was strengthened by state-level regulations in some fields limiting eligibility for professional licensing exams to graduates of accredited programs. The state and national calls for increased attention to academic outcomes noted above also have inspired new approaches to specialized accreditation, particularly in professional accrediting agencies where the participation of practicing professionals led to greater pressure for an emphasis on academic standards (Ewell 2008). For example, the American Board of Engineering and Training (ABET) has radically shifted its process for accrediting undergraduate engineering programs from an emphasis on curricular specifications to student learning outcomes and accountability. Accredited engineering programs now must publish specific goals for student learning and measure their achievement to demonstrate how well these objectives are being met.

Elaine El-Khawas analyzes one particularly innovative, specialized accrediting agency in the USA, the Teacher Education Accreditation Council (TEAC). TEAC is a voluntary agency founded in 1997 as an alternative approach to accreditation of teacher education. It combines an innovative focus on student learning and an evidence-based audit method that builds upon and extends the academic audit process first developed in the UK and adopted in a number of other countries (see the analysis by Massy in this volume). Its review process is supported by detailed protocols, well-defined visits, a strict separation between the auditor role and the summative evaluation role, and the application of scholarly standards for reliability and validity in the evaluation of quality assurance processes. Programs must meet standards that focus on student learning and its use for academic planning and improvement, but they select what evidence to use in making their claims. This approach has strengthened student assessment and helped programs gather detailed evidence that is both meaningful and useful for improvement.

The discussed instruments of UK external examining as well as institutional and specialized accreditation in the USA suggest the strengths and limitations of self-regulatory approaches to assuring academic standards in mass systems of higher education, evidence consistent with the experience with self-regulation in other professional fields and industries. Without the support and authority of the state, self-regulatory instruments are likely to be erratically enforced and/or insufficiently rigorous. In the case of higher education the rapid growth of enrollments, the proliferation of subject fields, and the increasing global rivalry for academic reputation have also created a new context in which a purely self-regulatory approach to assuring academic standards appears unlikely to effectively protect the public interest.

References

Baldwin, R., Cave, M. (1999). *Understanding regulation: Theory, strategy, and practice*. Oxford: Oxford University Press.

Brubacher, J. S., Rudy, W. (1968). *Higher education in transition: A history of american colleges and universities, 1636–1968*. New York: Harper & Row.

Ewell, P. T. (2008). *U.S. Accreditation and the future of quality assurance*. Washington DC: Council for Higher Education Accreditation.

Graham, P. A., Lyman, R. W., Trow, M. (1995). *Accountability of colleges and universities: An essay*. New York: Columbia University, Office of the Provost.

Nettles, M., Cole, J., Sharp, S. (1998). The Landscape: Tracking a subtle storm. assessment policies in higher education. *Change*, 30(2), 47–51.
Orlans, H. (1975). *Private accreditation and public eligibility*. Lexington, Mass. Lexington Books.
Schwarz, S., & Westerheijden, D. (2004). *Accreditation and evaluation in the European higher education area*. Dordrecht: Kluwer.
Selden, W. K. (1960). *Accreditation: A struggle over standards in higher education*. New York: Harper & Brothers.
United States Department of Education (2006). *A test of leadership: Charting the future of U.S. higher education*. Washington, DC: United States Department of Education (USDOE).
Warren Piper, D. J. (2004). *Are professors professional? The Organisation of university examinations*. London: Jessica Kingsley.

Chapter 2
External Examiner System in the UK: Fresh Challenges to an Old System

Richard Lewis

Introduction

In the mid-1970s I visited the United States for the first term to serve as a visiting professor in a large state university. I had no real knowledge of the US system of higher education and had no idea of the basis of grading. When I sought the advice of the chairman of the department, his words were to get out of town before the results were published.

I have no idea whether this was a typical attitude in the United States at the time or whether my chairman was untypically insensitive to the needs of his students. But what I did learn was the power of the individual instructor over the grading of students, a power which did, and probably still does, exist in a good number of other countries. This was to someone brought up in the British tradition an unpleasant shock. In the UK the grading of students is very much a collective exercise involving not only other faculty members from one's own institution but also faculty members from other institutions – the external examiners – the subject of this chapter.

The system is not unique to the UK as it is used in other Commonwealth countries, while aspects of it can also be found in a number of other European countries such as Denmark. This chapter will, however, be restricted to the UK experience. The external examiner system works at both research and taught degrees, but this chapter will concern itself with taught degrees.

In order to understand the role that external examiners play in assuring quality in the UK, it is necessary to understand the national system of external quality assurance.

R. Lewis (✉)
Centre for Higher Education Research and Information (CHERI), Open University, London, UK

D.D. Dill, M. Beerkens (eds.), *Public Policy for Academic Quality*, Higher Education Dynamics 30, DOI 10.1007/978-90-481-3754-1_2,

The UK System of External Quality Assurance in Higher Education

All higher education institutions in the UK in receipt of public funding are subject to the Quality Assurance Agency in Higher Education (QAA). Institutions are also subject to a range of professional associations who recognise or accredit certain programmes of study, mainly those of a professional or vocational nature, that are not unlike the specialised accreditation agencies in the United States. The QAA is owned by the organisations that represent the heads of UK universities and colleges (Universities UK, Universities Scotland, Higher Education Wales and the Standing Conference of Principals). It describes itself as being independent of UK governments (QAA 2005, p. 5), but this is, perhaps, not an entirely fair description.

The legislation that underpinned the creation of the four UK higher education funding bodies (covering England, Wales, Scotland and Northern Ireland) contains the provision that the funding bodies must ensure that there are adequate quality assurance systems in force covering the institutions which they fund. Prior to the establishment of the QAA in 1997, the funding councils were themselves directly involved in quality assurance, but they now contract the QAA to do the work on their behalf. Thus, the QAA is in a contractual relationship with a government agency, and it is clear that the government can influence QAA policy and even formulate their policy. This point was demonstrated in 2001 when there was pressure from the universities to lighten the very onerous quality regime that was then in place. The announcement that the system would be changed, the nature of which will be described later in this chapter, was made by the then Secretary of State for Education and Skills, David Blunkett, rather than the chairman or chief executive of the QAA (Brown 2004). As will be described later, the government continues to influence the development of quality assurance including the changing role of external examiners.

The QAA operates slightly differently in the four constituent countries of the UK (QAA 2005), but in the main it operates at institutional level.[1] In England it carries out what it describes as "institutional audits" which examine the internal quality assurance systems at institutional level. Institutional audits are undertaken every 6 years.

The examination of the institution's internal systems is done in the light of the QAA's *Code of Practice*. There are 10 sections to the Code, one of which, Section 4, deals with external examining. The code includes a number of precepts which are explained as expressing "key matters of principle that the higher education community has identified as important for the assurance of quality and academic standards" (QAA 2004, para 6). The QAA is anxious to stress that the precepts are not the same things as rules to which there must be rigid adherence:

[1] It carries out subject reviews in Further Education colleges and of heath-care programmes under a contract with the Department of Health.

> Institutions will not be asked about their adherence to the *Code of Practice* on a precept by precept basis. They will be expected to explain in their self-evaluation documents how they have addressed the intentions of the precepts, including any resulting changes to their practices. Any areas of difficulty that institutions have experienced should also be discussed in their self-evaluation documents. (QAA 2004)

The present system was not arrived at easily and differs markedly from the very onerous system that was introduced with the ending of the binary divide in 1992. One of the reasons why the new system was so onerous was that the external examiner system was not, at that time, held in high regard by senior civil servants and the officials of the funding councils.

The new system was far more intrusive than the polytechnics' experience in the last days of the Council for National Academic Awards (CNAA) and certainly far more demanding than the institutional audits conducted by the Academic Audit Unit (AAU) that had been established on a voluntary basis, possibly to forestall government intervention, by the Committee of Vice-Chancellors and Principals (CVCP) in 1990.

The new approach was twofold; the work of the AAU was taken over by the newly formed Higher Education Quality Council (HEQC), while the funding councils themselves conducted programme reviews that afterwards became to be called Teaching Quality Assessments (TQAs). It does appear that one of the reasons why the TQAs were introduced was the belief on the part of the government and its advisers that the external examiner system could no longer deliver on its prime purpose of ensuring comparability of standards. A related issue was that a number of the relevant staff of the funding councils had a school or further education background and appeared to be comfortable with an inspectorial, rather than a peer group, approach and, thus it was, that lectures at the old universities were officially inspected for the first time in their history.

The twin track was ended in 1997 when the QAA was set up to undertake the work formerly carried out by the HEQC and the funding councils; in the latter case the work was carried out under contract with the councils who retained the legal obligation to ensure that all the institutions they funded were subject to an effective quality assurance system.

The level of confidence in the external examiner system has ebbed and flowed over time, and 1992 was a time when confidence, particularly on the part of government, was low. For reasons that will be explained later, confidence in the external examiner system grew over the decade as reflected in one of the findings of the Dearing Committee that had undertaken the first major review of UK higher education since the Robins Committee whose report was published in 1963. The Dearing Report recommended the adoption of a lighter approach to quality assurance coupled with a strengthening of the external examiner system (Dearing 1997, p. 157).

These recommendations, not all of which have been implemented, will be discussed later in the chapter. The recommendation about lightening the external quality assurance was one of those which was not immediately implemented, and the fight continued until its successful conclusion, from the university's perspective, in 2001, with the Secretary of State's announcement referred to above. As a result of the change in policy, TQAs were ended,[2] but the government was not entirely reassured by the university's stance and insisted on a quid pro quo which involved placing greater reliance on the external examiner system. The way in which this is being done will be discussed later in the chapter, but in short it requires summaries of all external examiner's reports to be published on a website that is intended to provide information to potential students to help them in their choice of institution.

Policy Problem

Historical Background

The notion of importing examiners from other universities was first introduced by the University of Durham when it was established in 1832 (Silver 1994). These examiners were mostly drawn from the University of Oxford, and, while their main purpose was to increase the local "examining capacity", they also provided some evidence to the outside world of the acceptability of Durham's degrees. Thus, in the 18th century a policy concern was identified which has continued to concern the academic community – namely how to achieve comparability of standards across institutions.

It was, however, the creation of the Victoria University[3] in 1880 that gave birth to the system of external examiners that has existed to the present day. The charter of the Victoria University, as did the charter of the University of Birmingham and other universities founded after 1880, required that examinations be conducted by internal and external examiners with the role of the latter being explicitly recognised as providing public assurance that the standards of the new universities were comparable with those of their more ancient counterparts.

The two main traditional roles of external examiners were to ensure

- that degrees awarded in similar subjects are comparable in standard across higher education institutions and
- that students are dealt with fairly in the system of assessment and classification (Silver et al. 1995).

While these roles remain basically unchanged, the ways in which they are discharged has been subject to quite substantial change over the last two decades or so.

---

[2] The QAA was allowed to carry out so-called audit trails by which about 10% of the institution's programmes were reviewed as part of the assessment of the effectiveness of the institution's internal procedures. It now appears that the audit trails will be removed from the QAA's armoury (HEFCE 2005).

[3] The Victoria University had colleges in Manchester, Liverpool and Leeds. In 1905 the Victoria University divided into three components based in each of these cities.

The ways in which external examiners operated in the past can best be described in the context of the traditional degree programme, without continuous assessment and without any elements of modularity, where the degree award is based entirely on the student's performance at one, or perhaps two sessions of closed book, proctored, examinations.

In such a world the external examiner had three main tasks:

- to take part in the approval of the examination paper,
- to check the marking of the scripts usually by inspecting a sample,
- to attend the final meeting of the board of examiners.

In some cases the external examiner had the right of veto, in that the examination paper could not be approved or a degree awarded without his or her approval; in other cases the views of the external could be overruled. We will see later how the introduction of such elements as continuous assessment and modular structures has affected the way in which externals discharged their responsibilities.

In the early days, external examiners were usually drawn from the Universities of Cambridge, Oxford and London and from the older Scottish universities, but the 20th century saw the appointment of examiners drawn from a wider range of institutions. The next major development occurred in the 1950s with the establishment of the National Council for Technical Awards (NCTA) who awarded Diplomas of Technology in newly designated Colleges of Advanced Technology (CAT). The appointment of external examiners was one of the conditions imposed by the NCTA. Within 10 years the system changed: the CATs became universities and the diplomas degrees, and in place of the NCTA, the Council for National Academic Awards (CNAA) was created.

In the 1960s the binary system of higher education was created in the UK under which new forms of higher education instaurations were created. Those institutions, the polytechnics and colleges – the former generally being multi-faculty large institutions while the colleges were smaller and more specialised – were expected to focus on teaching vocationally related subjects. They were not encouraged to engage in pure research but instead concentrate on applied research and consultancy.

While the new institutions were able to provide courses and research opportunities leading to the whole range of academic qualifications including Ph.D.s and higher doctorates, they were not degree-awarding institutions; the qualifications were those of the CNAA which acted as a quality assurance body as well as a degree-awarding body.

As a quality assurance body the CNAA initially approved and reviewed, usually at 5-yearly intervals, the programmes of study leading to its awards. It also carried out institutional reviews, also normally every 5 years. By the time it came to its demise in 1992, the CNAA had given all the polytechnics and many of the colleges the right to approve and review programmes subject to continuing institutional review.

In terms of external examiners the CNAA continued the policy of the NCTA in that it required the institutions to appoint external examiners for all qualifications.

Under the CNAA system the institutions nominated external examiners, but their appointment was made by the Council. In contrast, the universities appointed their own external examiners without the intervention of an external agency.

From its inception the CNAA set out regulations covering the external examiner system, but the position in the universities was different. It was not until 1984, when the system was more than a century old, that the then Committee of Vice-Chancellors and Principals (CVCP), now called Universities UK, issued its own external examining recommendations. Up to that date although all universities, with the partial exceptions of Oxford and Cambridge, were fully committed to the external examining system, practices varied greatly. The issue of the CVCP recommendations initiated a process of standardisation that has continued to the present day, although there is evidence to suggest that this process still has some way to go.

The System Under Strain

The generally held view in the early 1970s in the UK was that the external examiner system was a valuable one that helped ensure reasonable comparability of standards across the country within disciplines. There was very little evidence that it helped ensure comparability of standards across disciplines, but this was not regarded as a matter of great concern.

The period 1975–1995 saw many changes in UK Higher Education that put great strains on the external examiner system. The most obvious was the move to a modular credit-based system whereby students were no longer assessed at only one or two stages in their programmes. Students now, very much in the US style, earned credits from separate units or modules, perhaps six to twelve units per academic year, with the final award being made when sufficient credits had been earned or accumulated.[4] A further complication was the introduction of continuous assessment so that, even within the module, there was no single examination on which the external examiner could focus. These changes put the traditional external examination system under considerable strain.

The other change was the very significant increase in the size of the system with full-time student numbers growing from just under 300,000 to close to 1 million (Dearing 1997, p. 15) There was a corresponding, but not proportionate, increase in the number of academics, particularly not of senior academics who were members of those disciplinary-based communities in which the members knew, or at the very least knew of, the other members. It was these, the senior members of the professoriate, who had, traditionally, served as external examiners but there were no longer

[4] The Open University used credit-based awards since its foundation in 1970. As a separate development, two polytechnics, Oxford Brookes and the City of London, introduced modular degree schemes in the 1970s. By the end of the century virtually all UK universities had introduced unit or credit-based systems.

enough of them to go around. The system had to change and these changes have largely occurred over the last decade.

The growing concerns about the external examiner system led to government support for an extensive survey of the system which was conducted by David Warren Piper (Warren Piper 2004) and which was the first of its type.

The survey indicated that only a minority of universities were following the recommendations of the CVCP guidelines that examination papers should be double marked and that all external examiners should produce reports addressed to the vice chancellor of the examining university. The report also indicated difficulties with multidisciplinary and joint degrees including a lack of effective academic contact between the departments contributing to a joint programme.

Another finding related to the choice of external examiners which was based on their reputed knowledge coupled with the fact that they were known to the senior members of the department and that consultation with those outside the department was rare.

The more recent changes that have taken place in the system are set out in Silver (2005), in which he compares the results of two investigations he carried out into external examining in 1994 and in 2004. The general view in 1994 was that the external examiner system needed to be retained but that the existing system, which was characterised as a voluntary, self-sacrificing, underpaid and time-consuming process in competition with other pressures, needed to be changed (Silver 1994). The 2004 study (Hannan and Silver 2004) described how the system had changed. The response to increased numbers was, what might be described as, the democratisation of the external examiner system. External examiner appointments were no longer largely made up of members of the senior professoriate. Many more appointments were drawn from the middle ranks of the academy, while the increased proportion of vocationally related programmes was reflected in the increased number of externals drawn from industry and the professions.

The other major change resulted from the changes in the structures of degree programmes. The fact that a students' final award was based on numerous instruments of assessment, and not on a set of end-of-year examinations, meant that the external could no longer be closely involved with all aspects of the assessment process. In consequence, external examiners now play little or no part in determining the grades of individual students but act more in as overall monitors or validators of the system (Silver 2005).

Content of the Policy Instrument

In one sense there is no "national system"; each university selects and recruits its own external examiners and each university has its own detailed regulations. There are no rules that specify who is entitled to act as an external examiner, nor is there a qualification that is necessary to possess before being appointed to the post. The nearest thing to regulations is the External Examiners section of the QAA *Code of Practice*.

While the precepts are not regulations which must be adhered to at all times, the institution does have to present evidence of a properly functioning external examiner system to the QAA as part of the institutional audit, and it is much easier to do so if it can be shown that the precepts are being followed.

The precepts will be summarised and discussed in this section. The core functions of the external examining process are set out in precept 1.

> An institution should ask its external examiners, in their expert judgement, to report on:
>
> i. whether the academic standards set for its awards, or part thereof, are appropriate;
> ii. the extent to which its assessment processes are rigorous, ensure equity of treatment for students and have been fairly conducted within institutional regulations and guidance;
> iii. the standards of student performance in the programmes or parts of programmes which they have been appointed to examine;
> iv. where appropriate, the comparability of the standards and student achievements with those in some other higher education institutions;
> v. good practice they have identified.

It can be seen that although the list of functions has been expanded, it still includes the traditional objectives described earlier in the chapter. The other precepts may be summarised as follows.

Roles and Powers

The institution should communicate to all concerned the roles, powers and responsibilities assigned to external examiners including the extent of their authority in examination/assessment boards (precept 2). Also, institutions should make known the programmes and awards, or parts of programmes, to which each external examiner has been appointed (precept 8).

The guidance notes to these precepts emphasise the discretion allowed to institutions. While the external may have a right of veto over the decision of an examining board, it is also pointed out that "In some institutions it is considered normal practice for a board of examiners to make collective decisions, with no individual having primacy" (precept 2).

Precept 2 also deals with the position in modular degrees where different external examiners may have different areas of responsibilities. A number of externals may, for example, be responsible for a number of modules while another might have oversight of the way in which the institution brings the credits together in order to make the award. In such complex cases it is not unusual for the external examiners to be constituted as a team under the leadership of a chief external examiner.

In the guidance accompanying precept 8 it is stated that the institution might decide to ensure that an external examiner is associated with the assessment of all student achievement that leads to an award but that it is acceptable for the external to be concerned only with the academic standards of a bachelor's programme and only implicitly endorse the levels and standards of a related interim qualification such as undergraduate certificates or diplomas which might count for credit against the bachelor's degree.

Appointment of Externals

Institutions should make every effort to ensure that external examiners are competent (precept 4), should have regulations governing their appointment and premature termination (precept 5) and should ensure that any potential conflicts of interest are resolved before the appointment is made (precept 6). Externals should be provided with sufficient information and support (precept 7). The subjects of training and of conflict of interest will be considered later in the section on current issues.

While the actual precepts do not specifically refer to the point, external examiner appointments are made for a limited duration of about 4 years in order to ensure that they can retain an element of independence from the institution.

Methodology

Prior to the confirmation of mark lists, external examiners are expected to endorse the outcomes of the assessments they have been appointed to scrutinise (precept 3). Institutions are expected to agree with external examiners the evidence that each external considers necessary to discharge their responsibilities (precept 9). The endorsement referred to in precept 3 is the agreement by the external that the assessment processes have been carried out in accordance with the policy and regulations of the institution. It appears from the guidance notes that the precept distinguishes between the external examiner's agreement that proper process had been carried, which if not obtained would suggest to the QAA that standards were at risk, and agreements with all the outcomes of the assessment, for as is pointed out in relation to precept 2, the institution's regulations might not give the external the right to overrule the collective view of the internal examiners.

External Examiner Reports

Externals should submit reports at regular intervals providing comments and judgements on the assessment process and the standards of student attainment (precept10) in a form and coverage specified by the institution (precept 11). The reports should be addressed to the head of the institution, or a person designated by the head of the institution and consideration should be given to the reports at both subject and institutional levels (precept 12). Full and serious consideration should be given to the reports, and the outcomes of those considerations, including any action taken, should be formally recorded (precept 13). The institution should provide the externals with feedback including information on any actions taken by the institution (precept 14).

The importance placed by the QAA on the reporting aspects of the external examiner's role is indicated not only by the number of precepts devoted to the subject but also by the fact that the reporting precepts allow very little discretion to the institution. An important consequence is that the external examining process has become far more open and represents a shift from one of the weaknesses in the way that the system operated in the past, which was often based on personal contacts. In the past it was too often the case that the external was effectively selected by the head of department and the senior staff of the department from among those that they knew or knew of. A further weakness was that the report, if there was one, was sent to the head of department and not necessarily widely circulated with a content that was more than occasionally devoted more to the quality of the external examiners' refreshments and the well-being of the head of department's spouse than the attainments of the students. In accordance with the QAA guidelines the institution must now have in place internal procedures for dealing with possible conflicts of interest, but it is impossible to remove entirely the point that the best people in the institution to judge the subject knowledge and experience of potential external examiner will be the staff of the department over which he or she will, in a sense, "be sitting in judgement".

Current Issues of Concern

While there have been considerable improvements in the way the external examiner system has worked in recent years, there continue to be a number of areas of concern and debate.

The Independence of External Examiners

The precepts deal with two aspects of the independence of external examiners. It deals with one in a reasonable fashion, but its approach to the other is more debatable.

The first aspect, covered by precept 6, covers such matters as ensuring that sufficient time has past before a former student or staff member of the institution is appointed as an external. The precept also warns against the danger of reciprocal, or mutual back scratching, appointments whereby institution A appoints an external from institution B and vice versa. The precept also warns against the replacement of an external by someone from the same institution.

A related issue, not strictly related to independence, is the desirability, set out in the guidance notes, of specifying a maximum number of appointments that the institution believes its appointees can hold.

All this seems reasonable enough, but there are suggestions that the *Code of Practice* deals less well of the other aspect of independence, independence from the design and operation of the assessment model on which judgement has to be passed.

External examiners in the traditional model had the rare privilege of being allowed to be wholly negative in their criticisms. It was argued that should the external examiner start suggesting ways of overcoming problems that they had identified and that if the proposals were implemented, then the external would no longer be able to bring an independent judgement to bear on the process to which they were party. The *Code of Practice* takes a far more "liberal" position in that although it points to the potential dangers of employing externals to perform other tasks in respect of the programme they are examining, it also concludes that "It may, in some circumstances, be appropriate for someone who has acted as an external referee or advisor in curriculum design or in the development of a new programme to become an external examiner subsequent to fulfilling the advisory role" (QAA 2004, Section 2). Not all would agree with this view.

Modular and Joint Degrees

This is a long-standing issue that was identified by Warren Piper in his 1988 study (Warren Piper 2004). An underlying issue that has been long recognised but largely accepted, if only by default, is the disciplinary differences that are found in grading. For example the Higher Education Statistics Agency (HESA) statistics reveal that in 2003/2004, 70% of graduates in Languages and History and Philosophy were awarded first or upper second class honours degrees, while the corresponding figures for Education and Business and Administration were 50% and 46%, respectively (HESA 2009).[5]

Thus, a student taking a joint honours degree in say Mathematics and Business, a not unusual combination, would have a final overall grade that is the aggregate of marks from disciplines with rather different approaches to the granting of good honours degrees. The external examiners who will be drawn from those disciplines are not in a position to compensate for these differences.

The problem is compounded in modular degree structures where students study different subjects in varying proportions. In such cases the final honours classification will usually be based on an arithmetical calculation based on the grades obtained from the individual modules, and in many institutions the work of the external examiner would be only at the level of the module. An external examiner might be involved at the final examination board where the overall grade is decided but in such cases they will, as stated earlier, play little or no part in determining the

---

[5] In most disciplines the majority of bachelor's degrees are honours degrees, which are awarded in four classes: first, upper second, lower second and third. Many employers recruit only graduates with a first or upper second class honours degree.

classification of individual students but instead act more as overall monitors of the system (Silver 2005).

The Training and Selection of External Examiners

As has already been pointed out, the most recent major enquiry into UK higher education, the National Committee of Inquiry into Higher Education under the chairmanship of Sir Ron (now Lord) Dearing, recommended the adoption of a lighter approach to quality assurance coupled with a strengthening of the external examiner system (Dearing 1997, p. 157). The proposed strengthening included the creation of a UK-wide pool of recognised academic staff from which universities and other degree-awarding bodies would have to select their external examiners. It was suggested that the pool be managed by the QAA who would also be responsible for selecting its members. In order to achieve consistency of approach it was suggested that external examiners would have to undergo a process including thorough familiarisation, training and preparation, including a trainee/apprenticeship model for new external examiners (Dearing 1997, p. 162).

While these specific recommendations did not come to pass, the period following the publication of the Dearing Report did see the increased recognition of the need to strengthen the external examining system.

Concerns were expressed about any move towards a compulsory period of training for external examiners for perhaps two overlapping reasons. One was the view that an experienced academic did not need any training to discharge the role. The other reason was that it was felt that the requirement to undergo a compulsory period of training would deter many potential external examiners, especially those with the most experience, from accepting appointments. The Teaching Quality Enhancement Committee (TQEC) which was established by HEFCE, Universities UK and the Standing Conference of Principals firmly rejected the principle of compulsory accreditation (Cooke 2003) but did make a number of proposals with the aim of strengthening the external examiner system. These included

- Improved induction of external examiners by the institution employing their services. Such institutional induction programmes might be submitted on a voluntary basis for external accreditation.
- Improved institutional preparation, i.e. through an apprenticeship programme for internal examiners, which prepares them to take on the role of external examining. Such institutional preparation might also be submitted for external accreditation on a voluntary basis.
- The availability of a national programme for the training of external examiners.

The system for voluntary accreditation of institutionally based programmes is not yet in place, but the newly established Higher Education Academy (HEA) is engaging in, and commissioning, research in a number of areas that would support the practice of external examining. It is, for example, finding out more about what external examiners actually do and what information and support would be helpful. In terms of training and preparation it is examining the idea of a voluntary continuing professional development programme that might be used to support external examiners and is considering how recognition might be given to institutional programmes for the preparation and development of external examiners.

The HEA also maintains two email discussion groups, one for external examiners themselves and one for the institutional administrators that support the system.

External Examiners' Fees

Silver (2005) identifies an interesting difference between the results of the two research projects, one carried out in 1994–1995 and one in 2004, with which he was associated. The respondents to the first survey complained they had been underpaid, but by the second survey the burden of the complaints had moved from the level of the fees to the apparent unjustified variations in the level of fees paid for what appeared to be the same work.

There is as yet no national survey of the level of fees paid although there is a good deal of demand for the publication of these figures. At a 2005 seminar on external examiners organised by the Centre for Higher Education Research and Information (CHERI) of the Open University, Professor Howard Colley, a senior advisor to the HEA, reported that a recent survey of external examiners' fees carried out in 2004 revealed that some institutions paid a fixed fee while others paid a variable fee depending on such matters as the number of students or the number of courses. The fees reported to the researchers ranged from £250 to £650.

When considering the fees paid to externals it must be remembered that, with a few exceptions, they are full-time academics paid by their employers. A rigid view, which would not be held by many academics, is that since they are doing the work in "office time" they are lucky to be paid anything. The more commonly held view among the external examiner community is that they already do more than they are paid for and hence they deserve to be paid as the external examining role adds to their existing overload.

A Fresh Challenge

In the earlier section of the chapter, reference was made to the intervention of the government, at the best of the leading universities, to reduce the burden of external quality assurance. This was achieved by the elimination of the system of programme or disciplinary-related reviews. A price, however, had to be paid, and part of that price was that institutions would need to publish more information about their quality and standards (Brown 2004, p. 133). The information required is set out in the

HEFCE report *Information on Quality and Standards in Higher Education* (HEFCE 2002) and on the website of the TQI.

As far as external examiners are concerned, institutions are required to publish summaries of the findings of external examiners, at programme or subject level, on an annual basis.

It is recommended that the summaries should be written by the external examiners themselves. Where it is more appropriate for the institution to prepare the summaries, the text should be approved by the relevant examiners. The template for the entry to the website includes the opportunity for an institution to respond to the findings of examiners.

The TQI website is, at the time of writing this chapter, still in the course of development, and no research has yet been done to test the notion of government ministers, and their advisers, that the publication of the summaries of external examiner reports will help to sustain standards and will be of significant help for a student deciding where to study.

Costs

It is impossible to obtain a precise cost of the system of external examining in the UK. One reason for this is that the cost falls on different shoulders, including the home institution of the external examiners and possibly the external his or her self if the fee does not provide a fair return for their services. The costs of the employing institution are difficult to estimate for, while the fees and expenses are easily measured, the other costs associated with providing the necessary information and support for the externals are less easy to measure with any degree of objectivity.

While a precise figure is difficult to obtain, it does seem possible to estimate its order of magnitude and to compare that estimate with the total cost of UK higher education. In their 1995 study, Silver et al. estimated that there were between 10,000 and 15,000 external examiners; since student numbers have increased by about 25% over the period, let us assume a proportionate increase in the number of external examiners and take a conservative upper estimate of 20,000. An alternative approach is to start with the figure of 2.2 million students in UK higher education in 2003/2004, and if one assumes an average of one external examiner for every 100 students, there would be 22,000 externals. In order to maintain the spirit of prudence, this figure will be rounded up to 25,000.

The 2004 study referred to earlier found that most fees fell in the range £250–£650; to be prudent an average of £500 will be assumed, to which £200 will be added for expenses. The direct costs of externals would thus be in the order of £17.5 million, and if one, again to be prudent, added an overhead percentage of 50%, the total cost would be about £26 million or £12 per student. The total expenditure of UK higher education institutions in 2003/2004 was £16,900 million, so the over-estimated cost of external examiners would be of the order of 0.15% of the total spend.

Conclusions

It is open to judgement whether these admittedly very crude calculations suggest that the external examiner system is an expensive one or not, but the general view is that it is an expensive system but worthwhile. To quote the words of the Teaching Quality Enhancement Committee,

> The committee notes that the UK has a nearly (sic) unique system of external examining. Most HE systems have no equivalent. The system of external examining represents a substantial cost to the sector but it also demonstrates a tangible commitment to standards and quality. External examiners are appointed by higher education institutions (HEIs) and are widely recognised by HEIs and other stakeholders to be a crucial component of the operational quality assurance arrangements within the sector and institutions. (Cooke 2003)

It is very difficult to present solid evidence of the impact of the UK external examiner system by virtue of the fact that in modern times it has always been with us and it is not possible to present a "before and after" picture or present the results of a controlled experiment comparing the results with and without external examiners. As stated earlier there are periods when confidence in the system has been relatively low, but on close examination the concerns are usually fundamentally about whether the methods of student assessment used by institutions are appropriate and rigorous rather than about the role of the external examiners (Brown 2004).

All the studies of the system made by either the great and the good such as the TCEC committee or the Dearing Committee, or of the actual practitioners, suggest that, despite its faults and weaknesses including those related to joint honours and modular programmes, the system is an effective way of helping to ensure comparability across institutions, if not across disciplines, and of ensuring fairness for students.

The overwhelming consensus is that while the system needs to be supported and cosseted from time to time, and occasionally given a good shake-up, it remains an essential part of the UK higher education system which is supported by the overwhelming majority of academics and students who would not wish to see it disappear.

References

Brown, R. (2004). *Quality assurance in higher education: The UK experience since 1992*. London: Routledge Falmer.

Cooke, R. (2003). *External examiners: Report from the Teaching Quality Enhancement Committee*, letter from Sir Ron Cooke, chairman of the TQEC to Sir Howard Newby, Chief Executive of HEFCE, 22 Jan 2003.

Dearing, R. (1997). *Higher education in the learning society*, Report of the National Committee. London: HMSO.

Hannan, A., Silver, H. (2004). *Enquiry in external examining: Final report*. Higher Education Academy. www.heacademy.ac.uk.

HEFCE (2002). *Information on quality and standards in higher education*. HEFCE 2002/15.

HEFCE (2005). *Review of the quality assurance framework*. HEFCE 2005/35.

HESA (2009). *Students and qualifiers data tables: Qualifications obtained 2003/04.* http://www.hesa.ac.uk/dox/dataTables/studentsAndQualifiers/download/quals0304.csv?v=1. Accessed 12 June 2009.

QAA (2004). *External examiners code of practice.* http://www.qaa.ac.uk/academicinfrastructure/codeOfPractice/fullintro.asp. Accessed 12 June 2009.

QAA (2005). *Quality assurance in UK higher education: A guide for international readers.* http://www.qaa.ac.uk/international/studentGuide/English_readers.asp. Accessed 22 January 2009.

Silver, H. (1994). External examining in the UK: how did it start? In HEQC *External Examining in Focus*. November 2004.

Silver, H. (2005). External examining – Ten years on. In CHERI, *Higher Education Digest*. Spring 2005.

Silver, H., Stennant, A., Williams, R. (1995). *The external examiner system: Possible futures.* London: QSC.

Warren Piper, D. J. (2004). *Are professors professional? The organisation of university examinations.* London: Jessica Kingsley.

Chapter 3
The Teacher Education Accreditation Council (TEAC) in the USA

Elaine El-Khawas

Introduction

The education of prospective teachers is important in all societies. Well-qualified teachers help ensure that a nation's youth acquire the skills and knowledge to find productive adult roles and meet future challenges. In recent decades, America's teaching force has come under scrutiny, out of concern that teachers are not well prepared and that teacher training is weak. These concerns have found expression in special commissions and reports, including a landmark report, *A Nation Prepared: Teachers for the 21st Century* (Carnegie 1986). A decade later, the National Commission on Teaching and America's Future issued a report, *What Matters Most: Teaching for American's Future* (NCTAF 1996) which, among its recommendations, called for accreditation of all programs of teacher education.

The preparation of teachers in the USA is primarily the responsibility of colleges and universities. The dominant pattern is one in which colleges and universities provide the content knowledge and practical experience (in cooperation with local schools), while state governments and professional associations exert external quality control. About 1,300 of the nation's 4,000 institutions of higher education offer teacher education programs. This includes private colleges that enroll fewer than 100 students in education and also large state universities where schools of education enroll 600 or more students.

States have two roles: they regulate programs and they license new teachers. All programs of teacher education must receive state "approval," by meeting the state's formal standards. Standards vary, with some states maintaining quite general requirements. Separately, each graduate must pass the state examination to become a licensed or "certified" teacher, required for employment by school systems. Again, states vary in the coverage and rigor of this exam and whether the license is required for private as well as public school teaching.

E. El-Khawas (✉)
George Washington University, Washington, DC, USA

D.D. Dill, M. Beerkens (eds.), *Public Policy for Academic Quality*, Higher Education Dynamics 30, DOI 10.1007/978-90-481-3754-1_3,

Each program also may seek accreditation, a voluntary, nongovernmental review organized by professional associations. Subject-specific accreditation is quite common in the USA. More than 60 professional agencies accredit academic programs in electrical engineering, art, music, nursing, social welfare, business, medicine, and other subjects. Multiple accrediting agencies exist in several fields, including business, nursing, law, and teacher education. In the past, many programs of teacher education did not seek accreditation, in part because they operated on the basis of state approval and their graduates mainly found employment within their state. Generally too, their institution had "regional" accreditation, a broader form of recognition. This pattern is changing today in teacher education, as programs grow and compete with each other and as some states now require teacher education programs to be accredited (Murray 2005).

Two external bodies provide national-level oversight of this system of decentralized accrediting agencies (El-Khawas 2001). A nongovernmental umbrella organization, the Council for Higher Education Accreditation (CHEA), evaluates and publicly attests to the legitimacy and appropriate standards of each of the accrediting agencies that it recognizes. In addition, an advisory council – The National Advisory Committee on Institutional Quality and Integrity (NACIQI), convened by the U.S. Department of Education and assisted by a small evaluation office within the department – applies its own evaluative process to recognize accrediting agencies. Because of the voluntary nature of program accreditation, some academic fields have very high rates at which programs are accredited, while other fields have much lower rates. In fields where most programs are accredited, or where the strongest programs choose to be accredited, accreditation status is seen as conferring status or prestige. Accreditation thus acquires an additional role – as a mark of status – beyond its quality control role of assuring the public that a program meets external standards of quality.

Teacher education is a field in which many programs have not sought accreditation, instead relying on state approval. In 2003, about 560 of the country's 1,300 teacher education programs (43%) were accredited by one of the long-established accrediting bodies, primarily NCATE (the National Council for the Accreditation of Teacher Education), and a small number by the Montessori Accreditation Council (which accredits free-standing Montessori schools). Accredited programs nevertheless account for about 80% of the annual supply of new teachers. In recent debates about the quality of school teachers, concerns have been voiced about whether nonaccredited programs are offering sufficient quality in their training.

The Policy Problem

The Teacher Education Accreditation Council was created in 1997 in a context of lively debate on ways to improve perceived shortcomings in teacher preparation in the USA. During this period, multiple initiatives, both governmental and nongovernmental, have been competing for support and legitimacy. Since 1987, for example, the National Board for Professional Teaching Standards has pioneered a system for certifying teachers. Also in 1987, a coalition of state education offices, education organizations, and institutions of higher education formed INTASC (the Interstate New Teacher Assessment and Support Consortium), dedicated to improving the licensing and continuing education of teachers. In 2004, the U.S. Department of Education provided financial support to the American Board for Certification of Teacher Excellence to develop and administer an online standardized test for teacher licensing. This test has already been accepted in Pennsylvania and Florida.

It also was a time when accrediting agencies were making significant changes, both to streamline requirements and to focus attention on student learning (El-Khawas 2001). Their actions, especially those directed toward student learning, were a response to criticisms that accreditation procedures were too heavily concentrated on indirect evidence about the "capacity" of a program or an institution (e.g., good faculty and facilities; responsible procedures for governance and administration) rather than actual accomplishments. Other accreditation reforms were spurred by a 1996 article by Dill et al. (1996) that described academic audit as an attractive approach. As Ewell has remarked, this period offered "...an unparalleled opportunity to respond to growing demands that accrediting bodies pay greater attention to student learning outcomes in their review processes" (Ewell 2001, p. 2).

While influenced by these trends, the founding of TEAC as a new accrediting agency was also a response to dissatisfaction with existing requirements for accreditation in teacher education. Several concerns came together, among them the need for an accreditation process more compatible with the characteristics of teacher preparation programs at relatively small colleges. Smaller institutions argued that NCATE's accreditation model included criteria for faculty credentials, research productivity, facilities, and governance that did not fit their circumstances. In 1996, a survey by the Council of Independent Colleges (CIC) among its membership (about 404 private colleges) found that most members were not satisfied with NCATE's approach. These results led to the creation of a broad-based committee to plan an alternative approach. While CIC was the sponsor, committee members included deans from three major public research institutions (the University of Michigan, Indiana University, and Iowa State University) and a public 4-year college (Millersville) as well as several presidents from private colleges (Ekman 2003).

The Teacher Education Accreditation Council thus was founded during a time of sharp debate, competing factions, and differing approaches to the reform of teacher preparation. Building on several reforms, it developed an alternative approach based on an innovative audit method that emphasizes verifiable evidence for student learning. TEAC's focus on student learning, while distinctive, is similar to recently adopted approaches of other US accreditation agencies. In the late 1990s, for example, WASC (the Western Association of Schools and Colleges) adopted two primary criteria for accreditation: educational effectiveness and institutional capacity.

Content of the Instrument

Within a framework used by most US accrediting agencies, TEAC developed a detailed process to evaluate programs and decide whether to accredit them. Table 3.1 offers an overview of the TEAC procedures. The following description is based on TEAC's Standards and Guidelines, available on the TEAC website. To be considered for accreditation, a program notifies TEAC of its interest and submits an eligibility application, attesting that it meets five eligibility requirements:

- the program is committed to TEAC's goals and quality principles;
- the program faculty understand that TEAC may disclose the member's accreditation status;
- the program faculty will provide any information that TEAC may require;
- the institution giving the program has regional accreditation or its equivalent;
- the program' graduates are eligible for the state's professional teaching license.

Table 3.1 TEAC's accreditation process at a glance

Stages in TEAC	Program Faculty Actions	TEAC Staff Actions
1. Application	Program faculty prepares and submits application and fee.	TEAC staff consults with the institution and program faculty; TEAC accepts or rejects application (on eligibility requirements) and accepts or returns fee accordingly.
2. Formative evaluation	1. Program faculty attends TEAC workshop on writing the Inquiry Brief or Inquiry Brief Proposal. 2. Program faculty submits working drafts or draft sections of Brief with *checklist*.	1. TEAC staff reviews draft Brief or sections for coverage, clarity, and auditability and returns drafts for revisions and resubmission as needed. 2. If appropriate, TEAC solicit s outside reviews on technical matters, claims, and rationale.
3. Inquiry Brief or Inquiry Brief Proposal	1. Program faculty responds to TEAC staff and reviewers' comments. 2. Program submits final Brief with *checklist*.	1. TEAC declares Brief auditable and instructs program to submit six copies of that final version of the Brief. 2. TEAC accepts Brief for audit and submits it to the Accreditation Panel chair for instructions to auditors.
4. Call for comment		TEAC places program on TEAC website's Call for Comment page and circulates Call for Comment page to program faculty and staff to forward to school superintendents, state board of education, teachers, principals, and employers.

Table 3.1 (continued)

Stages in TEAC	Program Faculty Actions	TEAC Staff Actions
5. Audit visit	1. Program faculty submits data for audit as requested. 2. Program faculty receives and hosts auditors during visit (2–3 days). 3. Program faculty responds to Audit Report (2 weeks).	1.TEAC schedules audit. 2. Panel chair formulates questions and instructions for auditors; auditors verify submitted data. 3. Auditors complete visit to campus. 4. Auditors prepare Audit Report and send to program faculty, TEAC, and Accreditation Panel chair. 5. TEAC staff responds to program faculty's comments about the draft Audit Report. 6. Final Audit Report prepared and distributed.
6. Staff analysis		1. TEAC completes staff analysis and sends to program and panel. 2. TEAC sends Brief, Audit Report, and faculty response to panel members; panel members complete worksheets.
7. Accreditation Panel	1. Program head attends meeting (optional). 2. Program faculty responds (within 2 weeks).	1. Panel meets and formulates Accreditation Report and sends report to program faculty. 2. Call for Comment announced via e-mail and website.
8. Accreditation Committee		1. TEAC sends Brief, reviewers' comments, Audit Report, accreditation report, staff analysis, and panel recommendation to Accreditation Committee for decision. 2. Accreditation Committee meets; TEAC sends Accreditation Committee's decision to program.
9. Acceptance or appeal	Program faculty accepts or appeals TEAC's action within 30 days.	If the decision is to accredit and the program accepts the decision, TEAC announces the decision and schedules the annual report. If the decision is not to accredit and the program appeals, TEAC initiates its appeal process.
10. Annual report	Program faculty submits annual report to TEAC by anniversary date of accreditation decision.	TEAC reviews annual reports for as many years as required by program's status with TEAC.

Source: TEAC 1998.

Once eligible, a program has "candidate" status for 5 years. It arranges a schedule with TEAC for preparation of a self-study (called an Inquiry Brief) and a review of the evidence in the Inquiry Brief through an expert visit (called an Audit).

Compared to other quality assurance models, TEAC's procedure is heavily grounded in explicit standards and detailed evidence that standards are met. The standards, called principles of quality, have three elements:

- Quality Principle I: Evidence of student learning, involving mastery of content knowledge and pedagogical skills;
- Quality Principle II: Valid assessment of student learning, involving evidence that the program's method(s) for assessing student learning are valid; and
- Quality Principle III: Institutional learning, involving evidence that the program undertakes continuous improvement and quality control based on its assessment of student learning.

All steps in the TEAC evaluation are based on these three principles of quality, and also on standards for demonstrating institutional "capacity," or the ability to sustain a program at acceptable levels of quality. Seven components of capacity, required by the U.S. Department of Education for all accrediting agencies, specify requirements related to curriculum; program faculty; facilities, equipment, and supplies; fiscal and administrative capacity; student support services; recruiting and admission practices; and student feedback. TEAC's requirements on each of these components are specified in the Standards and Guidelines.

Preparation of an Inquiry Brief is the first step in seeking TEAC accreditation. This self-study describes the program and its results, and provides documentation on its faculty, its requirements and standards, and its quality control system. Distinctive to TEAC's review, the program also must explain the evidentiary basis for its "claims" that it provides effective student learning, has effective methods of assessing learning, and actively uses assessment to make continuous improvement. It also must address whether the evidence is dependable, persuasive, and representative of the program. All claims must be supported by multiple forms of evidence that are mutually consistent.

There is flexibility in what evidence is presented. TEAC lists 20 possible forms of evidence to demonstrate Quality Principle I, on student learning. Box 3.1 lists the forms of evidence, including grades, evaluations of teaching skill, test scores, rates of student success, or studies of graduates or employers.

Similarly, for Quality Principle II, methods of assessing learning, evidence can take different forms, including interviews, surveys, and classroom observations. Evaluations by employers, analyses of graduates' scores on licensure exams, or case studies of the achievements of program graduates might be offered. Programs must use multiple measures, show data by meaningful subcategories of students, and present evidence for the reliability and validity of assessment results and their interpretations of data.

Box 3.1 Types of Evidence to Demonstrate Student Learning

Grades

1. Student grades and grade point averages in each component of *Quality Principle I*: subject matter, pedagogy, and teaching skill

Scores on Standardized Tests

2. Student scores on standardized license or board examinations in any of the areas of *Quality Principle I*
3. Student scores on admission tests for graduate study in the areas of *Quality Principle I*
4. Standardized scores and gains of the program graduates' own pupils

Ratings

5. Ratings of portfolios of academic accomplishment
6. Third-party rating of the program's students
7. Ratings of in-service, clinical, and PDS teaching
8. Ratings by cooperating teachers and college/university supervisors, of practice teachers' work samples

Rates

9. Rates of completion of courses and program
10. Graduates' career retention rates
11. Graduates' job placement rates
12. Rates of graduates' professional advanced study
13. Rates of graduates' leadership roles
14. Rates of graduates' professional service activities

Case Studies and Alumni Competence

15. Evaluations of graduates by their own pupils
16. Alumni self-assessment of their accomplishments
17. Third-party professional recognition of graduates (e.g., NBPTS)
18. Employers' evaluations of the program's graduates
19. Graduates' authoring of textbooks, curriculum materials, etc.
20. Case studies of the graduates' learning

Concrete evidence also must be presented with respect to Quality Principle III, active use of assessment to improve learning. A program might describe ways that past decisions to modify the program have been shaped by assessment results. Also expected are plans for further use of assessment results to improve the program.

Case studies on the use of assessment results can be offered, but they must include evidence that their methods are dependable and trustworthy.

Such in-depth attention to the evidence for student learning is distinctive to TEAC's accreditation criteria. Also exacting are its requirements with respect to program capacity, mentioned earlier. Evidence on program capacity might include data showing that the program is supported at a level that is in line with support given to other programs at the institution. Information might also be provided on what was learned through the program's internal audit of its quality control system. TEAC needs such evidence to fulfill requirements imposed by the U.S. Department of Education and by the Council for Higher Education Accreditation (CHEA). In addition, TEAC considers the evidence of capacity to be important for the program's own assessment of what contributes to its success or failure with student learning, and also as evidence of good procedures for monitoring and improving programs.

TEAC has an unusual requirement that each program audit its quality control system, following TEAC Guidelines. This internal audit, intended to show that systems function as intended, takes a sample of student records and follows an "audit-like" trail to ascertain whether appropriate procedures were followed as students were admitted and progressed through their studies.

Distinctive, too, is the assistance that TEAC provides to programs as they develop the Brief. TEAC offers workshops on how to develop the Brief. When a program is drafting its Brief, staff members who have been trained to examine program statements and claims will work with the Brief's authors to ensure that it is clear and complete and that evidence is in a form to be "auditable." Typical staff comments include requests for clarifications to strengthen the document's evidentiary basis, perhaps to be more specific about how program faculty are "involved" in planning, how often students meet with their advisors, or what is the basis for other general statements (Workshop... 2005). Once the Brief is ready for audit, it is formally submitted to TEAC and an audit visit is scheduled.

Next is an external audit process, designed to provide independent verification of the evidence in the Inquiry Brief. The on-site audit visit is conducted by at least two TEAC-trained auditors. It is a rigorous and systematic review, bounded by detailed protocols and checklists developed by TEAC. The audit team's role is to determine whether the evidence in the Inquiry Brief on quality and capacity is trustworthy and whether the evidence justifies a claim that the institution is committed to the program.

Audit team members probe whether the language of selected claims is accurate and precise. Fact checking is conducted through interviews, use of institutional data to recalculate statistics, and direct review of other primary sources that were used to support claims for each quality standard. Auditors might ask to see the actual course evaluations cited in the Brief, examine course syllabi for several different years, or ask questions about how reported problems in program management are being handled (Workshop... 2005). Consistency among multiple sources of evidence is sought, and there is specific attention to whether contradictory or disconfirming evidence exists. Auditors are trained to avoid offering opinions or informal advice.

The audit report, completed within 2 weeks of the visit, describes the audit tasks that were completed and gives the team's formal opinion on the accuracy of the evidence in the Brief. The report does not offer recommendations or make judgments, as it is designed only to vouch for the accuracy of the evidence in the Inquiry Brief, possibly modified by information gained during the audit visit. A draft of the Audit Report is sent to the program for corrections, revised if necessary, and then submitted to the TEAC Accreditation Panel.

When the audit report is complete and the auditors have attested that the evidence is accurate, TEAC procedure moves to two further, sequential steps. First, a review is conducted by a seven-member Accreditation Panel. The Panel is chosen from a pool of 12 educators appointed for 3-year terms, primarily for their skills in evaluating evidence. The Panel's task is to weigh and assess the documented evidence about a program's capacity and the results it has achieved.

Using evaluation worksheets, the Panel evaluates whether each TEAC standard is met, based on the evidence found in the Inquiry Brief, the audit report, and other materials. For Quality Principle I, it reviews whether the evidence is complete, consistent, sufficient, and precise about student learning. For Quality Principle II, (valid assessment of student learning), it reviews the rationale for completeness, strength, and faculty support and whether it is grounded in scholarship. It also reviews the assessment system in terms of its design, use of multiple measures, sufficiency, and precision. For Quality Principle III (use of evidence to improve programs), it reviews decision-making and the quality control system and looks for completeness, precision, and evidence of quality improvement. Evidence on capacity is reviewed for completeness, commitment, sufficiency, and precision.

The Panel makes a summative judgment, deciding whether there is *sufficient* evidence that the program is "above standard" on each quality component, considered separately. Evidence of student learning is the pivotal factor in recommending full accreditation. The Panel also assesses alternative interpretations for the evidence the program presented. If evidence is determined to be sufficient to support the program's claims, the Panel prepares a report with written justification for its recommendation on accreditation, based on a minimum of four affirmative votes.

The Panel recommends one of TEAC's categories of accreditation. The recommendation may identify weaknesses and may stipulate areas requiring remedy within a short period. Negative decisions include the following: Denial, where TEAC's standards are not met, and Adverse Action, where accreditation is revoked following a finding that a program no longer complies with TEAC standards. The Panel's report is transmitted to TEAC's president.

The Accreditation Panel meeting can be attended by a representative of the program being discussed, allowing the program to be aware of the reasoning of the Panel. There also is a point in the Panel's review when the program representative may respond briefly to questions seeking clarification on factual points. Once the Panel's report is transmitted to the TEAC president, it is sent to the program faculty, which has 2 weeks to respond in writing to the arguments and findings in the report.

The Accreditation Committee of TEAC's board of directors conducts the next step in the review. The Accreditation Committee examines all documentation to determine, first, that TEAC has correctly followed its own procedures and, second, that the Panel's report and accreditation recommendation are convincing and consistent with the Committee 's own readings of the materials. While the Committee's review method is closely connected to the Panel's method, it has a different, complementary task: it attempts to find evidence that would undermine a recommendation or finding. The committee then makes TEAC's final decision, with a majority vote to accept or reject the Panel's recommendation on accreditation status. If it rejects, it gives written reasons for its decision. Program faculty may appeal a decision, within 30 days, to a five-member Appeals Panel appointed by the chair of the TEAC board of directors.

In sum, TEAC has a three-step review, each focused on evidence provided by the program about its capacity and results. Each step has a different emphasis: the audit determines whether the program's claims are accurate and trustworthy, the Panel determines whether the evidence is sufficient to warrant accreditation, and the Committee determines whether it, independently, can support or undermine the Panel's recommendation.

Costs

TEAC, while still a start-up organization, has benefited from foundation support, most recently the Pew Charitable Trusts. Other funding has come from the Carnegie Corporation of New York, the Olin Foundation, Atlantic Philanthropies, and an anonymous donor. A FIPSE grant, which supported initial design and pilot testing, ended in December 2001. Overall, TEAC has had more than $3 million in external funding.

Following procedures of other US accrediting agencies, TEAC is primarily supported by fees (for audit visits) and by annual dues paid by members. No governmental funds, state or federal, are provided to TEAC to support its operations. TEAC's annual dues in 2005 were $2,000, with over 100 institutional and organizational members. An audit fee of $1,500 per auditor (for 2–4 auditors) is assessed for each program that begins the TEAC accrediting process. The program also pays for audit visit expenses. Auditors are paid a small honorarium and reimbursed for expenses. TEAC currently has a total staff of about seven persons.

In designing its procedures, TEAC sought to keep costs to a minimum. Data are often sent to TEAC prior to the visit to make more efficient use of the visit itself. The detailed review of an institution's Inquiry Brief before the visit also contributes to efficiency. The use of detailed protocols allows for smaller visiting teams. Programs understand, prior to the visit, what evidence to make available for the auditors. As one program reported, this significantly reduced the time and expenses of preparing for the visit, which was more intense and meaningful than other accrediting visits (Cohen 2003).

Implementation and Impact

In January 1998 the Fund for the Improvement of Postsecondary Education, a competitive-grants agency within the US Department of Education, awarded a grant to the Council of Independent Colleges to develop and test a new model for accrediting teacher education. The grant also supported a pilot test of the new accrediting process, which was carried out at three institutions: the University of Virginia, Fort Lewis College, and Texas Lutheran College.

TEAC's first 4 years were evaluated by Peter Ewell, a well-respected US assessment analyst, between summer 2000 and December 2001. He assessed the experiences of the three pilot accreditation reviews and compared TEAC's methods to other quality assurance approaches (Ewell 2001). He reported several strengths:

- Clear objectives, focused on high-priority issues, with an "... insistence that each program ... provide solid and direct evidence of student learning consistent with the program's goals." (p. 13)
- A structured and tightly focused approach, which makes the ground rules for the review extremely clear. (p. 15)
- Direct evaluation of the veracity of the submitted evidence on student achievement.
- Well-trained visiting teams, with extensive prior practice with audit techniques.
- Value to the programs, based on pilot-institution comments that the self-study's focus on student learning, by separating "relevant from irrelevant data," allowed them to gather data that were really useful for improvement. (p. 9)

Ewell concluded, "TEAC has ... succeeded in creating a review process that puts – and keeps – evidence of student learning at the center" (p. 14). In his view, TEAC's process matches best practice in other "leading-edge" quality assurance models. With respect to its evaluation method, he concluded, "... no other quality assurance agency in higher education ... engages in this level of detail when examining assessment-based evidence of student achievement" (p. 5).

Between 2001 and 2003, TEAC achieved two forms of recognition that are essential in establishing its legitimacy as an accrediting agency. In 2001, TEAC gained recognition from the Council for Higher Education Accreditation, signifying that it met its standards. In 2003, TEAC gained recognition from the U.S. Department of Education's National Advisory Committee on Institutional Quality and Integrity, which determined that TEAC met the department's standards for accrediting agencies. TEAC's petition for federal recognition had been endorsed by the American Council on Education and by other higher education associations.

More recently, TEAC accreditation has been included in the teacher licensing requirements of at least six states, with other states likely to follow suit (Murray 2005). In 2004, TEAC held discussions with 12 other states on linking TEAC accreditation to their licensing process. To some degree, this attention to building support among individual states represents a shift in tactics for TEAC. With state endorsement, TEAC may gain greater support among teacher education programs and, thereby, offset the effects of the continued rivalry between TEAC and NCATE.

By 2007, TEAC has gained experience and has built up its membership. One hundred and nineteen institutions of higher education are members of TEAC. Notably, TEAC has been able to broaden its base of support, as its 119 members include 30 public institutions, including 15 public doctoral universities. Thirty-four programs at 32 institutions have been accredited by TEAC as of early 2007. Another 11 institutions have satisfied TEAC eligibility standards and hold "candidate" status, the first step toward accreditation.

One surprise TEAC has encountered is that many programs have found it hard to use the logic of educational research for assessing their own programs. While education faculty are well trained in research inquiry – developing clear questions, designing and implementing steps to assemble evidence to answer those questions, and evaluating findings for their reliability and validity – they generally have not been systematically applying those methods to their academic courses and programs. The focus on evidence-based inquiry is a laudable part of TEAC's design, one that has been maintained, but it has taken considerable effort and rethinking for most program faculty to start with their own questions about their mission and to gather needed evidence.

TEAC has responded by designing and holding workshops and by assigning coaches to guide and assist institutions in preparing evidence for quality. TEAC now holds at least two workshops each year to assist participants with the tasks required by the TEAC accreditation process. Workshops emphasize ways to construct clear, concise, and defensible claims about programs and ways to select, evaluate, and use evidence to support those claims. TEAC now invests staff time in formative evaluation, assisting institutions and offering advice as they work on their Inquiry Briefs. Staff examine draft materials and provide feedback on areas in which the draft does not give sufficient detail or where descriptive language is imprecise. They point out where the evidence is inconsistent, where connections have not been demonstrated between program claims and the assessment evidence, or where no plans are shown for systematically relying on evidence for future improvement (Workshop... 2005, p. 101).

TEAC also has found that members need greater guidance on how to meet another requirement, conducting an internal audit of their quality control system. What TEAC has learned is that most programs may have a general sense that their quality control systems work as intended, but have conducted only occasional or limited inquiries. The systematic scrutiny required by TEAC, therefore, calls for a change in program practice. As one response, TEAC has developed a separate training manual on the internal audit.

Evidence of TEAC's overall impact is found in the experience of its members, especially the institutions that have achieved accreditation or are actively preparing for TEAC review. A consistent message emerges: the TEAC process forces attention to a program's own questions about its effectiveness and yields evidence that is readily used in improving the program. In other words, the TEAC process itself adds value: program improvement is embedded in the process of developing an Inquiry

Brief. Faculty members report that they have improved their program while conducting the TEAC review process – clarifying objectives or claims, identifying the extent to which evidence supports those claims, and making changes to better align their offerings with their objectives. Procedures are tightened up, better assessment tools are put to use, and faculty confront weaknesses and make changes.

Under the demanding TEAC model, many programs initially make slow progress, sometimes needing to revamp program features or undertake studies as part of their preparation. Participants generally attest that the process fostered program improvement, rather than being done as an administrative task useful only to an external agency. As reported by TEAC, members have commented that "...TEAC demands honest scrutiny of programs and resources" and that "...greater opportunity for program improvement is possible than with other accreditation systems." Another comment is that "...the process is one that is relevant and is a matter of documenting and clearly articulating what we as a teacher preparation program are actually doing." Another reported a specific result: TEAC "...enabled us to get assessment front and center ... and made faculty aware of the importance of assessment" (TEAC 2005).

Reports from institutions that have completed TEAC review support these judgments. The University of Virginia reported that the TEAC review showed that certain procedures were not functioning as intended. As a result, it strengthened the alignment between student portfolios and its learning goals and developed more systematic ways to learn from its graduates (Cohen 2003). Hollins University learned during its TEAC review of inconsistencies in the way that educational and theoretical concepts were taught, and the program faculty made changes to strengthen the teaching program. Because the internal audit conducted by the Hollins faculty identified areas where their quality control system needed improvement (primarily with missing information from student files), they took steps to ensure complete files. Also, they took action to strengthen teaching components related to technology skills and classroom management, based on the TEAC review (Hollins University 2002). At Rockhurst University, program faculty found important discrepancies between the ratings students were given by their own instructors and those given by teachers who observed student teaching. In their Inquiry Brief, they reported several corrective steps they took, including a new departmental policy to use a new database with evidence on student progress (Workshop... 2005). Other institutions have reported that, being dissatisfied with gaps in what they could document, they have developed novel ways of measuring learning. One institution conducted several analyses of data on their students in mathematics education, comparing their achievement to mathematics majors and also conducting correlation analyses to determine the extent to which measures of learning were coherent and consistent (F. Murray, personal communication, March 31, 2005).

During its short history, TEAC has weathered a certain amount of criticism (e.g., Darling-Hammond 2000). An early concern was that TEAC does not have sufficient standards or that its standards were not performance based. It is true that TEAC's standards were more flexible than the NCATE standards in effect throughout the 1990s, standards generally seen as too prescriptive. However, an independent comparison (AACTE 2003) has shown that TEAC standards are substantially parallel to NCATE's new standards, adopted in 2000, following its own standards revision.

While its unusual approach departs from other accreditation approaches, TEAC does have explicit and focused standards, calling for evidence of student learning, use of that evidence for program improvement, and the capacity to sustain a program of quality. These standards, and their detailed provisions, have been formally recognized by CHEA and the US government.

Some critics have said that TEAC's standards are softer than NCATE's because TEAC allows each program to decide what evidence to use to meet TEAC standards. TEAC has taken this approach for well-considered reasons, mainly to allow flexibility to programs. Thus, for example, where a program believes that its state exam is poorly designed, the program may use other evidence of student achievement, defending their choices. At present, some states have licensing exams where a "passing" score can be achieved with fewer than half of test items answered correctly. Even so, TEAC expects a persuasive rationale for the evidence that is chosen, strong evidence for each standard, and reporting of any negative evidence as well. In actual practice, most programs that have sought TEAC accreditation have used conventional evidence – grade averages and scores on state examinations – that are also used in meeting NCATE standards. To reduce the burden on programs, TEAC will accept evidence that was initially assembled for other accrediting organizations when programs are preparing their Inquiry Brief for TEAC.

Some criticism of TEAC may be related to its origins in the midst of controversy about whether teacher education programs offer good training. As one journalist characterized the conflict, "...the battle for turf involves money and politics as much as educational values" (Basinger 1998). In this climate, some educators may be reluctant to endorse a new, unproven approach to accreditation, preferring its competitor, NCATE, which is well known and has a long history. NCATE accreditation, held by relatively few programs, is considered prestigious. Some educators also expressed concern that, with two accreditation options, programs might "shop" for the easier process, which could cause quality to decline. For these and other reasons, TEAC has seen slow development as an alternative approach.

Both TEAC and NCATE, as voluntary accrediting bodies, face challenges from broader trends in the USA that have imposed more governmental requirements, both on accreditors and on programs of teacher education. TEAC has pledged its willingness to cooperate with other accrediting organizations, and many educators look to a unified accreditation process sometime in the future.

Comparisons

TEAC's accreditation model combines elements from several existing forms of quality assurance. It builds on recent reforms in US accreditation and adopts some emerging ideas about effective practice from international experience with academic audit. It also borrows from widely accepted norms of scholarly inquiry.

Over the last decade, a variety of US accreditation reforms have continued to use guidelines, self-study, and site visits but have added a requirement for evidence of student learning. This "outcomes" focus is strongest among those accreditation agencies that conduct subject assessments in such fields as business, engineering, medicine, or nursing. It emphasizes the quality of delivered performance, typically documented by data on the performance of graduates on licensing exams or in initial employment. TEAC's model adopts this "outcomes" emphasis, as well as the effort by accrediting agencies to streamline their requirements on institutional capacity.

International developments with academic audit are also reflected in TEAC's approach (Dill 2000; Massy 2003). Audit models are strongly evidence based, comparable to financial audits. Where accreditation models traditionally allow programs to prepare general reports, audits focus on institutional processes that support quality. After a decade's experience with audit in different countries, the audit approach is increasingly valued for its systematic, "evasion-resistant review methodology" that also spurs improvement. Audit processes are useful to programs because they require thoughtful and systematic attention by a program's faculty about how and why their program is meeting worthy objectives (Jennings 2003; Massy 2003, pp. 230–231).

TEAC has built on this international experience with audit. It borrowed the audit model's attention to institutional processes and its focus on evidence and introduced an innovative requirement that each program conduct its own internal audit. It goes further than many process-oriented audit models by requiring substantial attention to evidence of student learning and program results. Thus, where most audit models look at processes that support quality, TEAC adds a critical preliminary step: programs must explain and defend their learning objectives. They also must document the processes that assure them those goals are met and the processes that lead to quality control and improvement. This approach parallels the emphasis in Hong Kong's most recent approach, termed Quality Review, that looks to the design of curricula, methods of assessing learning outcomes, and adequate resources to offer a quality program (Massy 2003).

Unlike other approaches to quality assurance, TEAC has grounded its model in an evaluative philosophy based on scholarly research. Thus, it requires that claims be supported by evidence, expects valid assessment of student learning, and has rules for providing and evaluating evidence (consistency, reliability, validity, representativeness, etc.). Under TEAC's approach to assessing learning, for example, programs are expected to employ multiple measures to achieve a dependable finding and must provide evidence that the inferences they make conform to accepted research standards for reliability and validity.

Useful perspective on TEAC's assessment approach can be gained by comparing it to subject evaluations in Denmark (see Chapter 10, "Subject Assessments for Academic Quality in Denmark"). The Evaluation Institute's (referred to as EVA) system for assessment, which covers a range of academic subjects, has elements that can be considered as best practice in external quality assessment. EVA aims to have a "concrete, transparent and trustworthy process." Its approach is structured and systematic, relying on detailed protocols and specific criteria that guide the program's self-studies and site visits. Programs must provide quantitative data on their accomplishments (including data on completion rates and drop-outs), and EVA conducts surveys among recent graduates or employers to supplement the self-study evidence (Kristoffersen 2004, pp. 91, 95). The 1- or 2-day site visit is designed to validate information in the self-study through meetings with institutional representatives and students and to examine documentation more closely. EVA staff participate in each visit, partly to ensure that evaluations follow correct procedure. The evaluation report is prepared by EVA staff on the basis of input from the visiting team. After institutions have a chance to correct factual mistakes and comment on the evaluation process, the report is made public.

TEAC's approach is also highly structured and systematic. It uses detailed checklists and decision rubrics with specific criteria to guide its audits and the decisions of its Accreditation Panel and Accreditation Committee. Its entire process focuses on evidence for claims made about student achievement. Programs must give evidence about student achievement, then document the validity of that evidence, and, finally, show that such evidence is used for program improvement. As Ewell commented about TEAC's model, "...the amount and quality of evidence about student learning...easily exceeded that present in virtually any other extant accreditation process" (2001, p. 14).

As with EVA's visit, the TEAC site visitors limit their role to validating the evidence in the self-study. Areas for further probes are identified in advance, based on TEAC standards, and auditors have extensive training, including practice with audit techniques.

TEAC's decision process is exacting. Its Accreditation Panel and its Accreditation Committee conduct two separate reviews of the program's evidence and make a series of decisions covering all components of the accrediting standards. Unlike EVA, TEAC separates this summative role from the site-visit auditor role, seeking to avoid the issues of "blurred boundaries" identified by some analysts (Schwarz and Westerheijden 2004).

EVA's objectives have been described as dual, combining accountability and improvement (Kristoffersen 2004, p. 26). Under the accountability function, programs are required to conduct self-studies and to host site visits, and they know that EVA's reports will become public. Also contributing to accountability are EVA's independent surveys among employers and graduates. The improvement function is equally important, based on EVA's responsibility under its initial legislation to "...inspire and guide..." programs (EVA 2004).

TEAC also has a dual emphasis. Accountability goals are served by its focus on student learning and its stringent reliance on evidence to support claims about learning. Improvement goals are encouraged by TEAC's self-study requirements that call for detailed probing of a program's aims, accomplishments, and evidence that it uses assessment findings to improve its program.

A core issue for all systems of external evaluation is the role of precise, defined standards versus standards that allow flexibility. EVA's view largely mirrors TEAC's stance, in that both approaches try to balance the need for defined standards with support for improvement. As EVA recently noted, there is a risk that tightly defined standards based on a conservative concept of quality could impede innovation (EVA 2004).

EVA has followed a "fitness for purpose" model with an overall framework that allows each evaluation to be designed according to what makes sense for each specific subject being reviewed. There is active dialogue with institutions prior to the evaluation, and self-study guidelines encourage open discussion about teaching and learning issues. TEAC also allows flexibility within an overall framework. It begins with standards that require a focus on student learning and its use for academic planning and improvement. However, each institution decides what evidence it will use to show that it meets TEAC's standards.

Conclusions

The Teacher Education Accreditation Council was designed to offer an approach to accreditation that reflected some of the best contemporary thinking on external quality review. Unlike quality assurance agencies in many countries, it was not a product of a government ministry or a parliamentary decree. With this relative freedom but guided by the context in which it was developed, TEAC's founders developed an innovative approach worthy of attention by others who are interested in models for quality assurance.

Readers interested in program evaluation will find much of value in TEAC's approach. It requires that programs focus on important objectives related to student learning, but it allows them to demonstrate accomplishment of those objectives in varying ways. This approach does not stifle innovation, but it does insist on evidence that students are learning what is needed to be effective, well-qualified teachers. Elements of good research technique have been built into the evaluation design. Reviewers are taught to consider criteria of validity, reliability, and representativeness and to be alert to evidence that suggests alternative explanations.

Readers interested in uses of audit will find interesting refinements in audit technique. TEAC's audit is sharply defined and focuses heavily on evidence of student learning. Prior to the audit visit, the program's Inquiry Brief is examined closely and clarifications are sought to strengthen the document's evidentiary basis. Site visitors are given a narrow role and are well trained, including receiving guidance on not offering opinions. TEAC also requires programs to conduct internal audits of the administrative procedures that affect student progress; conducting this audit gives programs direct evidence of their strengths or weaknesses.

TEAC also has achieved a significant degree of transparency, attained through detailed, explicit, and closely-adhered-to procedures that consistently focus on learning outcomes. Programs have detailed guidelines for preparing the self-study and can expect site-visit auditors to maintain a disciplined focus on evidence for the program's statements. Reports from early participants indicate that TEAC's model is practical to implement and supportive of educational enhancement (Cohen 2003).

TEAC is still a "start-up" agency, but it has achieved formal recognition and has built up its membership. As a voluntary accrediting agency, it must continue to prove its worth to colleges and universities. It must also provide a voice for strengthening teacher preparation in the larger US debate over the quality of teaching in the nation's schools. Its innovative approach and focus on student learning afford it a position of strength for the years ahead.

References

AACTE (American Association of Colleges for Teacher Education) (2003). Comparison of NCATE and TEAC processes for accreditation of teacher education. www.aacte.org

Basinger, J. (1998). Fight intensifies over accreditation of teacher education programs. *The Chronicle of Higher Education,* October 9, A12 p.

Carnegie Corporation, Task Force on Teaching as a Profession (1986). *A nation prepared: Teachers for the 21st century*. New York: Carnegie Corporation.

Cohen, S. (2003). *Statement to the national advisory committee on institutional quality and integrity*. Washington, DC: US Department of Education. June 10, 2003. www.cic.org Accessed 6 October 2004.

Darling-Hammond, L. (2000). Teaching for America's future: National commissions and vested interests in an almost profession. *Educational Policy, 14*(1), 162–183.

Dill, D. D. (2000). Designing academic audit: Lessons learned in Europe and Asia. *Quality in Higher Education, 6*(3), 187–207.

Dill, D. D., Massy, W. F., Williams, P. R., Cook, C. M. (1996). Accreditation and academic quality assurance: Can we get there from here? *Change, 28*(5), 17–24.

Ekman, R. (2003). *Statement to the national advisory committee on institutional quality and integrity*. Washington, DC: US Department of Education. June 10, 2003. www.cic.org Accessed 6 October 2004.

El-Khawas, E. (2001). *Accreditation in the United States: Origins, developments and future prospects*. Paris, France: UNESCO/International Institute for Educational Planning.

EVA (2004). *Criteria-based evaluations: EVA's experience in evaluations based on criteria*. Copenhagen, Denmark: EVA.

Ewell, P. (2001). *Piloting a new approach to accreditation in teacher education: An evaluation of the TEAC/FIPSE project*. Boulder, Colorado: National Center for Higher Education Management Services.

Hollins University (2002). *Inquiry brief proposal*. www.teac.org. Accessed 1 March 2005.

Jennings, J. M. (2003). New Zealand – Higher Education. In *Educational evaluation around the world: An international anthology*. Copenhagen, Denmark: EVA.

Kristoffersen, D. (2004). Ministerial Approval and Improvement-Oriented Evaluation in Denmark: An Alternative to Accreditation? In S. Schwarz and D. F. Westerheijden (Eds.), *Accreditation and evaluation in the European higher education area* (pp. 87–100). Dordrecht, The Netherlands: Kluwer.

Massy, W. F. (2003). *Honoring the trust: Quality and cost containment in higher education*. Boston, MA: Anker Publishing.

Murray, F. B. (2005). On building a unified system of accreditation in teacher education. *Journal of Teacher Education, 56*(4), 307–317.

Murray, F. B. (n.d.). *On some differences between TEAC and NCATE*. Available at www.teac.org. Accessed 12 June 2009.

National Commission on Teaching and America's Future (1996). *What matters most: Teaching for America's future*. New York: NCTAF.

S. Schwarz and D. F. Westerheijden (Eds.) (2004). *Accreditation and evaluation in the European higher education area*. Dordrecht, The Netherlands: Kluwer.

Teacher Education Accreditation Council (TEAC) (1998). http://www.teac.org/index.php/accreditation/steps-to-accreditation. Accessed 19 December 2008.

TEAC (2005). Available at www.teac.org. Accessed 6 October 2004.

Workshop on the Inquiry Brief and Inquiry Brief Proposal (2005). Available from TEAC on request.

Part II
Market Regulation of Academic Quality

David D. Dill and Maarja Beerkens

It is obvious at the turn of the century that the world of higher education has become much more subject to market forces (Teixeira et al. 2004). In part this is because of globalization, particularly the growing international rivalry among universities for students, academic staff, and research reputation. In part it is because of the increasing role of profit-making institutions in massified systems of higher education. But in addition market forces have been promoted by national policies introducing so-called quasi-markets – mechanisms for competitively allocating state support for instruction and research – into public systems of higher education (Dill 1997).

Competition for students and government funds is widely believed to lead to quality and efficiency in the higher education sector (OECD 2008). Certain conditions, however, need to be met before competitive mechanisms will produce the expected benefits in systems of higher education. Adequate information about quality differences in the system is one of the most important conditions for a functioning academic market (Dill and Soo 2004) and the quality of goods and services can therefore sometimes be assured through the provision of appropriate consumer information (Gormley and Weimer 1999). One means of assuring the provision of necessary consumer information is via government action or regulation. However economists contend that if consumer information regarding a valued good or service is truly inadequate or insufficient, this will create a "secondary market" for the needed information that may be filled by commercial organizations without the need for government intervention (Weimer and Vining 2005). From this perspective the competitive market is potentially self-correcting and will essentially regulate itself. This, however, assumes that the cost of collecting and presenting adequate information is balanced by the consumers' willingness to pay for the information, that the consumers are sufficiently knowledgeable to request appropriate information, and that the information requested by private consumers corresponds to the full range of public interests in higher education.

It is certainly true that the world-wide massification of higher education has been accompanied by a global explosion of university league tables, rankings, and report cards designed to inform student consumers and the larger public about academic quality (Van Dyke 2005). But have these largely commercial efforts at information provision helped assure academic standards?

In a comparative analysis of the leading commercial league tables in Australia, Canada, the UK, and the USA, we investigated this latter question (Dill and Soo 2005). Our conclusions were that purely commercial university rankings and league tables have in fact undermined efforts to assure academic standards (see the similar analysis in Ewell's chapter to follow). The most appropriate information to aid student choice in all countries is valid and reliable knowledge about *subject fields*, including indicators of academic program performance and relevant information on the student educational experience.[1] But the challenge and cost of developing valid indicators of the quality of academic programs are significant and for-profit publications have little motivation to make such investments. Instead they enjoy substantial sales and influence among opinion leaders, high achieving students, and even university personnel by focusing on *institutional* rankings utilizing reputational surveys, input measures such as student test scores and financial resources, and indicators of research quality, all of which have questionable validity as predictors of effective student learning (Pascarella and Terenzini 2005). The signals of university reputation broadcast by these commercial league tables have drowned out the more relevant information on the quality of academic programs and have encouraged a costly, competitive pursuit of academic standing among all institutions that in turn has crowded out the educational activities necessary for the actual improvement of academic standards.

The failure of the commercial sector to adequately address the need for relevant knowledge on academic quality has motivated a number of non-profit initiatives to provide more valid and socially useful information. In this part we examine four of these innovative instruments of information provision.

Beerkens and Dill analyze the highly influential academic program rankings first developed in Germany by the Center for Higher Education (CHE) and now being implemented in a number of other EU countries as well as in Canada. While the CHE rankings were initiated and are maintained by a non-profit organization, the German government was nonetheless influential in their development. The rankings follow an economic logic in clearly being designed to inform new students' choices of academic programs. The rankings therefore present information on academic subjects rather than whole institutions, information that students truly consider when making their choice. The rankings were carefully developed by knowledgeable professionals utilizing existing research as well as surveys of German students. By employing multi-dimensional measures of program performance, systematic processes for collecting and validating data, and a presentation format that avoids ordinal rankings and the aggregation or weighting of indicators, the CHE rankings effectively respond to the noted conceptual and technical limitations of commercial rankings of academic quality. There is also some evidence that the CHE rankings

---

[1] In first-level US degrees the equivalent to the subject field would be a student's "major," which usually comprises about one-third of an undergraduate's academic program. In the US context, therefore, valid information on the quality of the overall undergraduate educational experience would also be needed, such as the National Survey of Student Engagement (NSSE) reported on in this volume.

have benefited student choice as well as assisted academic quality assurance in Germany. The CHE rankings thereby offer a potentially valuable benchmark for the design of information provision policies as part of a national framework on academic quality assurance.

Peter Ewell presents an analysis of the National Survey of Student Engagement (NSSE), a US survey of first and fourth-year students based on known "best practices" in teaching and learning. The survey is operated by an independent entity based at Indiana University and is supported by user fees paid by individual participating institutions. Since the survey's inception in 2000, nearly 1100 different institutions have participated with over 1,160,000 student responses compiled. To date, twelve state systems of higher education have participated in NSSE, with the resulting information used as part of state accountability requirements for public higher education. The instrument is currently being promoted, among others, by the U.S. Department of Education as a standard way to disclose quality information to potential students. NSSE information is also frequently featured by participating institutions in their required accreditation reviews.

The survey was originally developed to generate information on academic quality in response to the growing influence of annual media rankings of institutions. But voluntary participation and institutional desires to keep results confidential have limited its impact in this arena. Through an aggressive media effort of its own, however, NSSE has had a significant influence on public perceptions of how "quality" should be construed in higher education. At the same time, it has given participating institutions an important diagnostic tool for improving their instruction and their educational environments. NSSE's concept and approach is readily portable to other national higher education contexts, where more proactive government sponsorship or support might well result in greater policy impact.

Kerri-Lee Harris and Richard James analyze an information based policy tool in Australia. The Australian government conducts regularly two surveys of university graduates. The Graduate Destinations Survey (GDS) collects information on the labor market success of graduates and the Course Experience Questionnaire (CEQ) inquires about their satisfaction with the various aspects of their university studies. The findings are presented at the subject field level for each university and are thus easily comparable across universities. It is worth noting that the findings are also adjusted for personal factors that may affect labor market outcomes and student satisfaction, such as study field, age, the basis of entry, and enrolment type. The surveys serve two main purposes. Since 2006 the outcomes of the survey are linked to the Learning and Teaching Performance Fund (LTPF) which rewards universities that demonstrate excellence in teaching and learning. Secondly, the survey data is publicly available for prospective students and thus serves as a direct consumer information tool. The findings are further promoted by commercial university rankings (such as *Good University Guide*) that combine the results with other information for prospective students. Furthermore, universities can use the information to monitor their performance relative to their competitors and improve their teaching quality. While rigorous evidence about the impact of the initiatives has yet to be established, Harris and James believe that the information tool has raised awareness about teaching and learning in universities and helps institutions to focus on educational outputs instead of inputs.

David Breneman analyzes *Measuring Up*, the biennial Report Card prepared by the National Center for Public Policy and Higher Education, a non-profit, non-partisan organization located in San Jose, California. These reports focus on the fifty states within the U.S. and employ quantitative performance indicators to compare the states on six educational categories: preparation, participation, completion, affordability, benefits, and learning. Each state receives a letter grade (A through F) indicating its performance relative to top-performing states on each category and as such *Measuring Up* represents a classic benchmarking exercise.

The reports have garnered significant coverage in the media and can be used by state policymakers to examine the underlying policy issues that determine the grades. While the National Center does not advance explicit policy recommendations, the purpose of the Report Cards is to generate a conversation and further research within each state. Several states have extended the basic framework to the county-level, revealing how well (or how poorly) citizens of the state are served by their higher education systems. The Report Cards utilize nationally collected data that are available at the state level. The Center does not undertake its own data collection. The reports have generally been well received, although officials in a few states have raised technical objections to some of the indicators. Since 2006 *Measuring Up* has extended the methodology to include international comparisons.

These four analyses help reveal the limitations of information currently supplied by an unregulated commercial sector and provide valuable models for the development of more useful public policies on the provision of academic quality information. They also reveal the complexities of the competitive market for students and suggest that while appropriate quality information should be an important component of a national quality assurance policy, public provision of information for consumers alone cannot assure academic standards. Academic quality information must also play a more direct role in institutional efforts to assure and improve student learning. As will be discussed in the next part, for this to occur government policy needs to assure, in ways the market cannot, that more valid and reliable information on academic quality is not only produced by universities, but also utilized by the institutions to assure and improve their academic standards.

References

Dill, D. D. (1997). Higher education markets and public policy. *Higher Education Policy*, 10(3/4), 167–186.

Dill, D. D., & Soo, M. (2005). Academic quality, league tables, and public policy: A cross-national analysis of university ranking systems. *Higher Education*, 49(4), 495–533.

Dill, D. D., & Soo, M. (2004). Transparency and quality in higher education markets. In P. Teixeira, B. Jongbloed, D. D. Dill, A. Amaral (eds.), *Markets in Higher Education: Rhetoric or reality?* Higher Education Dynamics Douro Series (pp. 61–86). Dordrecht: Kluwer.

Gormley, W. T. Jr., Weimer, D. L. (1999). *Organizational report cards*. Cambridge, MA.: Harvard University Press.

OECD (2008). *Tertiary education for the knowledge society, vol 1*. P. Santiago, K. Tremblay, E. Basri, E. Arnal (Eds.). Paris: OECD.

Pascarella, E. T., Terenzini, P. T. (2005). *How college affects students: A third decade of research*. San Francisco: Jossey-Bass.

Teixeira, P., Jongbloed, B., Dill, D., Amaral, A. (Eds.) (2004). *Markets in Higher Education: Rhetoric or Reality?* Higher Education Dynamics Douro Series. Dordrecht: Kluwer.

Van Dyke, N. (2005). Twenty years of university report cards. *Higher Education in Europe*, 30(2), 103–125.

Weimer, D. L., Vining, A. R. (2005). *Policy analysis: Concepts and practice* (4th edition). Upper Saddle River, NJ: Pearson/Prentice-Hall.

Chapter 4
The CHE University Ranking in Germany

Maarja Beerkens and David D. Dill

Introduction

The idea of university rankings has gained remarkable popularity in the past 20 years (van Dyke 2005). The first academic rankings were created in the USA in 1925 by Raymond Hughes, a professor of chemistry and later a university president (Webster 1992). Responding to a request from the North Central Accrediting Association for a study of graduate schools, Hughes conducted a reputational survey and produced the first published academic rankings, which in this case were rankings of the quality of graduate-level degree programs in the primary academic disciplines in the USA. Hughes' rankings created an important precedent, and similar reputational rankings of the quality of graduate degree programs were repeated in 1934 by Hughes himself, in 1959 by Hayward Kinston, and in 1966 by Alan Carrter. This tradition of rankings of graduate degree programs based on reputational surveys has been continued in the USA by the National Research Council (NRC), a private, nonprofit institution that provides science, technology, and health policy advice to the federal government under a congressional charter. The inaugural *US News and World Report* (USNWR) rankings of 1983, usually cited as the first university rankings, therefore broke with this tradition, first by being produced by a commercial publisher rather than by members of the academy and second by providing rankings of whole institutions focused on the supposed quality of undergraduate or first-level education. The USNWR rankings inspired a new international industry in university "league tables,"[1] and now universities are regularly ranked in at least 20 countries (Merisotis 2006). The practice has been adopted in education systems as different as Germany, Canada, China, and Nigeria. Furthermore, university rankings have

M. Beerkens (✉)
Centre for Higher Education Policy Studies (CHEPS), University of Twente, Enschede, The Netherlands

[1] Outside the USA, university rankings are often described as ''league tables,'' reflecting the published rankings or standings of international football (i.e., soccer) teams. In this chapter, the term *league tables* will be used synonymously with university rankings.

D.D. Dill, M. Beerkens (eds.), *Public Policy for Academic Quality*, Higher Education Dynamics 30, DOI 10.1007/978-90-481-3754-1_4,

crossed national borders. As higher education has become more global and students more mobile, there is also a demand for cross-national and global rankings. The World University Rankings by *The Times HE Supplement* (THES) and the Shanghai Jiao Tong University (SJTU) ranking compare universities internationally and are well known among students and university administrators around the globe.

As university rankings have gained legitimacy and visibility in the world, rankings themselves have become a subject for academic research and international partnership. Several international workshops and conferences have been dedicated to the topic, special issues of higher education journals have been published, and numerous academic articles discuss the strengths and weaknesses of various rankings.[2] In 2002 UNESCO-CEPES began regular meetings of higher education experts, researchers, and practitioners who are involved in various university rankings in the world. The initial purpose of the meetings was to compare experiences, discuss methodological problems, and improve rankings, but now steps have been taken toward some professional self-regulation in the area. This initiative established an International Ranking Expert Group (IREG) and designed a set of principles of quality and good practice in university rankings (IREG 2006).

The existing university rankings are far from homogenous. They differ in terms of their stated purpose, the criteria that are used for evaluation, the methodology, and the nature of the producer. Following in the footsteps of Raymond Hughes, some rankings continue to be produced by academic researchers such as those at the University of Melbourne and the Shanghai Jiao Tong University or by not-for-profit organizations such as the Centre for Higher Education Development (CHE) in Germany or the National Academy of Sciences in the USA. But in the vast majority of cases, league tables are produced by commercial publications.

Most of the league tables suggest they are designed to inform student choice and as such provide rankings of entire universities and/or of academic degree programs, including first-degree programs, professional programs such as MBAs and law degrees, and research doctoral programs. This function of university rankings relies upon economists' argument that the provision of appropriate information in a competitive market can serve as a means of assuring the efficient price and quality of goods and services. Research on organizational rankings similar to university league tables suggests they can be a useful instrument for public accountability, supplying information to consumers and policymakers on measurable differences in service quality, while also providing an incentive to organizations for quality improvement (Gormley and Weimer 1999). However, if (university) rankings are to help assure (academic) quality, several linked behaviors need to occur (Gormley and Weimer 1999). First, rankings and related consumer information on academic quality need to utilize measures that closely approximate or are clearly linked to valued societal outcomes. Second, league tables must inform and influence student choice of university or encourage universities to act in anticipation of the potential effects

[2] See, e.g., the journal *Higher Education in Europe* 2002 vol. 27 no. 4 and 2005 vol. 30 no. 2, Dill and Soo (2005), and Usher and Savino (2006) for comparative analyses.

of published rankings. Third, universities must respond to student choices and/or to the potential effects of rankings by genuinely improving the educational benefits provided to students. In a separate analysis (Dill and Soo 2005) we examined the leading university rankings in Australia, Canada, the UK, and the USA and concluded that most commercial university rankings failed to meet these conditions and therefore were unlikely to be useful instruments for assuring academic quality.[3]

However, the ranking in Germany carried out by the CHE is explicit in its purpose, carefully designed, and acknowledged by a number of researchers as minimizing the conceptual and technical problems that often make university rankings controversial (Dill and Soo 2005; van Dyke 2005; Usher and Savino 2006). The CHE ranking is an independent, non-governmental initiative, and its primary purpose is to provide university applicants with reliable data on various aspects of German universities. The CHE began ranking German universities and universities of applied science (*Fachhochschulen*) in 1998. By 2007 the exercise had gone through three cycles and was well established in the German higher education environment.

Background

As previously noted, university rankings have had the longest tradition and highest visibility in the USA as well as in the UK. The rankings have been associated with specific characteristics of the related higher education systems: competition between universities for students, staff, and resources; competition between students to be admitted to a specific university; and a clear prestige hierarchy among universities. These principles were alien to the traditional German higher education system.

While German universities differ in their mission and size, they were traditionally seen as homogenous in terms of quality. Differences existed in the performance of individual academics, reflected for example in the resource allocations by the German Research Foundation, but these individual differences had not led to a formally stratified system of universities. Even though the equality in the system was regarded as more myth than reality (Müller-Böling and Federkeil 2007) there was no publicly obvious hierarchy of universities. The idea of homogeneity is grounded also in the traditional power structure of the German university system. On the one hand, universities were heavily regulated by the federal and Land governments; on the other hand, individual professors in the university had a high degree of autonomy. The university as an entity had quite limited power to plan its future and develop a unique identity. This traditional structure of German higher education, however, has been rapidly changing since the last decade. Lump-sum payments by the government have replaced the traditional rigid line item budgets and more decision-making power is delegated to universities. Also the idea of homogeneity across universities is under pressure as various evaluations demonstrate clear

---

[3] An important exception was the *Good University Guide* in Australia, which will be discussed below.

performance differences among universities (see, for example, Chapter 12, "The German System of Accreditation"). Furthermore, the idea of equality has been explicitly altered by the current higher education policy. In 2004 the federal government proposed the idea that Germany needs a band of elite universities in order to promote top-quality research and the international reputation of German universities (Kehm 2006). The first set of universities was selected for a special treatment in 2006.

The competition between students for access to certain universities is also alien to the German higher education system. Until recently, universities had no role in student admission and universities' revenue is not significantly affected by student demand. The number of students admitted to each academic program is decided centrally in each Land and the decision is based primarily on the teaching capacity of each university. In most cases all applicants will be admitted. If the number of student places is insufficient (e.g., in medicine), a central authority *Zentralstelle für die Vergabe von Studienplätzen* (ZVS) distributes applicants between universities. The decision used to be based primarily on school grades and the applicant's place of residence. However, since 2005 universities can participate in the selection process and can freely choose their own selection criteria. The final decision is now dependent on three components: the average grade in high school, waiting time after finishing high school, and success as judged by the selection criteria of the university.

Tuition fees remain a heated political topic in Germany. As elsewhere in Europe, higher education has been free of charge, but the idea of tuition fees has gained much popularity in the last decade. To protect free education in Germany, the federal parliament passed an amendment to the Higher Education Framework Law in 2002, which would guarantee free higher education to all students. Opponents argued that the law violated the constitutional autonomy of *Lände*r in educational and cultural issues and appealed to the Constitutional Court. In 2005 the Constitutional Court ruled that *Länder* have a right to decide on tuition fees. The first *Länder* (Bavaria, Baden-Württemberg, North Rhine-Westphalia) introduced tuition fees in 2006/2007.

German students are also affected by the "Bologna process." In 1998 an amendment was added to the Higher Education Framework Law, which introduced bachelor's and master's degrees on a trial basis, and in 2002 these degrees were instituted as regular degree programs. Currently, former Diplom and Magister degrees function in parallel to the new degrees.

The German higher education system is going through a significant transformation, and the CHE university ranking is both an expression of the new thinking and a contributor to the changes.

Policy Problem

The issues of competition, evaluation, and accountability in higher education were first raised in Germany in the 1980s. The German Science Council (*Wissenschaftsrat*), an advisory body on higher education, science, and research policy that is co-funded by the federal government and the governments of the 16 *Länder*, explored the possible advantages of introducing greater rivalry in German higher education and concluded that publicly available information on academic quality and performance was a necessary precondition for effective competition in the sector (Wissenschaftsrat 1985). The Council recommended that the German Rectors' Conference – an association of German universities – should collect and publish comparative information on all universities. The Rectors' Conference expressed little enthusiasm for the idea of university ranking, but did organize a symposium on performance assessment and cross-university comparisons and initiated publication of descriptive information on universities (Frackmann 1990). In 1988 the Science Council also started to publish comparative information on study duration, which was in those days a much discussed issue in Germany. The average duration of studies in German universities was 7 years, which was clearly above the international average and also above the length of study set out in university and examination regulations (Berchem 1991). It therefore became an important performance indicator in Germany. The information published by the Science Council was explicitly intended to inform students about differences in study time across the sector and emphasized institutional responsibility for the issue (Frackmann 1990).

As the public interest in the quality differences in German universities increased, several private actors began to publish university rankings. In 1989 the weekly news magazine *Der Spiegel* published a ranking of German universities. This ranking was based on a survey of students about their learning experience and satisfaction with the academic environment. Following this first initiative, several other university-level or subject-specific rankings were constructed and published (Ott 1999).

In the beginning of the 1990s, the German Rector's Conference developed a partnership with the Bertelsmann Foundation. The Bertelsmann Foundation is a private organization that promotes the idea that competition and civic engagement are essential for social progress. The foundation was seeking an agenda in higher education and discovered that its interests were aligned with those of the Rector's Conference. In 1994 the German Rectors' Conference and the Bertelsmann Foundation jointly founded the Center for Higher Education (CHE). Among the primary tasks of the center was to design and implement a university ranking. In 1998, after a 2-year preparatory phase, the CHE started to rank German universities.

The Nature of the Ranking

Most well-known commercial university rankings have been subject to serious criticism in recent years because of their conceptual and methodological weaknesses (Bowden 2000; Dill and Soo 2005; Ehrenberg 2003; Usher and Savino 2006; van Dyke 2005; Yorke 1997). One of the most fundamental weaknesses of rankings is the ambiguity of their purpose. When the purpose of the ranking is ill-defined, then the ranking obscures what exactly is being ranked: is it university prestige, research quality, teaching quality, or something else? This ambiguity leads to a further problem. When the purpose of the ranking is ill-defined, the choice of indicators remains arbitrary. There is often also a problem of aggregation. Most commercial rankings incorporate several indicators and each indicator is weighted for the aggregate rank. The problem is not so much the methodology for determining appropriate weights for each indicator, but the idea that there is a unique set of weights which reflects users' preferences. It is unlikely that all users value different aspects of a university identically and an aggregate ranking therefore fails to reflect personal preferences (Ehrenberg 2002). University-level rankings have an additional aggregation problem. While students have consistently demonstrated that the academic program is one of their major decision criteria (James et al. 1999; Connor et al. 1999), institutional-level aggregation hides differences between programs in one university. Therefore, a ranking of institutions does not show the best university for studying a specific field or doing research in a specific field. In sum, rankings that aggregate indicators and/or subject fields assume that there is a unique hierarchy of universities that is accurate for all purposes, for all users, and for all subject fields. However, this assumption is not justifiable.

The CHE ranking is different from most commercial rankings because these conceptual problems are carefully considered. Most importantly, the ranking defines clearly its purpose, and this definition serves as a strong foundation for selecting indicators and developing methodology. The CHE ranking is designed for students. Consistent with the research on effective organizational report cards (Gormley and Weimer 1999), the ranking is expected to inform the decisions of new students and of students who are planning to change their study programs. The format of the ranking and the selection of indicators therefore reflect the aspects that students consider in choosing a university.

The CHE ranking is built on four core principles that represent the conceptual foundation of the ranking.[4] These principles distinguish the ranking from most other commercial rankings in the world and address the major conceptual problems of university rankings.

The CHE ranking is a subject-level assessment, not a university-level assessment. The quality, performance, and expectations may vary significantly across fields of study in one university, and an average university-level evaluation does not provide helpful information for students who come to study a specific field. The ranking currently includes 35 subject fields which covers about 80% of all new students.

The CHE ranking is multidimensional. Unlike many other popular rankings, the CHE ranking does not attempt to construct a unique hierarchy of universities in each field. It is recognized that different aspects cannot be effectively combined into one aggregate measure – e.g., research measures or teaching measures can reveal very different hierarchies. It is also recognized that students have different priorities and they should have an opportunity to rank universities based on their own criteria. A universal ranking that would satisfy the needs of all students is therefore an impossible task.

---

[4] The CHE website www.che.de and the document CHE (2006) are used as primary source for the description below.

The CHE ranking combines different perspectives on performance. The ranking presents objective facts on program characteristics, but it also presents the opinion of students and professors about their satisfaction of the program and on the performance in general.

Instead of a linear ranking of universities, the CHE distributes universities into three groups – top group, middle group, and bottom group – with respect to each indicator. As a general rule, 25% of universities are included in the top group, 50% in the middle group, and 25% in the bottom group. Within each group universities are listed alphabetically. Such an approach is more accurate than ranking all universities sequentially, as do most university league tables. It has been demonstrated that differences between universities are often marginal and assigning ranks based on statistically insignificant differences is therefore unwarranted (e.g., Clarke 2002).

All subjects are reevaluated every 3 years. By now most of the subjects have gone through three evaluation rounds: 1998–2001, 2001–2003, and 2004–2006. One evaluation cycle takes 3 years. About one-third of subjects are evaluated in 1 year, and when the cycle is completed the next cycle starts again. This means that each year approximately one-third of results are updated in the ranking.

Indicators

When CHE started to design a framework for the ranking, the first step was to identify the criteria that students consider when they choose a university. The CHE engaged evaluation experts, members of professional and university associations, students, and school leavers in this preparatory process. From these discussions emerged a "decision model" – a list of indicators that students consider in their choice. Therefore, not all indicators are relative performance measures. The database includes also descriptive information on study programs, on the university in general, and on the location of the university, all of which students consider in choosing a university. Comparable indicators have been discovered to be influential on student choice in other industrial countries (Dill and Soo 2005). The indicators are also constantly revised and new indicators have been added if found helpful.

The decision model consists of nine components. The components are very different in their nature. Some of the components are more descriptive than evaluative – e.g., description of the location and general university characteristics. Other components are more evaluative – e.g., research output or overall assessment of students. Each of the components contains several specific indicators, some of them based on factual data, some on subjective assessment by students and professors. Individual indicators may vary somewhat between study fields depending on the specifics of the field. The list of components and most common individual indicators is presented in Box 4.1.

Since all subjects have gone through several rounds of evaluation, it is now possible to trace the development of a program over a number of years. Therefore, the recent ranking identifies also universities where performance has significantly improved or worsened, considering both relative and absolute change in performance.

Data Collection

The CHE ranking presents both objective factual data on study programs and universities as well as subjective assessment by students and professors. Most of the information is obtained from surveys. Four sets of questionnaires are sent out for each subject evaluation:

- survey of the subject area in each university,
- survey of universities,
- survey of students in the subject area,
- survey of professors in the subject area.

In the summer, before the results are to be published, CHE sends out a questionnaire to relevant faculties and/or departments. Questions concern general information about students and graduates, exam results (if applicable), financial and human resources, and research outcomes. The results are first reviewed and checked by the CHE and then returned to the units for final verification. At the same time another questionnaire is sent to the central administration of the university. The questionnaire asks about tuition fees, library hours, division of students across disciplines, and other university aspects. Since 2006 the questionnaire also includes questions about sports in the university – the number and nature of sports programs and courses. The response rate to the questionnaires is near 100%, both at the faculty level and at the university level (CHE 2006).

Box 4.1 The List of Components and Main Indicators

Component 1. Student Body

Total number of students, proportion of female students, number of first-year students, trend in student numbers over last years

Component 2. Student Outcomes

Average grade in final exams (if applicable), study duration, graduation rate, expected graduation time, the proportion of students who graduate in the expected time frame, documentation at the end of studies (transcript, diploma supplement, etc.)

Component 3. International Orientation

Double degree option, participation in European Credit Transfer System, field-specific foreign language courses, course programs in foreign languages, courses in foreign languages, mandatory experience abroad (study abroad semester or internship), international students, international visiting professors, primary foreign countries with whom the program has connections

Component 4. Teaching and Learning

Student–staff ratio, opportunity for a joint degree, the structure of the program, availability of professors (office hours and informal counseling), E-learning (Internet-based learning opportunities and virtual interaction environment), breadth of the program, transparency, interdisciplinarity, interaction with professors and other students

Component 5. Infrastructure

Library resources, IT infrastructure, laboratories and other field-specific facilities

Component 6. Research

Research funding per academic staff, patents, publications, citations, Ph.D.s supervised, research reputation as assessed by professors

Component 7. Labor Market

Internship requirements and opportunities, student assessment of career preparation

Component 8. Study Location and University

Location: population, proportion of students, primary mode of transportation, living conditions, rent. Universities: the number of students, primary study fields, student counseling, library hours, university sports, the year of foundation

Component 9. Overall Assessment by Students and Professors

Students' overall satisfaction with the program, professors' identification of five universities that they would recommend for undergraduate studies in their field and five leading research universities in their field

To get information on students' satisfaction with their studies, the CHE conducts a student survey in the subject area. Questions concern the organization of studies, the relevance of their studies, learning environment, student support and counseling, facilities, and their overall satisfaction with the program. Students are also asked about their living conditions (e.g., rent and mode of transportation). The survey is conducted online, and the list of students and their email addresses is obtained from the student office of each university. As a general rule, the questionnaire is sent out to all students in their 5th–12th (10th in *Fachhochschulen*) semesters in diploma programs and in 2nd–7th semester in bachelor's programs. When a program enrolls less than 500 students in this specified window (e.g., 5th–12th semester) then the questionnaire is sent to all students enrolled in the program regardless of their year of studies. Survey results are published only if at least 15 students in a program responded. In the last evaluation round the average response rate was around 20%, but the rate varied significantly across disciplines. For example, 35% of students in biochemistry programs returned the questionnaire, while only 9.5% of history students returned the questionnaire (CHE 2006).

The faculty survey includes all permanent academic staff in the subject field. The academic staff survey includes questions about working conditions: facilities, infrastructure, human resources, and support from the central administration. The survey also asks two questions on the quality and reputation of other universities. The first question asks: Which five universities in Germany, Austria, and Switzerland would you recommend for an undergraduate program, considering only the quality of the program? The second question asks: In your opinion, which five universities in Germany, Austria, and Switzerland are considered as leading universities in research in your field? In the last evaluation round ca. 31,000 questionnaires were distributed and about half of the academic staff responded. The response rate was the lowest in medicine (26%) and in mass communication (31%) and the highest in pharmacy (66%) and biochemistry (65%) (CHE 2006).

Research productivity is evaluated not only based on the faculty survey but also with objective, quantifiable data. The bibliometric analysis includes the number of publications and the number of citations. As a general rule, the count includes all publications in internationally recognized academic journals in the 3-year period. This information is extracted from discipline-specific databases: e.g., Science Citation Index (SCI) Expanded in sciences; MathSciNet in mathematics; and SOLIS, HWWA, ECONIS, and BLISS in economics and social sciences. Publishing behavior varies across disciplines and bibliometric measures are altered accordingly. In the field of sociology, for example, publications in what is considered as the top journals in the field were double counted. Publications in the field of economics, social sciences, and American studies were weighed based on their length and the number of authors. In mathematics, for example, only the number of publications, not citations, is presented because on average the number of citations is low and varies greatly between areas in the discipline. Scientific databases cannot be equally trusted in all fields. Because of methodological limitations, bibliometric analysis has not been utilized for a number of fields: computer science, German studies, geography and geology, architecture, political science, industrial engineering, and a few others. In engineering and some natural sciences, research output also includes the number of patents, and this data is obtained from the German patent office (PATDPA).

Where students have to pass a nationally recognized exam, the ranking also collects information on the exam results. Students of medicine, for example, must pass a qualification exam and the office in charge of the exam in each *Land* provides data on the number of exams taken, success rate, and grades.

The accuracy of data is essential for trustworthiness of a ranking. Data mistakes can have a fatal effect on the rank of the university as well as undermine the legitimacy of the entire ranking process. There is evidence of serious mistakes in rankings due to erroneous data; e.g., a leading business school in the USA was entirely dropped from a ranking because it was mistakenly associated with another school with a similar name (Economist 2007). The CHE has developed sound quality assurance procedures in order to ensure reliability of the data. Statistical tools are used for detecting outliers and inconsistencies in the data. As a preliminary assurance, all collected data will be sent back to universities for proofreading before indicators will be computed. As an additional assurance, each subject field that is being evaluated has its advisory board. The board advises about the choice of indicators and methodology, but it also analyzes the plausibility of results.

Dysfunctional effects of university rankings on universities' behavior are one of the biggest concerns of the rankings. In order to improve their position in the ranking, universities are likely to manipulate their performance data or even change their procedures without actually improving learning in the institution. Evidence from US universities suggests that dysfunctional effects are not rare. Some universities, for example, made standardized test scores (SAT) an optional application requirement in order to demonstrate a higher average SAT score for ranking purposes (Ehrengberg 2002). Other universities have excluded the scores of international students from the average SAT statistics in order to better position their institution in the ranking. Rankings that rely on self-reported data are particularly vulnerable to this kind of data manipulation. The CHE consciously avoids indicators that stimulate such a dysfunctional response from universities. While the CHE ranking is heavily based on self-reported data, the problem of data manipulation is somewhat alleviated by the nature of the ranking. Since the ranking refers only to single disciplines, not the entire institution, the academic community has a better overview of the situation in other universities and thereby functions as a social control mechanism. The CHE has encountered a few instances where departments tried to manipulate the student survey by telling their students that it is in their interest to graduate from a university that is ranked high. The CHE has a strict policy for occasions like this. In case there is any reason to suspect that a university has tried to manipulate the results, the universities are excluded from the ranking altogether and their scores are not published.

Publication

While the CHE alone is responsible for data, methodology, and design of the ranking, the publication, marketing, and distribution of the results take place in cooperation with a major national publisher. From 1999–2004 the results were published by the weekly news magazine *Der Stern*, and since 2005 the publishing partner is the weekly newspaper *Die Zeit*. *Die* Zeit is responsible for distribution and marketing.

The results are published in three formats: a summary in the regular issue of *Die Zeit*, a book, and an interactive website. The online version offers the most detailed information, the book provides program-level data on a few indicators, and *Die Zeit* summarizes the results from a more generic perspective.

The book *ZEIT-Studienführer* is published annually. For each subject field, universities are listed alphabetically and the score of the four or five most important indicators is presented. The selection of published indicators varies from subject to subject, but in most cases, information includes reputation as perceived by professors, research output, students' opinion of the learning environment, student support, and resources (e.g., libraries in the humanities and social sciences and laboratories in the natural sciences). The book also provides information on universities and general information on study fields.

More detailed information on individual subjects can be accessed in the online version. The website www.das-ranking.de is free of charge and requires only registration for some inquiries. As a first step the user has to choose a subject of interest. Then the site provides several options on how to approach data.

- *Concise Ranking* (Ranking Kompakt) presents an alphabetical listing of universities and gives scores (green, yellow, and red) for five main indicators. These five indicators are the same as the indicators in the book.
- *My Ranking* allows the user to identify important criteria based on personal priorities and then constructs an individualized ranking. The user can specify up to five indicators and must rank the indicators for the sorting sequence. The user can also specify whether he or she wants to consider only universities that are in the top group for *each* of the five indicators, the top and the middle group, or all universities.
- *Comparing universities*. After specifying the subject, the user can also select up to three universities of interest and compare them by an extensive list of indicators.

In addition to the subject-specific performance information, the website also provides a general description of each subject field and descriptive information on individual universities and their location.

DAAD (German Academic Exchange Service) has made the ranking available also in English and Spanish for international students.

Implementation

The ranking exercise started with the creation of the Center for Higher Education by the Rector's Conference and Bertelsmann Foundation. The founders believed that the ranking exercise should be conducted by an independent, non-political, not-for-profit organization. While the university ranking became the core task of the new center, the CHE's mission and contribution has expanded over the years. The CHE has become a respected party in public discussions on higher education issues, and it has developed expertise also in the areas of research policy, science policy, and internationalization of higher education. Today the CHE has a staff of 15 people. Its activities are guided and supervised by an advisory board which includes representations of the two founding organizations and other higher education and management experts.

The first round of ranking exercises started in 1998. The first ranking exercise was done in cooperation with a national foundation for testing goods and services, Stiftung Warentest, which provided both methodological and technical assistance in conducting surveys.

As most data is collected by organizational and individual surveys, the cooperation by universities and individual students and professors is crucial for success. The idea of evaluating universities and constructing a relative hierarchy of universities was not necessarily a popular idea among university leaders and professors. While the current rankings system is independent of the government, as noted earlier, it likely would not have been implemented without the pressure for comparative data on universities provided by the German Science Council (*Wissenschaftsrat*), an advisory body for political decision-makers and an instrument of cooperative federalism designed to promote scientific work in Germany, which is co-funded by the federal government and the governments of the 16 *Länder*. Within this supportive environment the CHE was able to develop the trust that has made the ranking effective. First, the ranking exercise gained the moral support and ownership of the Rector's Conference, which ensured the general compliance of university leaders. Second, the new head of the CHE was a former Rector of a German university and his academic background and managerial experience increased the trust for the CHE as a qualified partner for universities. Third, as other rankings were being published in Germany by private, for-profit actors, universities were open for a more systematic exercise even if not enthusiastic about the idea of ranking per se.

Participation in the ranking exercise is voluntary for universities and almost all universities choose to participate. The ranking results are not a part of the formal evaluation of universities. However, some *Länder* use the ranking results as a starting point for contracts between the Land and the university that outline needed areas of improvement.

Costs

The CHE annual budget is about € 3.2 million. A significant part of the budget is covered with an institutional grant by the Bertelsmann Stiftung. The rest of the funding is obtained on the project basis from various partners. The university ranking exercise takes circa €0.8 million per year. These funds pay for the data collection and project coordination in the CHE. The ranking is put together by a core group of five people with the help of a few student assistants. The data entry is contracted out to an outside partner.

Costs of marketing and publication of the ranking results are the responsibility of the newspaper *Die Zeit*. The paper copy of the student guide can be ordered online for €6. The online version is provided free of charge and the costs are covered with funds from advertisement and corporate sponsorship.

Impact

The primary purpose of the CHE ranking is to guide students in their decision as to which university to choose for their further studies. One success criteria is therefore the extent to which students indeed use the ranking in their decision-making. Evidence collected by the CHE indicates that approximately one-third of all entering students consult the rankings for general orientation among universities (Federkeil 2002). The number varies by study fields – e.g., 50% of engineering students consult the ranking, while only 19% of literature students use the ranking. The study also demonstrated that achievement-oriented students in particular use the ranking in their decision. The reported proportion and types of students who use ranking information for university choice making in Germany therefore correspond with research on the use of rankings in the USA and the UK (Dill and Soo 2005).

A survey of the first-year university students in 2005, ordered by the Federal Ministry of Education and Research, confirms the relevance of the university ranking among applicants. The survey shows that two-thirds of entering students consulted university rankings, and the use of rankings has increased 6% points in the past 2 years (Heine et al. 2007). On the other hand, only 13% of students recommend the rankings as the best information source for future students. The importance of ranking varies between fields: for law students the fame of the university and ranking results are the two most important decision criteria, while for arts students ranking results have only a marginal relevance. The report (Heine et al. 2007) also points out that the number of students that consider several alternative universities has increased, and the authors associate this trend with the availability of comparative information in the form of rankings.

A survey of high school students confirms the evidence of the relative importance of rankings in choosing a university (Heine and Willich 2006). Male students and students from more highly educated homes were more likely to use rankings. As noted, the fact that students from more highly educated homes are more likely to use the rankings seems to be universal (e.g., McDonough et al. 1998 in the USA).

There is also some evidence that the ranking results affect student demand. In the field of psychology the number of applications at the universities that scored well increased significantly. The number of applications rose on average 19% in universities that were recommended as excellent in research, and the number of applications increased 15% in universities that were recommended as efficient and supportive in teaching (Federkeil 2002).

University rankings can potentially be not only an information source for students, but also a quality assessment instrument. Federkeil (2003) points out that the quality assurance mechanisms are relatively weakly developed in Germany and since higher education is managed by individual states, there is no nation-wide government quality assurance instrument. The rankings compiled by private institutions are the only nation-wide initiatives in quality. Since 1998 Germany has been implementing a comprehensive assessment of teaching and learning in the format of program accreditation (see Chapter 12, "The German System of Accreditation"). However, the accreditation information suffers from the fragmentation of the accrediting process, and consequently its results are not easily comparable across the entire higher education system. Moreover, the accreditation procedure is relatively expensive for universities. In contrast, ranking, if rigorously designed, can serve as a tool for quality assessment.

The ranking can develop into the role of an assessment instrument in two ways – through either the market mechanism or central regulation. If performance as demonstrated in rankings proves to be an important influence on student demand and the attraction of other essential resources for universities, then the ranking will possibly encourage quality improvement in the system. Alternatively, ranking results may be monitored and utilized by government as a reliable means of performance assessment and improvement.

A university ranking has a larger impact on the performance of the higher education system only if universities consider their position in the ranking and are motivated to improve their performance. There is some indirect evidence in Germany that the universities use the rankings information to analyze their strengths and weaknesses. On the request by a university, the CHE provides a more detailed summary of student survey results to interested participating departments and universities increasingly use this opportunity. In 2007 the CHE created a for-profit agency, CHE Consult, that provides assistance to individual universities on means of improving their academic performance, designing university marketing strategies, and optimizing internal governance. As a fundamental principle, the university ranking and consulting activities within CHE are kept strictly separate and distinct, but the interest in the consultancy proves that universities are not ignorant to the way they look in public eyes.

The Future and Other Similar Initiatives

In the context of a common European higher education area and student mobility among nations, there is a greater need for internationally comparable data on European universities. The CHE ranking has been considered a potential foundation for such international endeavor and in recent years has been extended to other German-speaking countries. Since 2005 Austrian and Swiss universities are fully integrated into the general ranking procedure. In these countries the ranking is done in cooperation with local partners – the quality control agency in Austria *Qualitätssicherungsagentur* (AQA) and a private foundation swissUp in Switzerland.

In 2006 the European Commission funded a pilot project to explore whether the CHE ranking could be used as a basis for an international ranking. As an experiment the CHE ranking methodology was applied to universities in the Netherlands and Flanders and an international ranking of universities of five countries was produced. The pilot study pointed to a few serious challenges facing any international ranking. One of the strongest conclusions of the pilot was that outcomes of satisfaction surveys – a major indicator in the CHE ranking – are not internationally comparable (Westerheijden et al. 2008). For example, Dutch universities received consistently lower scores than German universities, even though some programs in the Netherlands attract German students because of their better quality. This indicates that the satisfaction scores can be interpreted only in a context as the scores are influenced by students' expectations and the perception of a "normal" grading scale. The problems have currently put a stop on further integration of Dutch and Flemish universities in the CHE ranking, but work on making the CHE ranking international continues. In 2009 the European Commission took an initiative to develop an international ranking that is more valid and accurate than commercial international rankings (RAPID 2008). The characteristics of the new ranking are similar to that of the CHE ranking (multidimensional, consumer friendly, etc.) and the CHE is one of the partners in developing the ranking.

The new *University Report Card Navigator* developed by the Canadian Educational Policy Institute was explicitly modeled on the CHE ranking.[5] Similar to the CHE ranking, the Navigator was designed and is maintained by a not-for-profit research institute, but is made available by a commercial publisher, *The Globe* and *Mail* newspapers. Like the CHE ranking the Navigator's focus is on student choice and provides multiple indicators of interest to students including relevant information on the student experience and measures of student engagement in higher education. The Navigator does not weight nor aggregate its indicators, but permits each student to choose the indicators and ranking appropriate to her or him. In the presentation of information, ordinal rankings of universities are similarly avoided and the Navigator relies primarily on non-university sources of data to circumvent the potential for institutional manipulation of ranking data.

---

[5] See: http://www.globecampus.ca/

As noted earlier, however, the CHE approach is not the dominant model of university rankings in the world. For example the Shanghai Jiao Tong University and *The Times HE Supplement* league tables both rank universities, not subjects, and provide an ordinal ranking of all universities. Both of these rankings are also heavily biased toward research reputation. The highest weight is assigned to the quality of staff, publication citations, and university resources. The available research evidence, however, suggests that research quality and university inputs do not significantly affect the quality of teaching in first-level degree programs (e.g., Terenzini and Pascarella 1994).

National rankings share the emphasis on research and reputation. From the measures utilized in university league tables we would infer that prominent research institutions give the best education, although it is more accurate to conclude that the listed performance indicators do a much better job in assessing the research quality of a university than its teaching quality (Yorke 1998). The league table rankings are heavily biased toward measures known to be associated with research performance including financial resources, numbers of faculty and research grants, as well as university reputation. Even the average faculty salary, which according to USNWR measures a school's commitment to instruction, more likely reflects faculty orientation to research and has been found to be negatively correlated with student learning in research studies in the USA (Astin 1996). An analysis of five rankings in Australia, Canada, the UK, and the USA demonstrated that the rankings are also heavily biased toward input measures – staff quality and student selectivity (Dill and Soo 2005). The importance of teaching output measures was significantly lower. The most commonly used output measures are student–staff ratio, graduation and retention rate, and some employability measure. However, unless output measures are controlled for the quality of the incoming students – which is rarely the case – even these limited teaching outcome measures will fail to capture the quality of the education process.

Furthermore, reputation biases university league tables not only via an emphasis on indicators of research, but also directly through reputation surveys. Many rankings include survey results about universities' reputation that have been collected from faculty members, administrators, or employers whose knowledge of the actual academic quality of a university is obviously limited.[6] Therefore, their opinions are likely to be influenced more by the existing reputation of the university (i.e., the "halo effect") than by actual knowledge of program quality (Clarke 2002). While the CHE ranking includes several reputational indicators among its components (see Box 4.1), unlike university league tables these indicators are presented separately and are not amalgamated into a weighted overall score or used to provide ordinal rankings of institutions or programs. Recent research (Berghoff and Federkeil 2006) on the reputational indicators used in the CHE ranking underscores the cited weaknesses of reputational measures in university league tables. Reputation was

---

[6] For example, reputational surveys provide 16% of the total score in the *Macleans* ranking, 25% in the USNWR ranking, and 50% in the THES World ranking (Berghoff and Federkeil 2006).

revealed to be a social judgment by particular groups or stakeholders, which was highly stable over time. Reputational judgments of the same universities or programs varied significantly among professors, students, and employers, indicating that their judgments were based on different perceptions and interests. Professors' reputational judgments, predictably, were correlated with research performance, but even among professors, reputational scores for different academic programs varied significantly within a single university, thereby confirming the meaninglessness of reputational scores for whole universities. Finally, the reputations of universities were found to vary systematically among survey respondents from Austria, Germany, and Switzerland, indicating that the regional distribution of a survey sample influences institutional reputation scores, a bias that will likely affect global rankings of universities as well.

More seriously, reputation-based rankings not only fail to indicate the quality of education in the university, but also encourage a socially costly "academic arms race" (Brewer et al. 2002). Research in the UK and the USA (Brewer et al. 2002; Rolfe 2003) suggests that to be competitive, universities are increasingly investing more financial resources and time in factors associated with reputation such as research doctoral programs, research facilities, and "star" researchers and investing less resources in activities known to be associated with improving student learning. Furthermore, because university ranking is a zero-sum game, investing in reputation is an endless process. Consequently, the dysfunctional effects of poorly designed university rankings are becoming a serious problem for society.

The CHE ranking has thus many advantages over most national and international commercial rankings. It has a well-developed conceptual foundation, follows a rigorous methodology, and provides rich data. Because of its design it also seems to avoid some of the potentially dysfunctional effects of rankings. There are other examples of rankings that aim to provide reliable data, rather than a simplified reputational hierarchy.

The Australian *Good University Guide* (GUG) (Dill and Soo 2005) has a number of similarities to the CHE ranking. The GUG focuses on subjects and academic programs, not whole institutions. The GUG also emphasizes indicators relevant to the preferences expressed by student consumers and offers a website in which prospective students can craft rankings individualized to meet their particular needs. Also similar to the CHE ranking, the GUG ranks academic programs in divisions or bands according to a variety of criteria, with no overall ranking for all institutions. The GUG measures reputation objectively using three indicators. While the GUG is a commercial publication, the source of almost all its data is government records, including the well-regarded and government-mandated Graduate Destination Survey and Course Experience Questionnaire, which yield generally reliable and educationally insightful information unavailable from commercial league tables in other countries (see Chapter 6, "The Course Experience Questionnaire, Graduate Destination Survey, and Learning and Teaching Performance Fund in Australia").

The most comparable rankings in the USA to those of the CHE are those conducted by the National Research Council (NRC), which has continued the practice of ranking research doctoral programs first initiated by Raymond Hughes (Dill

2009). While the NRC rankings are subsidized by US federal agencies including the National Institutes of Health and the National Science Foundation, similar to the CHE, the NRC is a private, nonprofit institution that also provides science, technology, and health policy advice to the federal government under a congressional charter. Because of its national stature, the NRC rankings are designed and carried out by some of the leading social scientists in the USA. The NRC rankings include objective data on measures that research has indicated are important determinants of academic quality in research doctoral programs. These include inputs, such as the number of faculty members and doctoral students in each program, and crucial process measures, such as student time to degree. The measures also include objective output measures such as the number of doctoral graduates each year and the number of faculty publications, as well as significant outcomes, such as the number of times faculty publications were cited and the number of distinguished awards received by the faculty.

The form of the NRC's rankings also has a number of similarities to the CHE rankings. Institutional ratings data are provided in the form of ranges rather than rankings to diminish the incentive for institutions to take actions designed purely to "move up in the rankings." Future NRC rankings will include student assessments of their educational experience, their personal research productivity, and their institutional and program environment in order to encourage a greater focus by programs on education in addition to research. Finally, the NRC ranking, similar to the CHE ranking, presents all its data in an unweighted form. Thus users of the assessment can apply their own preferences to the data and make their own comparative judgments, which is impossible with weighted measures.

The NRC rankings have also traditionally included reputational peer judgments of research doctoral programs. Following its last ranking exercise, however, the NRC commissioned a study by leading social scientists of the methodology used in that assessment (Ostriker and Kuh 2003). The NRC reputational measure had included two questions, one on the scholarly quality of the program faculty and a second on the effectiveness of the doctoral program in training scholars. While the reputational survey had been limited to members of the discipline being rated, nonetheless, the committee concluded – consistent with the research reported above – that the strong correlation between the two reputational measures in past NRC assessments "suggests that raters have little knowledge of educational programs independent from faculty lists" (Ostriker and Kuh 2003, p. 36). Therefore, while the reputational measure will be continued, it will be limited to scholarly reputation of the program faculty alone. Furthermore, the NRC committee determined that because more highly ranked programs were most visible, some measure of the rater's familiarity with the program should also be included.

Conclusion

Organizational rankings have been used in many other sectors in order to promote accountability (Gormley and Weimer 1999). Rankings have the potential to be a valuable instrument for academic quality assurance, because they can provide useful information to consumers and policymakers about quality differences in academic programs and can also provide an incentive for universities to improve their academic standards. This, however, requires that rankings are compiled rigorously and carefully and published in an appropriate and useful form. The Berlin Principles on Ranking is one attempt by an international expert group to outline the criteria for a good quality university ranking (IREG 2006). These principles specify the criteria for defining the purpose, selecting indicators, managing data collection, and presenting results.

After criticizing the legitimacy of the reputation rankings, it should be noted that the intention of all rankings is not necessarily information for consumers. Rankings clearly also have an "infotainment value." That is, they provide some information but are also a source of entertainment for those within and outside of higher education (Bowden 2000). Because of the entertainment value, rankings are a commercially valuable product. Overly simplified, institutional-level rankings, as opposed to more sophisticated evaluations of different aspects of universities, have a higher "infotainment value" and therefore greater popularity and visibility in society. For this reason there is little likelihood that commercial publishers will invest the time and/or the money necessary to create valid, reliable, and academically useful university league tables. If university rankings are to help assure rather than distort academic quality, they will need to be guided by appropriate public policies. While the CHE rankings are published commercially, their validity and reliability is supported by the nonprofit status of the organization, the private source of its funding, as well as by the indirect support of the German Federal and *Länder* governments exercised through the oversight of the German Science Council. The NRC in the USA is also a highly respected, nonprofit scientific organization, whose university rankings are subsidized by the federal government. The well-designed GUG in Australia is highly dependent on relevant data designed and subsidized by the national government as well as by the government's encouragement of university engagement in the process. In each of these cases, the positive contribution of the university rankings to academic quality assurance is influenced by government policy.

There is certainly a market for different kinds of university rankings. However, a ranking that has the ambition to be "the third leg of the quality-assurance stool, along with accreditation and government regulation and licensing" (Bollag 2006) must fulfill the criteria of conceptual and statistical rigor. The CHE ranking is in this respect a valuable model for policymakers.

Acknowledgement We would like to thank Gero Federkeil for providing us with valuable information on the CHE ranking system and for his helpful comments on an earlier draft of the chapter.

References

Astin, A. W. (1996). Involvement in learning revisited. *Journal of College Student Development*, 40(5), 587–597.

Berghoff, S., Federkeil, G. (2006). *Reputation indicators in university rankings*. Paper presented at the Consortium of Higher Education Researchers (CHER) 19th Annual Research Conference, Kassel, Germany.

Bollag, B. (2006). *International group endorses principles for ranking of higher-education institutions. Chronicle of Higher Education*, June 1. http://chronicle.com/daily/2006/06/2006060103n.htm. Accessed 12 June 2009.

Berchem, T. (1991). The internationalization of higher education: The German perspective. *Higher Education, 21*(2), 297–304.

Bowden, R. (2000). Fantasy higher education: University and college league tables. *Quality in Higher Education, 6*(1), 41–60.

CHE. (2006). *CHE HochschulRanking 2006: Vorgehensweise und Indikatoren*. http://www.che.de/downloads/Methoden_2006.pdf. Accessed 12 June 2009.

Clarke, M. (2002). Some guidelines for academic quality rankings. *Higher Education in Europe, 27*(4), 443–459.

Connor, H., Burton, R., Pearson, R., Pollard, E., Regan, J. (1999). *Making the right choice: How students choose universities and colleges*. London: Universities UK.

Dill, D. D. (2009). Convergence and diversity: The role and influence of university rankings. In B. M. Kehm, B. Stensaker (Eds.), *University rankings, diversity, and the new landscape of higher education*. Rotterdam, The Netherlands: Sense Publishers.

Dill, D. D., Soo, M. (2005). Academic quality, league tables, and public policy: A cross-national analysis of university ranking systems. *Higher Education, 49*(4), 495–533.

The Economist (2007). News from the schools. March 16, 2007. http://www.economist.com/business/globalexecutive/education/displaystory.cfm?story_id=8814515. Accessed 30 March 2007.

Ehrenberg, R. G. (2002). *Tuition rising: Why college costs so much*. Cambridge, MA. Harvard University Press.

Ehrenberg, R. G. (2003). *Method or madness? Inside the USNWR college rankings*. Paper presented at the Wisconsin Center for the Advancement of Postsecondary Education Forum on The Abuse of College Rankings, Madison, Wisconsin, 20–21 November.

Federkeil, G. (2002). *The CHE-/Stern-ranking*. Presentation at the International Roundtable on Statistical Indicators for the Quality Assessment of Higher/Tertiary Education Institutions, Warsaw, Poland, 13–15 June 2002.

Federkeil, G. (2003). The lack of a national policy regime of quality assurance in Germany: Implications and alternatives. *Higher Education Management and Policy, 15*(2), 63–71.

Frackmann, W. (1990). Resistance to change or no need for change? The survival of German higher education in the 1990s. *European Journal of Education, 25*(2), 187–202.

Gormley, W. T. Jr., Weimer, D. L. (1999). *Organizational report cards*. Cambridge, Mass: Harvard University Press.

Heine, C., Kerst, C., Sommer, D. (2007). *Studienanfänger im Wintersemester 2005/06 Wege zum Studium, Studien- und Hochschulwahl, Situation bei Studienbeginn*. Forum Hochschule 1/2007. HIS Hochschul-Informations-System GmbH.

Heine, C., Willich, J. (2006). *Informationsverhalten und Entscheidungsfindung bei der Studien- und Ausbildungswahl Studienberechtigte 2005 ein halbes Jahr vor dem Erwerb der Hochschulreife*. Forum Hochschule 3/2006. HIS Hochschul-Informations-System GmbH.

IREG (2006). *Berlin principles on ranking of higher education institutions*. http://www.ihep.org/assets/files/publications/a-f/BerlinPrinciplesRanking.pdf. Accessed 12 June 2009.

James, R., Baldwin, G., McInnis, C. (1999). *Which university? The factor influencing the choices of prospective undergraduates*. Canberra, ACT: Australian Government Publishing Service.

Kehm, B. M. (2006). The German "Initiative for Excellence" and the issue of ranking. *International Higher Education, 44*(Summer). http://www.bc.edu/bc_org/avp/soe/cihe/newsletter/Number44/p20_Kehm.htm. Accessed 12 June 2009.

McDonough, P. M., Antonio, A. L., Perez, L. X. (1998). College rankings: Democratized college knowledge for whom? *Research in Higher Education, 39*(5), 513–537.

Merisotis, J. (2006). *Introductory remarks. CHEA International Commission Meeting, 25–26 January 2006*. http://www.chea.org/international/commission2006/JMerisotis_IntroRemarks_IC012606.pdf. Accessed March 30, 2007.

Müller-Böling, M., Federkeil, G. (2007). The CHE-ranking of German, Swiss, and Austrian universities. In J. Sadlak, L. N. Cai (Eds.), *The world-class university and ranking: Aiming beyond status* (pp. 189–203). Bucharest, Romania: Cluj University Press.

Ostriker, J. P., Kuh, C. (2003). *Assessing research-doctorate programs: A methodological study*. Washington, DC: National Academies Press.

Ott, R. (1999). Darstellung und Beurteilung von Hochschul-Rankings in Deutchland. *Beiträge zur Hochschulforschung*, *4*, 309–322.

RAPID (2008). *Ranking Europe's universities*. IP/08/1942 (December 11, 2008). http://europa.eu/rapid/pressReleasesAction.do?reference=IP/08/1942&format=HTML&aged=0&language=EN&guiLanguage=en. Accessed 12 June 2009.

Rolfe, H. (2003). University strategy in an age of uncertainty: The effect of higher education funding on old and new universities. *Higher Education Quarterly*, *57*(1), 24–47.

Smith, J., McKnight, A., Naylor, R. (2000). Graduate employability: Policy and performance in higher education in the UK. *The Economic Journal*, *110*(464), 382–411.

Terenzini, P. T., Pascarella, E. T. (1994). Living with myths: Undergraduate education in America. *Change*, *26*(1), 28–32.

Usher, A., Savino, M. (2006). *A world of difference: A global survey of university league tables*. Canadian Education Report Series. The Educational Policy Institute. http://www.educationalpolicy.org/pdf/World-of-Difference-200602162.pdf. Accessed 30 March, 2007.

Westerheijden, D. F., Federkeil, G., Cremonini, L., Kaiser, F., Soo, M. (2008). *Piloting the CHE ranking of study programmes in Flanders and the Netherlands*. Enschede, the Netherlands: CHEPS.

Wissenschaftsrat (1985). *Empfehlungen zum Wettbewerb im Deutschen Hochschulsystem*. Cologne, Germany: Wissenschaftsrat.

Van Dyke, N. (2005). Twenty years of university report cards. *Higher Education in Europe*, *30*(2), 103–125.

Webster, D. S. (1992). Reputational rankings of colleges, universities, and individual disciplines and fields of study, from their beginnings to the present. In J. C. Smart (Ed.), *Higher Education: Handbook of theory and research, Vol. VIII* (pp. 234–304). New York: Agathon Press.

Yorke, M. (1997). A good league table guide? *Quality Assurance in Education*, *5*(2), 61–72.

Chapter 5
The US National Survey of Student Engagement (NSSE)

Peter T. Ewell

The National Survey of Student Engagement (NSSE) was originally conceived outside government and is operated by an independent third-party agency.[1] Its origins, however, are rooted in policy because its original intent was to provide a means for students, citizens, elected officials, and taxpayers to better assess the quality of higher education institutions. A secondary objective was to create a tool for the oversight bodies of publicly funded institutions to monitor the quality of higher education provision at the institutions for which they are responsible. While the NSSE has to date not completely fulfilled this first rather lofty original expectation, its design and implementation hold many lessons for other non-governmental approaches to quality assurance. At the same time, growing state-level use of NSSE in publicly supported colleges and universities provides useful lessons about how to create and manage systems based on more general public reporting of institutional performance.

Higher Education Context

The policy environment for higher education in the United States within which NSSE evolved is complex, decentralized, and in many ways unique. There is no national ministry governing public higher education. The role of the federal government in quality assurance is indirect, operating through approved accrediting organizations that certify institutional worthiness to receive federal funds (see below). Governance and support of public higher education is instead a responsibility of 50 individual states, which differ markedly in how they approach the task. Some are organized as systems with centralized policies with respect to administration, finance, and curriculum. Others comprise individual institutions that are connected to one another only in that they receive public funds. Complicating

P.T. Ewell (✉)
National Center for Higher Education Management Systems (NCHEMS), Boulder, CO, USA

[1] The NSSE website is http://nsse.iub.edu/index.cfm.

D.D. Dill, M. Beerkens (eds.), *Public Policy for Academic Quality*, Higher Education Dynamics 30, DOI 10.1007/978-90-481-3754-1_5,

the picture, about half of all degree-granting colleges and universities are private, nonprofit entities governed by independent lay boards and funded largely by income from tuition and fees. Although these institutions tend to be small – enrolling only about 20 percent of the total undergraduate student population in the country – their numbers (and, to a certain extent, their prestige) make them a formidable presence in the nation's higher education system. Despite their independence, moreover, private institutions receive a good deal of indirect public support through federal student scholarship and loan programs and, in many cases, through parallel state financial aid programs.

Partly as a result of these conditions, the US higher education system is strongly conditioned by classic market forces. Private institutions charge what the market will bear, and the 200 or so of the most selective and prestigious "national institutions" can command tuition "sticker prices" of over $40,000 per year. Private institutions farther down the prestige ladder still have tuition charges of over $15,000 per year. For these institutions, moreover, maintaining tuition revenue is a matter of life or death, as typically over 90 percent of their costs must be covered through tuition charges. For public institutions, market forces are buffered somewhat by public subsidy, but tuition still accounts for about a third of institutional revenues. Recently, due to substantial shortfalls in state revenues, public institutions in most states have engaged in record-breaking increases in tuition. All of this means that the behaviors of colleges and universities in the United States are shaped substantially by the market. Attracting a sufficient number of students to pay the bills is fundamental for most. And for those whose endowments and reputations allow them a bit more flexibility, attracting an ever-more-selective student body remains a priority.

Under these conditions, the factors that influence student choices about where to attend affect institutions far more than government regulation or steering. To be sure, public institutions in the United States are subject to various forms of regulatory control in realms such as finance and procedural accountability. More recently, most of the states have established systems of performance reporting, based largely on efficiency measures, and in a substantial minority, institutional performance on such indicators is consequential (Burke 2002, Burke and Minnassians 2003). As noted, a purportedly voluntary institutional accreditation system, loosely regulated by the federal government, requires all institutions – both public and private – to undergo a periodic comprehensive review that examines resources, organizational structures, instructional processes, and (most recently) student learning outcomes. But all of these public and quasi-public government regulatory and steering mechanisms operate on the margins of an enterprise that is shaped heavily by the marketplace of student choice. The factors influencing this marketplace, therefore, are fundamental to higher education policy in the United States.

Policy Problem

A major force shaping institutional behavior in this environment is media rankings of institutions. The *U.S. News and World Report* annual survey of "America's Best Colleges" was the first such venture in the world, with its inaugural edition launched in 1983. Since that time, additional media rankings have emerged in the United States, including a "value for money" review by *Money Magazine* and a burgeoning industry in college guides. And for better or worse, much of the world seems to be following this "league table" phenomenon with examples ranging from *McLean's Magazine* in Canada, through *The Times* in the United Kingdom, to *Der Spiegel* in Germany. Research in the United States suggests that such publications exert very little leverage over actual student choices, although they can sometimes noticeably affect admissions markets in the short term for institutions in the most selective tier (McDonough et al. 1997). More important are their indirect – and often substantial – effects on institutional behaviors, which have been repeatedly documented (Machung 1998). The institutions whose admissions pools might actually be affected by the changes in fashion reported by the national media quite naturally attempt to improve their rankings. And the vast majority, although their admissions markets are local, follow these leaders in a continuing attempt to move up the ladder of prestige.

All of this might be considered beneficial if the metrics of "quality" underlying media rankings faithfully represented institutional capacity and performance. Market forces, as in any other field, would automatically induce institutions toward ever-increasing "quality" (at least as perceived by the customer), thus serving public purposes. Indeed, this popular policy logic has been used increasingly to steer institutional behavior in the United States since at least 1989 when Congress first required colleges and universities to disclose graduation rates to prospective students. And it was a particular theme of the most recent report on reforming higher education by a national commission appointed by the US Secretary of Education (USDOE 2006). But the problem with media rankings in the eyes of most critics is that they are based on a badly flawed metric of "quality" driven essentially by institutional resources and reputation. The *U.S. News* measures, for instance, began as a reputational ranking done by college presidents and only gradually added such measures as dollars spent per student, admissions selectivity, and alumni loyalty as measured by financial contributions. While these factors produce a familiar list of "winners" drawn from the nation's best-known colleges every year, they say nothing about the question really being asked: what do institutions *do* to enhance student learning and how *well* do they do it?

In framing the policy problem that NSSE was originally intended to address, moreover, it is important to emphasize that moving the metrics of quality away from resources and reputation toward student outcomes was part of a larger undergraduate reform movement and a consequent change in the way governments approached accountability for higher education in the United States. Part of the impetus for this arose from the academy itself, stimulated by reformers worried about growing lack of coherence in the undergraduate curriculum (AACU 1985). But part of it also came from state governments, reflecting a new view of higher education as a "public good" connected directly with such statewide benefits as economic development and functional citizenship (NGA 1986; Ewell 1997, 2002). By the late 1980s, many states had enacted requirements for institutions to assess student learning and report publicly on the results, and by 1989, the federal government mandated institutional accrediting bodies to adopt such requirements as well. The following year, state and federal actors came together to proclaim a set of "National Education Goals" to guide educational policy for the coming decade. Although mostly about elementary/secondary education, these goals included an explicit commitment to "increase substantially" the ability of the nation's college graduates to "think critically, communicate effectively, and solve problems." The implied promise to develop the metrics needed to track progress on these elusive qualities was one of the many roots of NSSE because it stimulated thinking about how to examine them *indirectly* by looking at what institutions did to promote them (Ewell and Jones 1993). Not only would such an approach be less intrusive and expensive than launching a massive national testing program, but it could also be built on a solid tradition of research about effective student learning environments in the United States using the proven technology of survey research.

With this as a backdrop, the Pew Charitable Trusts – a charitable foundation with considerable visibility and influence in the United States – launched a multifaceted program to stimulate quality improvement in undergraduate education in the mid-1990s. The bulk of this effort comprised grants to individual colleges and universities intended to support promising innovations in teaching and learning. But some of it was designed to influence institutional behavior indirectly by reshaping the structure of regulations and incentives within which colleges and universities must operate. And a prominent negative element in this environment, at least for those at the Pew Trusts, were media rankings that rewarded institutions for the wrong things and reinforced the public's image that institutional "quality" was simply a matter of money and selectivity. To attack this perceived problem, Pew convened a meeting of higher education leaders concerned about the rankings in the spring of 1998. One conclusion was that the Trusts should underwrite a new survey of college student perceptions and behaviors, based on the kinds of indirect indicators of "good practice" suggested earlier as an approach to assessing the National Education Goals.

Design and Development

The NSSE is a national survey that focuses on specific undergraduate student experiences and features of the educational environment (Kuh 2001, 2003). The concept of "engagement" that constitutes its core reflects the results of at least two decades of research in the United States, identifying specific factors of both experiences and environment that are associated with high learning gain (e.g., Astin 1978, 1993; Pace 1979; Pascarella and Terenzini 1991, 2005). These factors are embodied in the five "benchmarks" scales around which NSSE results are reported:

- Level of academic challenge, consisting of items on the amount of time students spend on academic work and the kind of assignments and exercises expected of them.
- Active and collaborative learning, consisting of items on student participation in group work, and active participation in learning activities in and out of class.

- Student–faculty interaction, consisting of items on various kinds of contact between faculty and students in and out of class.
- Enriching educational experiences, consisting of items on particular curricular and experiential features of the educational environment including service learning, study abroad, or senior capstone projects and other independent work.
- Supportive campus environment, consisting of items on the availability and use of various academic support services as well as the general atmosphere of support for student achievement generated by faculty, staff, and other students.

Items on the survey were specifically selected for inclusion only if there was a clear empirical case in the literature on college student learning and development that the factor represented could be associated with learning gain. Indeed, given the initiative's origins, documenting the relationship between the "engagement" concept and actual learning has been crucial to its implementation. This has been accomplished in several ways. First, those responsible for NSSE stressed this relationship from the outset through documents describing the instrument's conceptual and empirical foundations.[2] Second, the survey was extensively validated through two major field tests (see below) which involved student focus group work to both refine item content and collect external evidence of links between particular item responses and actual student experiences (Ouimet et al. 2004). Finally, NSSE has engaged in ongoing attempts to directly validate the link between survey items and direct measures of student learning through the cross-administration of NSSE with a number of cognitive assessment measures (Kuh et al. 2006; Carini and Kuh 2004; NCHEMS 2003). This unusual level of conceptual and empirical documentation was seen by NSSE's founders as important in gaining public credibility for an indirect approach to examining academic quality and is frequently cited as a factor in its success.

NSSE was developed entirely through non-governmental means. The Pew Charitable Trusts, which initiated the effort, is a private foundation with a strong interest in education and education policy issues. The National Center for Higher Education Management Systems (NCHEMS), an independent nonprofit research center, was contracted by the Trusts to design and pilot the survey. The NSSE itself is a self-supporting entity housed in the Center for Postsecondary Research at Indiana University – a public research university.

NCHEMS began the task of designing the survey by convening a team of recognized experts on college student survey research and higher education quality. With an initial design in place, a successful pilot study involving 12 institutions was undertaken in the spring of 1999 to test the instrument itself. This was followed in the fall by a 60-institution field study to test survey administration procedures at different kinds of institutions. Both pilots were administered under subcontract to

[2] The NSSE is built on the foundation of past and current research on college student development and student learning. Information on the conceptual framework, psychometric properties, and other NSSE research-related issues is available at: http://nsse.iub.edu/html/researchers.cfm.

Indiana University's Survey Research Center, which was then chosen to house and administer the survey under a competitive RFP-based selection process. NSSE was launched on a national basis in the spring of 2000 supported by a 3-year grant from the Pew Trusts with the understanding that the survey would be self-supporting via user fees by the end of this period – a goal which has since been accomplished. A "sister" survey targeted at 2-year institutions, the Community College Survey of Student Engagement (CCSSE), was launched in 2003, also underwritten by the Pew Charitable Trusts, and is housed at the University of Texas Austin.

How the Survey Works

The NSSE is administered to samples of students at the end of their 1 year of study and just before they are expected to receive a baccalaureate degree. Sample sizes are based on the size of the institution and range from 450 to 1000 students (or up to 3000 in Web-based administration). A substantial advantage is the fact that the survey is administered to students at all institutions directly by Indiana University's Center for Survey Research using state-of-the art survey research techniques which require little work by participating campuses. This approach not only helps maximize response rates, but also helps ensure that data are comparable across campuses because administration procedures are standardized. Participating institutions are asked to send an electronic list of all students qualified to be chosen as part of the designated sample. NSSE staff then select a random group of students to be surveyed from this list and administer the survey directly. The survey is available both in Web-administered form and as a paper questionnaire sent through the mail, with the proportion between these two modes of completion shifting markedly toward the former: in recent administrations, almost 80 percent of all respondents completed the survey online. Response rates for both versions have averaged around 37 percent, and while there is individual variation in response rates across institutions, this national average response rate has been maintained within two or three percentage points for 9 years (and was also obtained by the pilots).

Volume of Participation

Institutional participation is voluntary and the enterprise is at this point entirely supported by user fees. Nevertheless, numbers have steadily increased over 5 years, with the latest administration involving 752 institutions and over 350,000 students. The total number of institutions that have participated in NSSE since its launch in the spring of 2000 is about 1200, with well over a million students responding. Looking at volume another way, institutions that have participated in NSSE now represent more than 71 percent of the total number of students enrolled at 4-year institutions in the United States.

Costs

The original development of NSSE, including the design of the survey and two field tests, was underwritten by grants from the Pew Charitable Trusts totaling about \$650,000. Pew further supported the first 3 years of implementation of the survey with a grant to Indiana University totaling \$3.3 million. Currently, NSSE is self-supporting with an operating budget exceeding \$2 million. Based on 4 years of operation, survey costs approximate \$6–7 per student surveyed. Direct costs to institutions in the form of user fees vary by size and range from \$1500 to \$7500. Other institutional costs are almost exclusively incurred in the form of staff time to compile the lists of qualified students and their associated contact information to create the sample pool that is sent to Indiana University.

Reporting

NSSE results are reported nationally in the form of an annual report issued by Indiana University, summarized in terms of the five "benchmark" indicators, which represent reliable aggregates of individual survey questions.[3] Although institutions typically choose to participate individually, to this point some 12 state systems of public colleges and universities have administered NSSE on a statewide basis with data reported publicly. NSSE does not release scores or benchmark results for individual institutions without their consent, but CCSSE results for individual institutions are posted publicly on the survey's website.[4] Individual institutions receive comprehensive reports on the responses of their own students, as well as electronic files to support further local analyses of the survey data. These institutional reports contain two features not typically found in standard survey reports. Benchmark scores are reported comparatively in the form of deciles, so an institution can see immediately where it stands comparatively with respect to other institutions of its type. At the same time, institutions are provided with assessments of the statistical significance of these differences and effect sizes, where appropriate, which is unusual in higher education survey reporting. Finally, for an additional modest cost, institutions can receive specially constructed peer comparison reports that show their results against other groups of institutions that they select. Feedback from institutions suggest that the information supplied by NSSE reports is far superior in form and content to commercial or other national surveys in which they have participated.

---

[3] See http://nsse.iub.edu/html/annual_reports.cfm.

[4] See http://www.ccsse.org/.

Implementation Issues

In order to fulfill the Pew Trusts' original notion of mounting a direct challenge to media rankings, it was originally envisioned that comparative NSSE results for institutions would be reported publicly – and perhaps even be incorporated in some way into the *U.S. News* ranking methodology. NSSE's sponsors expected the higher education community to welcome this proposal – both because institutional leaders visibly and vocally opposed existing media rankings and because the content of the survey addressed precisely the elements of academic quality that faculty felt such media reports lacked. But NSSE's leaders were appropriately cautious about immediately proposing such a position and engaged in a number of deliberate efforts to tap community opinion on this matter. One line of inquiry consisted of open meetings held at major national meetings representing various institutional constituencies. Another was the full-scale field test at 60 institutions held in the fall of 1999, which was consciously structured to include participation by various types of institutions – small independent colleges, large public research universities, and regional public universities – chosen in consultation with the national associations that represent them.

As these conversations unfolded, institutional opposition to the planned public release of survey results became increasingly vocal. This was especially the case among the high-prestige independent colleges that were currently at the top of the media rankings. Such institutions, though they were intrigued by NSSE's content and continued to maintain a public posture of vehement criticism of actors like *U.S. News*, were not ready to risk losing their dominant position as "America's Best" by participating in a process whose results were uncertain. Public institutions, already prominent in the public eye by this point due to numerous public-reporting mandates imposed by states, were less uncomfortable with the notion of public reporting. But early opposition by an influential sector of institutions quickly revealed the policy dilemma embedded in the enterprise from the outset. To leverage quality through the market, NSSE needed everybody to participate. But why would institutions pay for a voluntary data-gathering effort that might yield adverse public information about their own performances and undermine their markets?

After some consideration, the architects of NSSE decided to implement the survey on a confidential basis. The resulting data, according to a carefully worded institutional participation agreement, would be the joint property of the institution and NSSE, and the agreement stated explicitly that "results specific to each institution and identified as such will not be made public except by mutual agreement." While this wording protected institutional confidentiality where it was desired, it also did not preclude the release of comparative results for any public institutions that participated in NSSE as a system, in which case the "client" was a state governing board instead of an individual institution. As a result, some 12 state systems of higher education have administered NSSE at all of their campuses, with results reported publicly by institution. Significant examples here have included Kentucky, North Carolina, and South Dakota. Kentucky's use of the survey is typical, in that a report on survey results was used to generate cross-campus discussions about how to improve quality, and campus-by-campus results were posted on the state agency's website. Kentucky now administers NSSE at all public institutions on a biennial basis. Other states, for example, Virginia and Missouri, have encouraged institutions to adopt NSSE items as part of their response to state-mandated performance measures where institutions are allowed to choose the array of statistics to report. In these cases, NSSE has given participating states a cost-effective policy tool to monitor the quality of teaching and learning environments that they would otherwise have had to invent and run themselves.

Given the prominent position of highly selective independent institutions in the United States, though, the confidentiality provision was essential to obtain widespread and broadly representative institutional participation in NSSE early in the implementation process. Were this decision not taken, the initiative might have failed to achieve self-sufficiency altogether or be confined almost exclusively to public institutions. But this necessary decision forced a change of NSSE's main strategy from direct to indirect with respect to inducing quality improvement. Instead of creating a highly visible and authoritative competitor to the media rankings that might directly influence institutional behavior, NSSE would instead provide institutions with a tool that could be used to guide their own internal improvement efforts and would continue to supply states with a readymade quality measure should they choose to participate. At the same time, NSSE would work consistently through the media to change perceptions of institutional quality among opinion leaders and the public at large. Opportunities to pursue this latter "quasi-public" agenda were exploited from the outset and kept NSSE a *policy* initiative instead of just another consortium-based institutional data-gathering effort.

A first such opportunity arose in the summer of 2000 at the point when NSSE was to issue its first national report. Since the effort was new and had been much heralded in the higher education press, this report was the first significant chance to promulgate the "engagement" view of quality in a public forum. NSSE leaders knew that the report's punch (and its consequent press appeal) would be greatly enhanced if specific colleges and universities that exemplified particular characteristics were explicitly named. Naming institutions, all recognized, would have to be confined to top performers – both because such institutions would be unlikely to withhold permission to be named and because illustrating "best performance" was an effective way to communicate NSSE's more basic message about the nature of institutional quality. But naming institutions publicly, even with their permission, might upset the delicate perceptual balance embodied in the institutional participation agreement. By implication, *not* naming other participating institutions automatically implied they were of lesser quality. And losing institutional trust at this early stage – even if the letter of agreement with respect to the particular institutions being named was honored – would be a significant setback.

The decision was made to go forward with this strategy, and four institutions were named in the national report as top performers on all five benchmarks. All four were selective private liberal arts colleges – though they were not necessarily the top performers in the *U.S. News* ranking. Although this action produced some grumbling about "broken promises," it did not noticeably affect NSSE's institutional participation rate in subsequent years (although it is not known whether or not it influenced the decisions of particular institutions – or even institutions of particular types – to participate). This episode also illustrated the double-edged nature of participants' self-imposed ban on disclosing individual results. Some top performers, after discovering that they were such, quickly proclaimed this result in their recruitment materials and websites – illustrating precisely but unintentionally the ability to leverage institutional behaviors through student choice that NSSE's sponsors originally hoped for. And by implication, those not named either might have been spurred toward further efforts to improve engagement or, equally plausibly, might have decided not to risk another encounter. The relative numbers making each decision, of course, are unknown. But after 5 years of operation, two things are clear. First, with only a few exceptions, institutions placing in the top 50 in the *U.S. News* rankings have not participated in NSSE. And second, NSSE has up to now refrained from naming "top performers" in its national report, choosing instead to illustrate findings (with permission) drawn from the entire range of institutional types.

A second revealing incident in NSSE's early implementation resulted from the survey's systematic and highly proactive media strategy. Working with a nationally known media consultant, NSSE quickly became positioned as a premier press source about both college quality and other higher education matters. But part of the way this was accomplished was by continually questioning the *U.S. News* rankings or, more accurately, allowing others to take on *U.S. News* themselves. The most prominent of these early attacks occurred in 2000 when the *Washington Monthly* published a lengthy criticism of both the magazine and its methods, co-authored by a former *U.S. News* insider. In it, NSSE was prominently featured as an example of the kind of data source that *U.S. News* could use if it was serious and sincere. Subsequently picked up by *The New York Times*, this attack quickly got the attention of the editors of *U.S. News*. *U.S. News*, of course, cannot use NSSE data in its rankings even if it wanted to because not all institutions participate in the survey – a fact that its editors have repeatedly (and appropriately) claimed in their defense. But they needed some kind of response to what had become a considerable body of negative press about their methods.

Their answer was to try to publish NSSE data in some form in the upcoming college rankings issue for the fall of 2002. When approached directly by *U.S. News* to provide the needed data, NSSE staff honored the institutional participation agreement and chose not to do so. NSSE staff then took the further step of declining to help *U.S. News* choose a subset of survey items that the magazine could ask participating institutions to supply directly, on the grounds that any substantive contact with the magazine might be seen by participants as collusion. In a letter to all participants, NSSE made its position clear: NSSE had nothing to do with the *U.S. News* request (though the magazine had a perfect right to make it); it was up to each institution to decide how to respond. More than a hundred institutions chose to supply *U.S. News* with requested data about selected questions on the survey (about a quarter of then-current NSSE participants). This request was repeated in 2004 and subsequently, each year to date, with a gradually rising level of response.

Again, participation rates in NSSE seemed unaffected as a record number of more than 750 institutions chose to participate in the survey in its most recent year. But this episode illustrates the delicacy of using a voluntary reporting process – regardless of government or private sponsorship – to serve public accountability purposes.

A later and more serious effort to use NSSE results as public consumer information emerged in the fall of 2006. In summer 2005 the US Secretary of Education, Margaret Spellings, empanelled a "Commission on the Future of Higher Education" to critically examine higher education in the United States and make recommendations on how to improve it. From the outset, the commission proved interested in comparative measures of institutional performance and how these could be used to hold institutions accountable and inform potential students and stakeholders where institutions stood. Partly because of its media prominence, the NSSE was one of several measures that the commission wanted to know more about, and NSSE was one of only three instruments noted by name as exemplary in its final report (USDOE 2006). In the aftermath of the commission, the Secretary promoted NSSE in two ways. First, she made a proposal that scores on NSSE and similar comparative performance statistics be made public as part of the federal government's regular statistical reporting system, participating in which is mandatory for an institution to receive federal funds. This might be done through a common reporting template or through a link to the institution's website. Second, she pushed hard for accreditation agencies to require institutions to use instruments like NSSE as part of the accreditation process. Both of these initiatives ended without accomplishing the Secretary's goals. But the threat of government action of this kind stimulated institutions to act themselves to make more data about performance publicly available. The most prominent of these initiatives – the Voluntary System of Accountability (VSA) – is being undertaken by the two major associations of 4-year public universities in the United States and includes the results of student surveys, most prominently, NSSE.

Impact

From the standpoint of providing information useful to institutions to guide internal improvement, NSSE has clearly had a major positive impact. The nine national reports issued to date list scores of examples of how individual institutions have harnessed NSSE data for local planning and quality-enhancement purposes. Further examples quoted in these reports strongly demonstrate the utility of NSSE information in informing "soft" accountability processes that rely on institution-chosen measures of effectiveness (Kuh 2003). For example, some institutions have made indicators constructed from the NSSE survey the centerpiece of their response to state requests for information about performance. Many more have featured NSSE data in their institutional accreditation reviews (with no small number reporting such data on publicly accessible websites). In fact, this latter use of the NSSE data has become so prominent that NSSE constructed an "accreditation toolkit" to guide institutions in preparing and displaying data explicitly for this purpose.[5] More explicitly, though admittedly far less frequently, NSSE data have been displayed in a format that allows direct comparisons of performance as originally envisioned. And, as noted earlier, some 12 states have to date arranged for all public institutions to administer the survey as part of their accountability programs, with states like North Carolina and Kentucky featuring comparative institutional results on their websites. And as noted earlier, the Voluntary System of Accountability (VSA) developed in reaction to the Secretary of Education's Commission on the Future of Higher Education is making a steadily increasing volume of NSSE data available to the public in comparative form.

With respect to the NSSE's sponsors' original intentions, however, neither the survey nor its associated view of quality has directly displaced media rankings as public markers of quality. And the lack of participation by highly selective "national" institutions means that its impact in shaping the marketplace of college choice has so far been limited. But NSSE's "semi-public" strategy of aggressive media relations – together with its growing use as a state reporting tool as well as rising levels of participation in VSA – continues to attract considerable press and political attention. Seeing a potential opportunity to compete with *U.S. News* in the lucrative college choice market, for example, both *Newsweek* and *Time* magazines regularly publish fall issues featuring stories based on NSSE. In the fall of 2003, the prestigious *Atlantic Monthly* magazine parodied the rankings mania and highlighted NSSE as posing the kinds of questions that parents and prospective students ought to be asking about the colleges they are considering. Finally, *USA Today*, an influential national daily newspaper, successfully negotiated with NSSE for a "preferred" relationship with respect to reporting periodic statistics from the survey and has run two major stories as a result. Meanwhile, NSSE continues its campaign to affect public perceptions of quality through its own initiatives. One of the most interesting of these is the production of a pocket-sized brochure on student engagement entitled *College: What You Need to Know Before You Go* that outlines the questions that prospective students should ask when they visit a college. Although impacts of this campaign are not documented, several hundred thousand of these brochures have to date been distributed to high school guidance counselors and others charged with helping students select colleges to attend.

Finally, NSSE has clearly had an impact on campuses themselves in stimulating conversations about academic quality improvement among faculties. Indeed, a faculty version of the instrument, the Faculty Survey of Student Engagement (FSSE), was developed precisely to help stimulate such conversations. Teaching staff complete this survey, which consists of parallel items on the NSSE, before they know what their students' responses look like in order to provide a point of contextual comparison. Many institutions have used FSSE and NSSE together as

[5] See http://nsse.iub.edu/institute/index.cfm?view=tools/accred_index.

part of their required accreditation reviews, and this appears to be a process that could be employed effectively in other quality review settings based on academic audit or similar mechanisms.

Comparison with Related Policies

The closest counterpart to NSSE as a policy instrument is the Australian Student Course Experience Questionnaire (SCEQ), which was first administered as a census to all enrolled undergraduate students in 1999 and which has subsequently been administered annually to stratified random samples of undergraduate students. Questions on this survey address various aspects of the student experience including multiple items on the quality of teaching and learning. The survey is administered centrally by the Teaching Evaluation Enhancement Service and results are disseminated publicly as part of an institutional quality indicators system (see Chapter 6, this volume). The SCEQ was preceded by the Course Experience Questionnaire (CEQ), which asks similar questions of university graduates and has been administered since 1993. Both surveys are based on the Quality in Courses survey originally developed in the UK by Paul Ramsden. The primary difference between the Australian approach and NSSE is the fact that both Australian surveys are administered by a public agency, participation is mandatory, and results are presented publicly. As a growing number of US states participate in NSSE and as institutional participation in reporting initiatives like the VSA continues to grow, it will be interesting to see if the leverage on institutional behavior they obtain parallels the Australian experience.

Conclusion

Although NSSE has fallen short of the direct impacts on student choice and institutional behavior that the Pew Trusts originally hoped for, the survey's indirect impact on public perceptions of quality and on institutional actions to improve practice appears to be substantial and growing. Continued media presence and widespread voluntary acceptance by institutions are indeed gradually "changing the public conversation about quality." But the continuing problems that surround public release of NSSE results illustrate the difficulty of using voluntary processes to decisively steer institutions in desired directions if government stands on the sidelines. This difficulty is faced by institutional accreditation mechanisms in the United States as well. In a different public policy environment, a well-conceived and constructed survey like NSSE that addresses the quality of instructional provision directly might have a notably different impact, as is beginning to be demonstrated by growing state-level participation in NSSE and institutional participation in the VSA.

Yet even when governments decide to use such information to leverage institutional behavior, policymakers face a major dilemma about how to proceed. If, as typically the case for survey-generated information, they display it as a performance indicator and expect institutions to strive toward improvement simply to improve their public posture, impacts will likely be limited because institutional consequences are equally limited. But choosing to make performance on such measures consequential through approaches like performance funding may be equally inappropriate because of the very nature of survey-based evidence. Unlike direct measures of performance such as enrollments or completion rates, survey-derived statistics are subject to multiple uncertainties because of factors like sampling error and response bias. As a result, they usually lack sufficient precision to appropriately ground formal performance-funding schemes or other high-stakes decisions (Ewell 1999). As a result, government's best bet may be to either advance such statistics more aggressively as aids to student choice – thus harnessing the market – or to use them proactively as benchmarks to trigger deeper questions about teaching and learning for institutions whose performance remains consistently low. Both lines of action, of course, demand universal institutional participation and active government advocacy.

References

Association of American Colleges (1985). *Integrity in the college curriculum: A report to the academic community*. Washington, DC: Association of American Colleges (AACU).

Astin, A. W. (1978). *Four critical years: Effects of college on beliefs, attitudes, and knowledge*. San Francisco, CA: Jossey-Bass.

Astin, A. W. (1993). *What matters in college: Four critical years revisited*. San Francisco, CA: Jossey-Bass.

Burke, J. C., Associates (2002). *Funding public colleges and universities for performance: Popularity, problems, and prospects*. Albany, NY: Rockefeller Institute of Government, State University of New York.

Burke, J. C., Minnassians, H. (2003). *Performance reporting: "Real" accountability or accountability "lite"*? Albany, NY: Rockefeller Institute of Government, State University of New York.

Carini, R., & Kuh, G. D. (2004). *2002 NSSE-RAND cross validation study: Some insights into the role of student engagement to student learning*. Bloomington, IN: Indiana University, Center for Postsecondary Research.

Ewell, P. T. (1997). Accountability and assessment in a second decade: New looks or same old story. In *Assessing impact: Evidence and action* (7–21). Washington, DC: American Association for Higher Education (AAHE).

Ewell, P. (1999). Linking performance measures to resource allocation: Exploring unmapped terrain. *Quality in Higher Education*, 5(3), 191–209.

Ewell, P. T. (2002). An emerging scholarship: A brief history of assessment. In T. W. Banta, & Associates, *Building a Scholarship of Assessment* (3–25). San Francisco, CA: Jossey-Bass.

Ewell, P. T., Jones, D. P. (1993). Actions matter: The case for indirect measures in assessing higher education's progress on the National Education Goals. *Journal of General Education*, 42(2), 123–148.

Kuh, G. D. (2001). Assessing what really matters to student learning: Inside the National Survey of Student Engagement. *Change Magazine*, 33(3), 10–17.

Kuh, G. D. (2003). What we're learning about student engagement from NSSE. *Change Magazine*, 35 (2), 24–32.

Kuh, G. D., Kinzie, J., Cruce, T., Shoup, R., Gonyea, R. M. (2006). *Connecting the dots: Multi-faceted analyses of the relationships between student engagement results from the NSSE and the institutional practices and conditions that foster student success*. Final Report to the Lumina Foundation. Bloomington, IN: Indiana University.

Machung, A. (1998). The rankings game. *Change Magazine*, 30 (4), 12–16.

McDonough, P. M., Antonio, A. L., Walpole, M. B., Perez, L. (1997). *College rankings: Who uses them and with what impact?* Los Angeles, CA: Graduate School of Education and Information Studies, The University of California at Los Angeles (UCLA).

National Center for Higher Education Management Systems (2003). *An analysis of relationships between NSSE and selected student outcomes measures for seniors attending public institutions in South Dakota*. Boulder, CO: National Center for Higher Education Management Systems (NCHEMS).

National Governors Association (1986). *Time for results: The governors' 1991 report on education*. Washington, DC: National Governors Association (NGA).

Ouimet, J., Bunnage, B., Carini, R. M., Kuh, G. D., Kennedy, J. (2004). Using focus groups to establish the validity and reliability of a college student survey. *Research in Higher Education*, 45, 233–50.

Pace, C. R. (1979). *Measuring outcomes of college: Fifty years of findings and recommendations*. San Francisco, CA: Jossey-Bass.

Pascarella, E. T., Terenzini, P. T. (2005). *How college affects students: A third decade of research*. San Francisco, CA: Jossey-Bass.

Pascarella, E. T., Terenzini, P. T. (1991). *How college affects students: Findings and insights from twenty years of research*. San Francisco, CA: Jossey-Bass.

United States Department of Education (2006). *A test of leadership: Charting the future of US. higher education* Report of the Commission Appointed by Secretary of Education Margaret Spellings. Washington, DC: United States Department of Education (USDOE).

Chapter 6
The Course Experience Questionnaire, Graduate Destination Survey, and Learning and Teaching Performance Fund in Australia

Kerri-Lee Harris and Richard James

The Australian higher education system has developed a distinctive national approach to the evaluation of higher education that involves the use of the *Course Experience Questionnaire* (CEQ) and *Graduate Destination Survey* (GDS) in a national survey of all university graduates in the year following graduation. To appreciate the significance of this survey and the application of comparative data for performance incentive funding for Australian universities through the Learning and Teaching Performance Fund, it is useful to begin with a description of the Australian higher education context.

The Context: An Overview of the Australian Higher Education System

Higher education in Australia has developed as a mainly public university system, with 37 public universities. There are a small number of private universities and other providers of higher education, but these enrol only a small fraction of the nation's students. The public universities are somewhat differentiated in mission, size, and activities, but most are large, comprehensive universities offering programs across most of the major fields of study. All of the universities are research institutions, though there are considerable differences in the intensity of the research activity.

Most Australian universities are established under State (or Territory) legislation; however, the federal government is responsible for funding and regulating higher education, exercised through the Department of Education, Employment and Workplace Relations (DEEWR). There has been a steady decline in the proportion of university revenue provided by government, with universities on average now receiving less than half of their annual revenue from public funding. Within the national regulatory and reporting framework, universities have considerable

K.-L. Harris (✉)
Centre for the Study of Higher Education, University of Melbourne, Parkville, VIC, Australia

D.D. Dill, M. Beerkens (eds.), *Public Policy for Academic Quality*, Higher Education Dynamics 30, DOI 10.1007/978-90-481-3754-1_6,

autonomy over their affairs, including over expenditure and programs, and are self-accrediting and largely responsible for their own academic standards. The government has retained the capacity to indirectly influence university activities through strategic funding and other mechanisms, despite a recent trend toward some deregulation of university activities. The federal government plays a key role in quality assurance and quality improvement through the establishment of independent bodies such as the Australian Universities Quality Agency and the Australian Learning and Teaching Council.

The number of students in the Australian university system expanded rapidly during the 1980s and early and mid-1990s. International students, predominantly from South East Asia and studying in the fields of business, computing, and engineering, have provided an important source of revenue in the face of the decline in per capita public funding and have helped create cosmopolitan campuses. The rate of growth of fee-paying international student enrolments is probably the most significant trend in Australian higher education of the past decade and universities have become heavily reliant on the fee revenue from these students.

The Policy Issues: Quality Assurance, Continuous Improvement, and Performance Indicators for Higher Education

Many of the contemporary issues for the Australian higher education system are familiar ones in developed nations. Federal policy for higher education has gradually changed direction: from the 1980s goals of participation expansion and equity to the 1990s objectives of market diversity and student-led choice. Since the late 1980s the government has expected universities to play a stronger role in national economic development. Federal policy has encouraged universities to be entrepreneurial and to seek alternative sources of revenue. A domestic national market in higher education has been strongly encouraged, on the assumption that students, if they have adequate information and act as rational consumers, will help stimulate diversity, efficiency, and quality in the system (James et al. 1999).

Australian higher education has developed an international reputation for quality and innovation. This has been made possible by a number of factors, including the relatively small size of the system, the centralization of policy development, and the imperative of remaining competitive in a rapidly internationalizing and globalizing higher education environment.

During 2008 a major national review of higher education was undertaken by the federal government, dubbed the 'Bradley review' after Professor Denise Bradley, the chair of the review panel. The final report titled *Review of Australian Higher Education* (RAHE) (Bradley et al. 2008) was released in December 2008 and its recommendations call for, among other things, renewed attention to widening participation and equity, a thoroughly restructured approach to institutional accreditation and quality assurance, and a new emphasis on the measurement of outcomes. Overall, the RAHE reports the need to renew key policy settings to create a higher education system responsive to a more volatile international environment.

Significantly, the report includes key recommendations for enhancement of the measurement of teaching and learning performance and suggests more emphasis be given to outcomes-based indicators.

Australia already has a national system of data collection in key performance areas in which all public universities participate by agreement. The initial introduction of a national measurement framework coincided with the relaxation of direct control mechanisms over universities and the government priority of stimulating an informed market in undergraduate education. In addition, because of the small size of the Australian higher education system, its growing reliance on international student revenue, and concern about the possible effects of internationalization and globalization on the student market, the original interest in performance measurement was underpinned by a desire to ensure standards and to protect and demonstrate the quality of Australian universities.

In the late 1980s, the federal government department engaged in a major project to develop quantitative indicators of the quality and diversity of the higher education system. This work occurred over a lengthy period and included the research carried out in 1991 by the Performance Indicators Research Group (Linke 1991) to trial a broad range of quantitative indicators suitable for evaluating performance in higher education. Some of the indicators were to be based on the existing data collections mechanisms at the time (such as the GDS), while others required the development of new data sources (such as the CEQ or new statistical data collected from students at the time of enrolment).

As a result of this work, Australia has developed a long-standing system of national data collection and reporting. Universities are required annually to provide the government with statistics on various areas of institutional operations and performance. Some of this information, such as student demographics, is collected routinely by student questionnaire on enrolment, in accordance with federal data definitional requirements. Other information, such as completion rates, requires extensive internal institutional data collection and analysis.

The government publishes this information in a comparative format,[1] listing well over 200 indicators. Indicators are reported annually for:

- Students (e.g., numbers overall, and by field of study, enrolment type, fee type, age, gender, basis of admission, and equity group membership)
- Staff (e.g., numbers overall, and by field, age, gender, and qualification level)
- Finances (e.g., operating revenue by source, operating expenses by category, assets by category)
- Student outcomes (e.g., retention rates, progression rates, graduate employment, and graduate satisfaction with courses and teaching)

[1] The Australian government department concerned with higher education publishes these statistics annually. An archive of reports, commencing 1992, is available at http://www.dest.gov.au/archive/highered/statistics/characteristics/contents.htm while more recent reports (from 2001) are available at http://www.dest.gov.au/sectors/higher_education/publications_resources/statistics/publications_higher_education_statistics_collections.htm. Accessed 15 January 2009.

The crude data for student outcomes are statistically adjusted on the basis of the student demographics of each institution and other factors known to influence outcomes in order to give an indication of 'value-added' effects. Both crude and adjusted data are reported. For the past 5 years, selected data from this data set have been used to rank universities for the purposes of incentive funding. This chapter critically examines this policy and its effects, after an explanation of the collection of the data on which it relies via the Australian Graduate Survey.

The Australian Graduate Survey: Graduate Destination Survey and Course Experience Questionnaire

The GDS and CEQ are national, coordinated surveys of recent university graduates. Distributed together, they are collectively known as the *Australian Graduate Survey*. Both domestic and international students are included in the surveys, with the exception of students studying at offshore campuses. Conducted annually and administered by individual institutions, data are collated and analyzed by independent agencies (Graduate Careers Australia, GCA; Australian Council for Educational Research, ACER). The results of both surveys are reported in a variety of aggregations and levels of detail by universities, GCA, ACER, and the government.

Graduate Destination Survey

The GDS has been conducted since 1971 by GCA (formerly Graduate Careers Council of Australia). This survey collects information on the employment (including industry, occupation, and salary level) and further study (including level and field) patterns of recent graduates. Administered by individual institutions, the GDS was traditionally a means by which institutions initiated a relationship with alumni. National employment outcomes of graduates in various fields of study were, and still are, made publicly available for the information of prospective and current students and careers advisors (see Graduate Careers nd). The published GDS data reveal a detailed picture of the employment and further study outcomes by institution and field of study, including trends in the labour market.

The GDS is a relatively straightforward instrument with questions sufficiently broad to accommodate the diversity of Australian higher education institutions and graduates from all types of programs. Graduates are asked for the following information:

- highest qualification, including field and mode of study;
- demographic and equity group information including age, sex, and language group;
- whether or not they held paid employment during their final year of study;
- their work status at the time of the survey, including whether full- or part-time, business type, duties, and annual salary;

- whether or not they were studying at the time of the survey and, if so, details of the level, field, and institution; and
- their strategies for seeking employment.

Course Experience Questionnaire

Since 1992 the CEQ, a survey questionnaire developed by Paul Ramsden (Ramsden 1991; Wilson et al. 1997), has been included in the graduate survey. The CEQ focuses on graduates' overall perceptions, looking back, of their course or program of study. The CEQ is a survey of all graduates, with the exception of higher degree research students. A variation of the CEQ, the Postgraduate Research Experience Questionnaire (PREQ), is used to survey PhD and masters by research graduates. The focus of this chapter is on the CEQ, as this involves the largest data set and is arguably the most influential of the two surveys, having been co-opted for the allocation of performance incentive funding since 2006.

Questionnaire Items

The questionnaire items comprising the CEQ are statements to which graduates respond on a five-point scale of 'strongly disagree' to 'strongly agree' (Fig. 6.1). Provision is made for respondents to distinguish between two fields of study, or 'majors', within their course. Graduates who have completed a course with a single

		MAJOR 1					MAJOR 2				
		Strongly disagree				Strongly agree	Strongly disagree				Strongly agree
13	Overall, my university experience was worthwhile	1	2	3	4	5	1	2	3	4	5
14	The course sharpened my analytical skills	1	2	3	4	5	1	2	3	4	5
15	My lecturers were extremely good at explaining things	1	2	3	4	5	1	2	3	4	5
16	The teaching staff worked hard to make their subjects interesting	1	2	3	4	5	1	2	3	4	5
18	I felt part of a group of students and staff committed to learning	1	2	3	4	5	1	2	3	4	5
20	Students' ideas and suggestions were used during the course	1	2	3	4	5	1	2	3	4	5
22	I learned to explore ideas confidently with other people	1	2	3	4	5	1	2	3	4	5
23	The course developed my problem-solving skills	1	2	3	4	5	1	2	3	4	5
27	The staff made a real effort to understand difficulties I might be having with my work	1	2	3	4	5	1	2	3	4	5
28	I usually had a clear idea of where I was going and what was expected of me in this course	1	2	3	4	5	1	2	3	4	5
31	I felt like I belonged to the university community	1	2	3	4	5	1	2	3	4	5

Fig. 6.1 Reproduction of an excerpt from the CEQ section of the Australian Graduate Survey questionnaire prepared by the University of Melbourne for 2005

major respond to each item only once, whereas those whose course was either a combined degree (e.g., bachelor of arts/law) or a double major (e.g., bachelor of science, with majors in both computer science and mathematics) answer separately for each field of study.

The CEQ currently comprises 10 groups of between 3 and 6 items, or scales, and a single item, the 'Overall Satisfaction Index' (Appendix). The CEQ scales as originally developed by Ramsden were:

- Good Teaching Scale (GTS) (basic elements of effective teaching, e.g., clear explanations, provision of feedback on progress, interest shown in student progress)
- Clear Goals Scale (CGS) (students know what is expected of them and what will be rewarded)
- Appropriate Assessment Scale (AAS) (assessment that rewards 'deep' approaches to learning rather than rote or surface reproduction)
- Appropriate Workload Scale (AWS) (manageable study load)
- Overall Satisfaction Item (OSI) (a single item, 'Overall, I was satisfied with the quality of this course')

An additional scale, the Generic Skills Scale (GSS) (e.g., written and oral communication, teamwork skills, capacity for critical thinking), was added for the purposes of the national survey. The Generic Skills Scale differs conceptually from the other CEQ scales insofar as it seeks student self-reports of their skill development.

In 2002, the government funded a project to expand the CEQ to measure broader dimensions of the student experience (McInnis et al. 2001; Griffin et al. 2003) in response to concern about the coverage of the instrument. These dimensions include the social environment for learning, the degree of challenge and stimulation, and the quality of learning resources. This project led to the creation of the following scales:

- Student support scale
- Learning resources scale
- Learning community scale
- Graduate qualities scale
- Intellectual motivation scale

Universities are able to customize the CEQ by selecting a subset of scales for inclusion in the survey of their graduates. The GTS, GSS, and OSI are a core requirement, while all other scales are optional.

Conceptually, the CEQ is designed to be a proxy measure of student learning outcomes, with items designed to probe student perceptions of the characteristics of the teaching and the educational climate that are believed to be associated with effective student learning. The CEQ is therefore conceptually different from instruments such as the National Survey of Student Engagement (NSSE), used in the USA, and the locally developed Australasian Survey of Student Engagement (AUSSE), which seek to measure the extent of students' study and related activities believed to be related directly to student learning.

Survey Methodology

System-wide participation in the Australian Graduate Survey is the result of a collaborative agreement on the part of Australian universities, coordinated through the nation's council of university presidents, Universities Australia (formerly the Australian Vice-Chancellors' Committee, AVCC). Universities Australia and GCA have published methodology recommendations (AVCC-GCCA 2005) and a *Code of Practice* (AVCC-GCCA 2001) that provide guidelines for the collection and interpretation of survey data as well as the public disclosure of that information.

The GDS and CEQ are administered together, with the questionnaire items incorporated into a single document. While the Australian Graduate Survey is a national survey in the sense that most higher education institutions participate,[2] it is not centrally administered, and as a consequence both the questionnaire layout and the survey method vary between institutions. This variation is encouraged, in part, in keeping with the voluntary nature of the survey and the recognized diversity in the missions and priorities of individual institutions. However, sufficient standardization is achieved to ensure the integrity of the data set overall through monitoring by a Survey Reference Group (SRG), which includes GCA, DEEWR, and Universities Australia representatives.

Questionnaire Production and Distribution

Individual institutions design and produce their own versions of the paper-based questionnaire, including the coding of forms so that graduates' responses can be matched to particular faculties or schools within institutions. Some institutions choose to 'brand' their documents with institution-specific cover pages and explanatory notes. Those institutions that choose to format their own questionnaire do so in consultation with GCA and the SRG. Other institutions elect to have GCA prepare the questionnaire for them. GCA also provides the questionnaire in an online format.

Graduates are surveyed approximately 4 months after completing their course. As course completion dates vary between institutions and courses, and as mid-year completion is common, there are typically two survey distributions, each with follow-up of non-respondents, for each annual cycle. To standardize the information gathered, however, specified 'census' dates are used – October 31 for mid-year graduates and April 30 for end-of-year graduates.

---

[2] In 2005 there were 43 participating higher education institutions, including several private providers such as Bond University.

Most institutions manage their own questionnaire distribution, although some utilize the services of GCA. The initial distribution may be specifically timed to coincide with the census date, or may be in combination with other mail-outs to graduates, such as graduation information, and so only loosely aligned with the census date. Either way, GCA recommends that institutions follow up non-respondents, twice if necessary. Institutions choose various forms of follow-up, including a second mail-out of the questionnaire, mail-out of a reminder notice only, or email. Some encourage initial non-respondents to use of the online version. While some institutions contact non-respondents by telephone, the GCA discourages institutions from completing the CEQ in this way due to the unknown influence of verbal completion upon the psychometric properties of the questionnaire.

Response Rates

Institutional response rates vary widely. In 2005, for example, the response rates for the CEQ ranged from 27 percent to 71 percent, with a median around 50 percent. The GDS response rate is slightly higher for most institutions, perhaps due to the order and arrangement of items on the questionnaire. The variation in the institutional response rates is partly a result of the different procedures used to follow up non-respondents and the differing intensity with which universities opt to do so.

Data Collection and Collation

Institutions are also responsible for collection of completed questionnaires, coding of responses, and data entry. To support standardized practice, GCA provide coding instructions and spreadsheets for data entry. GCA collects this data from each institution, cleans and analyzes it, and provides institutions with the complete data set in return for a modest fee. Originally, each institution received a data set in which other institutions were de-identified. However, this is no longer the case and data sets are transparent inasmuch as scores are reported for each institution by name.

Reporting and Interpretation of the Data

GDS and CEQ data are annually reported for each university on a field-of-study basis. As students' perceptions of teaching are influenced by their field of study, Ramsden (1991) has argued that it is only appropriate to compare like courses or fields of study and that comparisons of institutional aggregates should be avoided. Accordingly, the reporting of CEQ data has focussed on course/field of study, allowing comparison of the quality of teaching in similar courses across different universities.

CEQ data are used for public information and as a guide for quality review and improvement at course level. The commercial guidebook the *Good Universities Guide* draws on the data to develop ratings of universities and courses for prospective students. The universities themselves occasionally use the data in their own marketing and recruitment campaigns, but the principal institutional purpose is for self-review and continuous improvement. It is common for institutions to further analyze their own data and to produce internal reports for quality monitoring and development.

For these purposes, the CEQ is believed to be a valid, reliable, and stable instrument. Studies using Rasch modelling have suggested that it measures a single dimension; however, the original scales have face-validity and satisfactory construct-validity according to factor analysis (J. Ainley, 2005, Access Economics, personal communication). Field of study is the single greatest influence on CEQ scores. Age is also an influence, with older graduates rating their courses more highly than younger graduates.

After many years of use of the CEQ, some clear patterns have emerged. Typically, the differences between comparable courses across institutions are small on most of the CEQ scales. Further, the results are relatively stable year to year, though there is evidence of slight improvement in some fields of study. As noted earlier, the responses of students are highly discipline specific: students in the liberal arts and sciences report the greatest levels of satisfaction, whereas students in professional and business courses such as engineering and economics report lower satisfaction levels (Ainley, personal communication).

Small campuses appear to produce a student satisfaction 'halo effect'. The highest CEQ-ranking universities/campuses are generally small and tend to have lower entry score requirements. Campus size and the 'intimacy' of the experience (small classes, ease of access to staff) appear to be significant factors in student satisfaction ratings. The effects of students' academic ability on their expectations and rating of teaching are not well understood.

The annual CEQ report prepared by the GCA avoids overt rankings, but such comparisons can be made by third parties – the *Good Universities Guide*, for example, converts CEQ data into five-star ratings of the quality of teaching. There is little evidence to date that these ratings have significantly influenced the prospective student market.

Costs

While the Australian government funds GCA for their involvement, the cost to each institution is significant. Precise costs are difficult to calculate, given that institutional systems for GDS/CEQ administration are typically interwoven with their other planning and evaluation processes. However, a broad-brush estimate is total institutional costs of $6–10 million ($150,000–250,000 per institution) and an additional $0.5 million of government funding.

Issues and Concerns with the Graduate Survey

Various concerns and criticisms have been made about the national graduate survey over the years, most of which are methodological but some might be better described as political. Generally, the concerns have focused on the conflation of objectives of continuous improvement and accountability, the possible misuse of the data for marketing in a competitive system, and the potential for a standardized process to stifle diversity and innovation.

The CEQ has borne the most criticism. The GDS has been less controversial, for the design of the GDS instrument does not depend on the depth of theorizing that the CEQ does. Methodologically, some concern has been expressed about the narrow conception of 'good teaching' embodied in the CEQ items and scales and the expectation that students can make an 'average' judgement across an entire degree program. These are significant criticisms indeed, for they are at the heart of the validity of the measurement that is being purported. Other concerns are more procedural than conceptual, with the variations in survey methodologies among institutions and the variations in response rates raising questions about the appropriateness of comparisons between institutions, the acknowledged difficulty in accommodating demographic differences in the student population, and other contextual factors (despite the employment of econometric measures in an effort to adjust raw data) (DEST 2000).

Impact

There has been no systematic research into the influences on institutional policies and practices of the GDS and CEQ. The analysis to follow is therefore based in the main part on the authors' observations as higher education researchers. First, and most obviously, the GDS and CEQ have been effective in drawing attention to the teaching function of higher education in a context in which many of the common quantitative indicators of performance are focused on research. Related to this important influence has been the creation of an explicit focus on teaching and learning outcomes rather than inputs or processes. In this regard, the comparative dimension of the data set has been particularly important, allowing institutions – notwithstanding differing contexts and missions – to place their performance on a field-of-study basis under some objective scrutiny.

Second, the GDS and CEQ data have established important market information for use by prospective students. The data set as a whole, and the willingness of universities to collaboratively self-evaluate, have been valuable in assuring international markets of the quality of the Australian higher education system. At the level of individual institution or field of study, the data have been available to prospective students and have been presented in modified form in commercial publications such as the *Good Universities Guide*. However, the use of the data in institutional marketing has not been particularly prominent. Further, the influence of the data on student decision-making is not altogether clear, with student choice patterns following well-established institutional reputations and institutional 'positional-status' in the market.

Third, the CEQ has possibly stimulated or encouraged greater management intervention in the evaluation of the quality of teaching and learning within units, subjects, and courses. Most institutions have comprehensive internal systems for semi-standardized or fully-standardized evaluation for the purposes of accountability and continuous improvement. Arguably, the evaluation of teaching and learning has been to some extent 'de-professionalized' as top-down management requirements have tended to dominate the bottom-up initiatives of the academic community itself. The CEQ and its local institutional counterparts are therefore sometimes viewed as symptoms of the managerialism seen by some to be pervading Australian universities.

Finally, the CEQ has had a deep influence on conceptions of effective teaching. Understandably, universities have tended to mirror the CEQ items in their questionnaires for internal unit and course evaluation. The conception of good or effective teaching embodied in the CEQ design has therefore become the dominant paradigm, at least from a management point of view. On the one hand, this has been a positive outcome, for the CEQ contains items that indisputably relate to well-established good teaching practices. Equally, however, the CEQ is necessarily a broad, generic instrument that might be criticized for being bland, superficial, and unlikely to detect important nuances of the educational environment in specific contexts. There have been concerns, for example, that the CEQ is not an appropriate instrument for measuring the quality of problem-based or enquiry-based learning environments.

The Learning and Teaching Performance Fund

The introduction in 2006 of the national *Learning and Teaching Performance Fund* (LTPF) was a development directly associated with the CEQ and GDS. Immediately upon announcement, this initiative was contentious, since for the first time the data from the graduate survey would be used for funding purposes.

The background to the LTPF lies in the 2002 review of higher education, one of the irregular but frequent national reviews that take place in Australian higher education. *Higher Education at the Crossroads* was a wide-ranging review of higher education undertaken by the Australian government (DEST 2002). One of the objectives was to review quality assurance mechanisms, in particular the assessment of teaching and learning quality. The premise was that existing quality assurance relied too heavily on assessment of teaching 'inputs' – institutional reports of teaching approaches and internal processes – rather than student learning outcomes.

In response to this review, a package of reforms was announced in May 2003 (DEST nd) to be implemented over 5 years. These reforms included national strategies for the promotion of teaching excellence through performance-based rewards and incentives. The *National Institute for Learning and Teaching in Higher Education* was formed (later renamed the *Carrick Institute for Learning and*

Teaching in Higher Education, and then *the Australian Learning and Teaching Council*), the existing program of national Awards for University Teaching was greatly expanded, and the LTPF was announced.

The aim of the LTPF is to reward universities for demonstrating excellence in teaching. Universities are ranked on the basis of existing data sets, including the CEQ and GDS. The most highly ranked institutions each receive a financial 'performance bonus'. The LTPF is an incentive fund and not used for the allocation of core government funding. Compared with the overall government expenditure on higher education, the allocation to the fund is modest; however, the status outcomes are significant. For example in 2006, after 1 year of implementation, $54 million was distributed. The five universities most highly ranked among the 38 eligible institutions shared $30 million, and the remaining money was shared between nine universities in the second band. Twenty-four participating institutions received no funds from the 2006 round. Funding in the 2007 and 2008 rounds was increased to $83 million and more broadly distributed across institutions.

The LTPF employs seven performance indicators. Three are derived from the CEQ (GTS, GSS, and OSI) and two from the GDS (graduate progression to full-time employment and graduate progression to further full-time study). The remaining two are based on institutional statistics collected annually by the government (student progression through their studies and student retention).

Development of the LTPF Policy

The development of the LTPF provides an interesting case study of the consultative approach used to develop national policy of this kind. Following the announcement of the LTPF in 2003, the government initiated a process of consultation with the higher education sector. Input from Universities Australia informed the preparation of an issues paper, released by the government in April 2004 (DEST 2004) and inviting submissions from across the sector. In addition, a series of meetings with nominated representatives from the nation's universities were held around the country. Universities Australia subsequently released its response to the issues paper (AVCC 2004), including recommendations and a proposal that the organization continue to work with the government in the development of a model that would be acceptable to the sector.

Unsurprisingly, the commitment of significant funding allocated on the basis of performance data, and the prospect of 'league ladders', generated a wide range of methodological and political concerns within the sector. Formerly, the CEQ and GDS methodologies and data had not been subject to the levels of scrutiny that followed the announcement of the LTPF. The specific issues proposed for consideration in the government's issues paper were as follows:

- *Benchmarking or ranking?* Two alternative approaches were proposed for the allocation of funds: to all universities meeting or surpassing a set 'threshold' level of performance or through a competitive process involving institutional rankings.

- *How many indicators should be used?* The issues paper supported the adoption of multiple indicators for two reasons: no single measure is likely to adequately reflect the complexity of teaching and learning and a single indicator is likely to encourage 'inappropriate manipulation' by institutions.
- *Should the data be adjusted for context?* Differences in the characteristics of institutions, particularly in the characteristics of their student populations and fields of study, were recognized by the government as factors influencing a range of university statistics – including the proposed indicators for the LTPF. Earlier work by the government (DETYA 1998; reviewed in Access Economics 2005) had produced a formula designed to 'correct for' student characteristics such as gender, age, field of study, enrolment type, and basis of entry to university and for field-of-study differences. It was proposed that crude percentage scores for each of the indicators be adjusted, using such a formula.
- *Evaluation of whole institutions or of disciplinary elements?* By drawing comparisons at the whole of institution level, much of the diagnostic potential of the graduate survey data is lost. The rankings become less useful for prospective students choosing courses and for institutions seeking to highlight areas for improvement. An alternative approach tabled for consideration in the issues paper was to assess the performance of institutions within defined fields of study.
- *The target student cohort.* The outcomes for only undergraduate, domestic students are included. This group comprised approximately 56 percent of all higher education students in Australia at the time.[3] The government proposed that postgraduate students be excluded as postgraduate student data were factored into university research funding, while the difficulties of collecting representative data from overseas students was cited as justification for their exclusion.

Implementation of the LTPF in 2006

In order to be eligible for performance assessment under the fund in 2006, institutions were first required to demonstrate that learning and teaching was a high priority within their institutional policy and planning framework. For example, they were required to show that teaching was recognized in staff appraisal processes and that the results from student evaluation of subjects were publicly available. Thirty-eight universities applied, and all were deemed eligible.

In August 2005, the government publicly released the percentage scores for each of the participating institutions against each of the seven performance indicators (see DEEWR nd). For the CEQ indicators, the scores represented the percentage of students to either 'agree' or 'strongly agree' with the questionnaire items. The reporting was transparent inasmuch as both the raw scores and the adjusted scores were reported. The fact that the adjustments were tailored for each performance

[3] Based on DEST statistics for 2004.

indicator was reflected in the effect upon individual institutions – raw scores were raised for some indicators and lowered for others.

In calculating the overall score for each institution, the ordinal rank number within each of the seven tables was used. This information was sent to institutions, and they were each offered a 'right of reply' – an opportunity to present a case for special consideration on any of a number of specified grounds.[4]

While the government did not publish a ranked set of results at any stage, the calculations were readily performed using the publicly available data. Unsurprisingly, league tables quickly appeared and discussion was animated, both in the Australian press and across the nation's higher education sector. The topic of discussion was predominantly focused on rankings, as the nature of the funding allocation had yet to be announced. Even the institutions leading in the rankings did not know how much funding they would receive, if any.

In November 2005, the government minister responsible for higher education announced the fund's allocation to a higher education audience gathered to celebrate the 2005 national university teaching awards, reinvigorating sector-wide discussion of the fund and the performance indicators employed. Fourteen universities received funding, allocated as a $1 million base grant plus an additional amount on a per capita basis according to domestic undergraduate student load. An upper band of five institutions shared $30 million, with more than $10 million awarded to one large research-intensive university. The remaining $14 million was distributed between the nine universities forming the second band.

Through the subsequent rounds in 2007 and 2008, the LTPF retained the same seven indicators and its characteristic focus on rewarding excellence. Each year there were some changes to the processes for calculating scores and allocating funds, the most significant being a shift in 2007 from a whole-of-institution rating to assessment against four broad field of study categories (science related; business and law related; humanities, arts, and education; and health).

In 2009 a more fundamental change occurred with the introduction of an 'improvement' component alongside the approach of rewarding excellence. Based on the same indicators, the improvement model compares institutions in terms of their relative degrees of improvement over 2 years, rather than on the basis of their absolute scores alone. This significant philosophical shift was introduced following the election of a left-wing Labour government in December 2007.

A Critical Analysis of the Australian Experience

The concerns with the LTPF and the value issues associated with the merit of such a scheme are a long way from resolved. The LTPF is a contentious policy within the Australian higher education sector, and the future shape of this policy initiative and the current funding model are not clear, particularly in light of

---

[4] Six criteria were specified, including 'Evidence that the statistical year was atypical', and 'Evidence that a high result against one indicator directly contributed to a lower result against another indicator' (DEST 2005).

the government's plan to re-examine funding arrangements for quality teaching in response to the 2008 Review of Australian Higher Education. Nonetheless, the government appears committed to outcomes-based funding or incentives, at least in part. The Australian Research Council is presently developing the methodology for the Excellence in Research for Australia initiative, previously known as the Research Quality Framework, which will evaluate research in Australian universities at a discipline level (see ARC 2008).

The Australian experience shows, once again, that once quantitative indicator information is available there is a tendency for it to be used for purposes for which it was not designed (Cuenin 1988; Cave et al. 1997). Performance indicators at institutional level provide commercially sensitive information, especially in an increasingly market-oriented higher education system. From the government perspective, there is an understandable desire to be assured of the quality of Australian universities and a belief that external pressure is needed to stimulate enhancement efforts. These objectives are awkwardly juxtaposed with the imperative to communicate to domestic and international stakeholders the high quality of the system as whole and are possibly incommensurable within the current LTPF policy model with its inevitable rankings and the implications of poor performance for the lower ranked institutions.

In the past, the willingness of some universities to participate in the CEQ and GDS has occasionally been tested. The LTPF exercise has opened up further cracks in an already fragile policy framework. Nonetheless, it is now quite difficult for a university to opt to withdraw from the agreement to participate in the CEQ and GDS data collection. To do so might imply concerns about performance and would make the possibility for performance comparison limited (though institutions could of course conduct independent surveys and compare the findings with the publicly available national data).

The ultimate test for the effectiveness of the GDS and CEQ is whether there has been a positive impact on the quality of university teaching and learning. As noted earlier, this is a very difficult assessment to make with confidence, especially given the large number of variables at play in a system undergoing ongoing growth and change. Clearly the CEQ data have provided a hitherto unavailable objective, external reference point for the quality of teaching against which institutions can judge their performance, and this has been valuable.

The national survey of graduates has also resulted in greater internal evaluation. Most institutions have invested heavily in evaluative activity, and the quantitative information base on teaching quality is now considerable in most universities, including time series trend data. In addition, many internal approaches to evaluation at institutional level are now based on or mimic CEQ items or scales – in this regard the influence on the prevailing conceptions of effective teaching has been significant.

Overall, from a quality improvement perspective, never before have Australian universities had so much information on their performance available to them. However, the influence of these activities on improvement in undergraduate education is far from clear. Despite the attention which has been drawn to the quality of teaching for the best part of a decade or more, the evidence of improvement is at best modest. In fact, the patterns in the CEQ data are notable for their consistency over time. There appear to be at least four interwoven reasons for this situation (James 2000). First, the feedback on performance provided by the CEQ is often ambiguous. Much of what is valued in university teaching is difficult to measure, given the often high levels of abstraction of learning outcomes and graduate qualities, and important information is lost when it is quantified and codified at a blunt aggregate level. Further, the quantitative performance indicators being used in Australia for measuring teaching quality have a highly subjective element to them. This means that performance indicator data are contestable and are open to various interpretations.

The second problem for effective 'knowledge management' is the inherent lag time involved in the measurement process. There is a lengthy period between the collection of graduate CEQ responses and the teaching and learning activities on which the students are reporting. Internal university feedback loops are tighter, but even so the delays in feedback cause adjustments to be made well before evidence is available on the outcomes of previous actions. This creates a learning loop with the potential for wildly erratic fluctuations, and intelligent anticipation takes over. However, awareness of the delay dilemma may also create subtle climates of complacency or fatalism, for it is easy to be dismissive of lagged measurement data that is perceived to be of questionable relevance in an altered context (James 2000).

Third, causal links between actions and outcomes are often unclear. The relationship between actions and outcomes in higher education is not fully understood. Precisely what universities do to and for graduates, and how they do it – the value-adding effect of higher education – is not yet fully mapped (see Pascarella and Terenzini 1991). As a consequence, it can be difficult to identify the precise reasons for apparently high performance of courses (or low performance for that matter). Further, the performance indicators do not in themselves identify the actions that might be taken to lead to improvement.

Finally, quality assurance and the use of performance data have been perceived by many academic staff as an intrusion on academic work, part of a creeping 'managerialism'. The history of quality assurance in Australia is largely one of management intervention. The teaching staff who ultimately make the day-to-day decisions and actions that lead to quality improvement may find little personal or professional meaning in data collected as a management requirement using a highly generic instrument.

These observations suggest some natural limits on the capacity of universities to reliably and validly detect and measure their teaching performance for the purposes of continuous improvement. Overall, however, despite some concerns such as we have indicated, the CEQ and GDS are now widely viewed as useful policy developments. The findings are of value to the sector, and both surveys appear firmly embedded for the foreseeable future. The CEQ/GDS policy framework has been made possible by a small higher education system (in international terms) and a reasonably homogenous one. In addition, the institutional agreement to participate in the exercise has hinged on a shared commitment to quality and quality improvement and a collective interest in the international marketing of Australian higher education. Almost paradoxically, while Australian universities operate in a quite competitive environment, there is also acknowledgment of the need to act in the interests of the sector as a whole. The competition–collaboration nexus is therefore played out in interesting ways, especially with regard to quality assurance and performance indicators.

The concerns with the LTPF are likely to lead to ongoing adjustments of the funding model and the performance data on which it is based. One future possibility is the use of a national graduate examination, for which the scene is partially set. The Graduate Skills Assessment (GSA), developed by ACER with government funding, has been designed to assess the so-called generic skills of university graduates (ACER nd). The test has the following components: critical thinking, problem-solving, interpersonal understanding, and written communication. The GSA was piloted in early 2000 and is used on a voluntary basis by some universities. It has been suggested that at university entry level, the test could be used by institutions to assess areas in which students might need assistance. At exit level the results could be used by institutions to determine entry into graduate courses and by employers to assess generic skills for employment purposes. More significantly perhaps, the GSA could also be used to gain insights into the 'value-added' across institutions that enrol cohorts of differing academic achievement level at entry point. The GSA has appeal for the Australian government as it offers the promise of a direct indicator of student learning rather than a proxy. Graduate examination of any kind, however, has not received strong support from Australian universities.

International developments will influence Australian policies and practices in the measurement of teaching and learning performance in the mid to long term. Australia is a participant in the OECD's feasibility study Assessing Higher Education Learning Outcomes (AHELO), which is examining the possibility of measuring learning outcomes across nations, languages, and cultures and includes a strand investigating the potential for measuring value-added (OECD nd). The AHELO process and the conclusions drawn from it may dramatically refocus energies on outcomes-based indicators of performance in Australia. This would be especially true if AHELO-derived indicators became routinely factored into university rankings calculations. *The Times Higher Education Supplement* university rankings and the Shanghai Jiao Tong Index have had significant influence on the Australian higher education sector, in part because of the effects these rankings may have on international student markets. These ranking systems are presently heavily skewed toward institutional research performance and reputation. The inclusion of valid and reliable data on teaching and learning outcomes would substantially influence university behaviors.

In the more immediate future, the more widespread measurement of student engagement for current students is likely. The approach reduces the problem of the lag time associated with the CEQ graduate survey, as well as providing a superior proxy measure that is one step closer to student learning outcomes – that is, a shift from measuring student perception of teaching behaviors to measuring the learning activities of students themselves. The AUSSE instrument developed specifically for the Australian context has developed a strong profile within the higher education sector in the past few years and provides a 'leading indicator' for diagnostic purposes. Institutions presently participate in the AUSSE on a voluntary basis; however, national agreement to use this survey in a coordinated approach such as for the GDS and CEQ may well be a logical development.

Overall, the policies described in this chapter illustrate one nation's response to a more complex operating environment for universities, one in which a premium is placed on new levels of performance monitoring, quality assurance, and evaluation. In part these systems are a flow-on from external requirements for public accountability, but they are also driven by an internal institutional commitment to understand performance against external reference points. Australia has made considerable progress in establishing a quality assurance framework and in building a data set of performance indicators at system, institution, and discipline levels. Until now, indicator data has been intended principally for use within universities to aid self-review, rather than as information to aid public choice or as data for performance-based funding allocation. The trend appears to be toward more stringent approaches to outcomes measurement, greater transparency in the publication of comparative performance data, performance-based funding allocation, and the legitimization of university rankings schema of various kinds. The pressure to collect data on teaching and learning performance and outcomes in higher education can only rise.

Appendix: Course Experience Questionnaire Scales and Items

Scale	Item
Good teaching	The staff put a lot of time into commenting on my work
	The teaching staff normally gave me helpful feedback on how I was going
	The teaching staff of this course motivated me to do my best work
	My lecturers were extremely good at explaining things
	The teaching staff worked hard to make their subjects interesting
	The staff made a real effort to understand difficulties I might be having with my work
Generic skills	The course helped me to develop my ability to work as a team member
	The course sharpened my analytic skills
	The course developed my problem-solving skills
	The course improved my skills in written communication
	As a result of my course, I feel confident about tackling unfamiliar problems
	My course helped me develop the ability to plan my own work
Overall satisfaction	Overall, I was satisfied with the quality of this course
Appropriate assessment	To do well in this course all you really needed was a good memory
	The staff seemed more interested in testing what I had memorized than what I had understood
	Too many staff asked me questions just about facts

Scale	Item
Appropriate workload	I was generally given enough time to understand the things I had to learn The sheer volume of work to be got through in this course meant it couldn't all be thoroughly comprehended The workload was too heavy There was a lot of pressure on me as a student in this course
Clear goals and standards	It was always easy to know the standard of work expected I usually had a clear idea of where I was going and what was expected of me in this course It was often hard to discover what was expected of me in this course The staff made it clear right from the start what they expected from students
Grad qualities	The course provided me with a broad overview of my field of knowledge The course developed my confidence to investigate new ideas University stimulated my enthusiasm for further learning I learned to apply principles from this course to new situations I consider what I learned valuable for my future My university experience encouraged me to value perspectives other than my own
Intellectual motivation	I found my studies intellectually stimulating I found the course motivating Overall, my university experience was worthwhile The course has stimulated my interest in the field of study
Learning community	I felt part of a group of students and staff committed to learning Students' ideas and suggestions were used during the course I learned to explore ideas confidently with other people I felt I belonged to the university community I was able to explore academic interests with staff and students
Learning resources	The library resources were appropriate for my needs The study materials were clear and concise It was made clear what resources were available to help me learn Course materials were relevant and up to date Where it was used, the information technology in teaching and learning was effective
Student support	I was able to access information technology resources when I needed them Relevant learning resources were accessible when I needed them Health, welfare and counselling services met my requirements The library services were readily accessible I was satisfied with the course and careers advice provided

Note that when incorporated into the Graduate Survey, the items are not presented in scale groups.

References

Access Economics Pty Ltd. (2005). *Review of higher education outcome performance indicators*. Report prepared for DEST. June 2005. http://www.dest.gov.au/sectors/higher_education/publications_resources/profiles/review_highered_outcome_perf_indicators.htm. Acccessed 15 January 2009.

Australian Council for Educational Research (ACER). (nd). Graduate skills assessment – GSA. http://www.acer.edu.au/gsa-uni/. Accessed 15 January 2009.

Australian Research Council (ARC) (2008). ERA indicator principles. Canberra: ARC. http://www.arc. gov.au/pdf/ERA_Indicator_Principles.pdf. Accessed 19 January 2009.

Australian Vice-Chancellors' Committee (AVCC) (2004). Assessing learning and teaching: The AVCC proposal for the learning and teaching performance fund. AVCC. August 2004. http://www.universitiesaustralia.edu.au/documents/publications/policy/statements/AVCC-response_ALTPF.pdf. Accessed 15 January 2009.

Australian Vice-Chancellors' Committee and Graduate Careers Council of Australia (AVCC-GCCA). (2001). Code of Practice for the public disclosure of data from the Graduate Careers Council of Australia's Graduate Destination Survey, Course Experience Questionnaire and Postgraduate Research Experience Questionnaire. January 2001. Canberra. http://www.universitiesaustralia.edu.au/documents/policies_programs/graduates/2003/COP_2001.pdf. Accessed 15 January 2009.

Australian Vice-Chancellors' Committee and Graduate Careers Council of Australia (AVCC-GCCA). (2005). *Standard recommended methodology for the Graduate Destination Survey, Course Experience Questionnaire and Postgraduate Research Experience Questionnaire*. May 2005. http://www.graduatecareers.com.au/content/download/2611/10493/file/ Std_Recommended_Methodology_May_2005.doc. Accessed 15 January 2009.

Bradley, D., Noonan, P., Nugent, H., Scales, B. (2008). *Review of Australian higher education: Final report*. Commonwealth of Australia. http://www.deewr.gov.au/he_review_finalreport. Accessed 15 January 2009.

Cave, M., Hanney, S., Henkel, M., Kogan, M. (1997). *The use of performance indicators in higher education: The challenge of the quality movement* (3rd ed.). London: Jessica Kingsley.

Cuenin, S. (1988). *Performance indicators in higher education: A study of their development and use in 15 OECD countries*. Paris: OECD.

Department of Education, Employment and Workplace Relations (DEEWR) (nd). Student outcome indicators for the Learning and Teaching Performance Fund. Technical Note No. 2. http://www.dest.gov.au/sectors/higher_education/policy_issues_reviews/key_issues/assuring_; quality _in_higher_education/technical_note_2.htm. Accessed 15 January 2009.

Department of Education, Science and Training (DEST) (2000). *Characteristics and performance indicators of Australian Higher Education Institutions, 2000*. Occasional Paper Series. December 2001. http://www.dest.gov.au/sectors/higher_education/publications_resources/statistics/publications_higher_education_statistics_collections.htm#studpubs. Accessed 15 January 2009.

Department of Education, Science and Training (DEST) (2002). *Higher education at the crossroads: An overview paper*. DEST. April 2002. http://www.backingaustraliasfuture.gov.au/publications/crossroads/default.htm. Accessed 15 January 2009.

Department of Education, Science and Training (DEST) (nd). *Our universities: Backing Australia's future*. http://www.backingaustraliasfuture.gov.au/Default.htm. Accessed 15 January 2009.

Department of Education, Science and Training (DEST) (2004). *Learning and Teaching Performance Fund: Issues paper*. April 2004. http://www.dest.gov.au/sectors/higher_education/policy_issues_reviews/key_issues/learning_teaching/ltpf/data_and_ref_links.htm. Accessed 15 January 2009.

Department of Education, Science and Training (DEST) (2005). Learning and Teaching Performance Fund 2006: Administrative information for providers. http://www.dest.gov.au/

sectors/higher_education/policy_issues_reviews/key_issues/learning_teaching/ltpf/2006ltpf.htm. Accessed 15 January 2009.

Department of Education, Training and Youth Affairs (DETYA) (1998). *The characteristics and performance indicators of Australian institutions*. Occasional Paper Series 98-A. December 1998. Canberra: Higher Education Division, Department of Education, Training and Youth Affairs.

Graduate Careers (nd). GradStats & GradFiles. http://www.graduatecareers.com.au/content/view/full/24. Accessed 15 January 2009.

Griffin, P., Coates, H., McInnis, C., James, R. (2003). The development of an extended course experience questionnaire. *Quality in Higher Education,* 9(3), 259–266.

James, R., Baldwin, G., McInnis. C. (1999). *Which university? The factors influencing the choices of prospective undergraduates*. Canberra, ACT: Australian Government Publishing Service.

James, R. (2000). Quality assurance and the growing puzzle of managing organisational knowledge in universities. *Higher Education Management*, 12(3), 41–59.

Linke, R. (1991). *Performance indicators in higher education*. Canberra, ACT: Australian Government Publishing Service.

McInnis, C., Griffin, P., James, R., Coates, H. (2001). *Development of the Course Experience Questionnaire (CEQ)*. Canberra, ACT: Australian Government Publishing Service.

Organisation for Economic Co-operation and Development (OECD) (nd). OECD feasibility study for the International Assessment of Higher Education Learning Outcomes (AHELO). http://www.oecd.org/edu/ahelo. Accessed 15 January 2009.

Pascarella, E. T., Terenzini, P.T. (1991). *How college affects students: Findings and insights from twenty years of research*. San Francisco, CA: Jossey-Bass.

Ramsden, P. (1991). A performance indicator of teaching quality in higher education: The course experience questionnaire. *Studies in Higher Education*, 16(2), 129–150.

Wilson, K., Lizzio, A., Ramsden, P. (1997). The development, validation and application of the course experience questionnaire. *Studies in Higher Education*, 22(1), 33–53.

Chapter 7
National Report Card on Higher Education in the USA

David W. Breneman

Introduction

The National Report Card is a benchmarking exercise, focusing on the 50 states in the USA as the units of analysis. The Report Card assesses with quantitative data the educational performance of each of the states on six dimensions: preparation for higher education, participation, completion, affordability, benefits, and learning. The focus is solely on undergraduate education, as no indicators are included for either graduate education or research. Aggregate data from all forms of postsecondary education are incorporated, including public 2-year and 4-year institutions, private non-profit institutions, and (where the data allow) private for-profit institutions. No data for individual colleges or universities are reported, however; the measures are collective state data. Multiple indicators are used for each of the six categories, with each indicator weighted by its importance, and combined into a single numerical measure for each category. The result is a set of performance measures that are used to compare each state to the best-performing state on each measure, a classic benchmarking technique. Grades are assigned to each measure (A through F), allowing the report to be used by state-level policymakers to judge how well a given state is performing relative to the other 49.

The Report Card is the product of the National Center for Public Policy and Higher Education, a non-profit, non-affiliated, and non-partisan private organization located in San Jose, California. Founded in 1998, the Center is fully supported by private foundation grants, with core support from the Pew Charitable Trusts, The Atlantic Philanthropies, and The Ford Foundation; no state or federal governmental funds are involved. In that sense, the Center has no official governmental status. It provides the biennial reports on state performance as a public service, and no state agency is obligated to respond to, or even read, the reports. It is assumed, however, that the quality of the work and the salience of the measures provide useful information to policymakers as they consider policies that support and govern higher

---

D.W. Breneman (✉)
University of Virginia, VA, USA

D.D. Dill, M. Beerkens (eds.), *Public Policy for Academic Quality*, Higher Education Dynamics 30, DOI 10.1007/978-90-481-3754-1_7,

education in each state. (Information on the extent to which the states have used the reports will be discussed later.) The National Center has an appointed board of directors, a small professional staff, and a number of advisory committees that guide its work.[1]

Policy Problem

The National Center was created and supported to be an independent, non-partisan voice in the debates about higher education policy. The decision to create a National Report Card emerged from discussions that the leadership held around the country in 1998 and 1999, seeking advice and comments on how a small organization could speak to issues of higher education policy in all 50 states. It was noted in several of those meetings that report cards (or similar benchmarking techniques) have operated effectively in areas such as medical care and social welfare, and thus the first step was to determine whether a 50-state report card on higher education was feasible. It was decided early on that such a report card would have to be constructed from existing data sources, as the Center did not have sufficient funds or staff to undertake independent data collection. In 1999, a small advisory group met several times with Center staff to design and implement a pilot project, using 10 states and relying heavily on nationally collected data from such agencies as the National Center for Education Statistics and the Census Bureau. The pilot studies indicated that sufficient data sources existed to make the project a reality, and the first Report Card was issued in the year 2000. The central problem was to locate relevant and comparable data at the state level; many data sources report information on the national level, but the underlying surveys often are not large enough to provide adequate data at the state level.[2] The feasibility committee also had to consider such issues as the interstate migration of students, and the differences among states in the structures of their higher education systems, in particular the extent to which states differ in their reliance on private institutions and community colleges in achieving educational opportunities. The pilot study indicated that, while some of these problems would remain, a meaningful set of measures, as designed by the committee, could be assessed quantitatively at the state level, and the board of directors authorized the Center to undertake the Report Card as its central project. Reports have been issued subsequently in 2002, 2004, and 2006.

It should be noted that the National Center is not intended to continue in perpetuity; while no definite date has been determined when it will shut its doors, at some point that will happen. Whether another organization will decide to continue producing the Report Card is unknown at this time, although the developmental

[1] Membership of these groups can be found at the National Center's website: www.highereducation.org.

[2] This problem continues to plague aspects of the project, which will be noted accordingly in the text.

work has been completed and the Report Card is designed now for publication on a website, precluding the need to produce expensive, book-length products. If it has demonstrated its value, one hopes that another agency will pick up the project and continue Web-based publication in future years.

Content of the Policy Instrument

As noted earlier, the Report Card assesses six measures of educational performance – preparation, participation, completion, affordability, benefits, and learning. Each measure is made up of several underlying quantitative indicators that are weighted and aggregated into a score for each state on each measure. In identifying indicators for each measure, the Center was limited to those data that are available at the state level, which constrained the choices one might have made in an ideal world. Nonetheless, a good number of reasonable and relevant indicators were found, and while additional data could improve the quality of the Report Card (and some of those data elements will be discussed subsequently), the result has clearly passed the test of face validity and plausibility within the policy community.

Preparation. The indicators that make up the measure of preparation for postsecondary education (and the weights in parentheses) include:

- High School Completion (20%)
- K-12 Course Taking (35%)
  - 9th to 12th graders taking at least one upper-level math course
  - 9th to 12th graders taking at least one upper-level science course
  - 8th grade students taking algebra
  - 12th graders taking at least one upper-level math course
- K-12 Student Achievement (35%)
  - 8th graders scoring at or above "proficient" on the national assessment exam in math, in reading, in science, and in writing
  - Low-income 8th graders scoring at or above "proficient" on the national assessment exam in math
  - Number of scores in the top 20% nationally on SAT/ACT college entrance exam per 1,000 high school graduates
  - Number of scores that are 3 or higher on an Advanced Placement subject test per 1,000 high school juniors and seniors
- Teacher Quality (10%)
  - 7th to 12th graders taught by teachers with a major in their subject

Participation. The indicators that make up the measure of participation in postsecondary education (and the weights in parentheses) include:

- Young Adults (60%)
  - Chance for college by age 19
  - 18- to 24-year-olds enrolled in college
- Working-Age Adults (40%)
  - 25- to 49-year-olds enrolled part-time in any type of postsecondary education

Completion. The indicators that make up the measure of completion of postsecondary education (and the weights in parentheses) include:

- Persistence (20%)
  - 1st-year community college students returning their second year
  - Freshmen at 4-year colleges/universities returning their sophomore year
- Completion (80%)
  - First-time, full-time students completing a bachelor's degree within 6 years of college entrance
  - Certificates, degrees, and diplomas awarded at all colleges and universities per 100 undergraduate students

Affordability. The indicators that make up the measure of affordability for postsecondary education (and the weights in parentheses) include:

- Family Ability to Pay (50%)
  - Percentage of income (average of all income groups) needed to pay for college expenses minus financial aid at community colleges, at public 4-year colleges and universities, and at private 4-year colleges and universities
- Strategies for Affordability (40%)
  - State investment in need-based financial aid as compared to the federal investment
  - At lowest-priced colleges, the share of income that the poorest families need to pay for tuition
- Reliance on Loans (10%)
  - Average loan amount that undergraduate students borrow each year

Benefits. The indicators that make up the measure of benefits from postsecondary education (and the weights in parentheses) include:

- Educational Achievement (37.5%)
  - Population aged 25–65 with a bachelor's degree or higher

- Economic Benefits (31.25%)
  - Increase in total personal income as a result of the percentage of the population holding a bachelor's degree
  - Increase in total personal income as a result of the percentage of the population with some college degree (including an associate's degree), but not a bachelor's degree
- Civic Benefits (31.25%)
  - Residents voting in national elections
  - Of those who itemize on federal income taxes, the percentage declaring charitable gifts
  - Increase in volunteering rate as a result of college education
- Adult Skills (0%)[3]
  - Adults demonstrating high-level literacy skills, quantitative, prose, and document

Learning. This category was graded as incomplete in both 2000 and 2002, as the country has no well-defined indicators at the state level that measure college-level learning. With support from the Pew Charitable Trusts, staff of the National Center and outside consultants began exploring this topic to determine whether some collection of existing tests and new instruments might be assembled to provide indicators for the learning category. A National Forum on College-Level Learning was organized under the auspices of the National Center in November 2001, and a broad group of participants, including educators, business people, and state and federal policymakers, reviewed the initial work and agreed that the topic was sufficiently important that it should be pursued to a pilot phase. That work was accomplished, and the Report Card for 2004 did contain results from the pilot effort in five states – Illinois, Kentucky, Nevada, Oklahoma, and South Carolina.

Essentially, the pilot effort developed three broad types of indicators of learning – literacy levels of the state population, graduates ready for practice, and performance of the college educated. For the first indicator, the Center used the 1992 National Adult Literacy Survey (NALS) for residents ages 25–64, updated using the 2000 census, and weighted at 25%. The second indicator involved the use of various licensure examinations, competitive admissions tests, such as the Graduate Records Examination and the Medical School Admissions Test, and measures of teacher preparation. Finally, for the third indicator, the WorkKeys assessment administered by the American College Testing Service was used for 2-year institutions, and the Collegiate Learning Assessment (CLA) developed by an offshoot of the Rand

[3] This indicator is an example of a data problem, in that the underlying National Adult Literacy Survey, was last conducted in 1992, and although those data were used in the 2000 and 2002 Report Cards, they were deemed too far out of date to use in 2004. A new National Assessment of Adult Literacy has been recently conducted, but may prove difficult to extend to the state level.

Corporation, was used for graduates of 4-year colleges. For both WorkKeys and CLA, the Center engaged a sample of colleges and universities in the five states and arranged for the tests to be administered to a representative sample of students. Given that the Center's general operating policy is not to collect original data itself, future efforts to measure learning will have to be done by the states themselves, but the Center has demonstrated the way forward. A 2005 publication of the National Center, *Measuring Up on College-Level Learning*, by Margaret A. Miller and Peter T. Ewell, provides considerable information on the results and issues confronted in the five pilot states and will be a valuable resource for anyone seeking further information in this area.

Implementation

As noted, the Center has now had experience producing and disseminating three Report Cards, for the years 2000, 2002, 2004, and 2006. Each time one has been published, a new set of issues and complications have been encountered. Most of the issues have surrounded the fact that the first Report Card was a one-time snapshot, while each successive version opens up the potential for longitudinal analyses, i.e., comparisons of 2002–2000 within the same state, rather than simply comparing states against each other. Any group undertaking a similar exercise would encounter the same issues, but it is fair to say that many of the issues raised by subsequent reports were not originally foreseen. The result has been a series of hard decisions that have been made as each new project is underway.

For the first Report Card, the key tasks were assembling the data, dealing with missing observations, and refining the aggregation techniques. Each indicator was tabulated, and the weights that had been determined by consensus and best judgment were then applied to each indicator – as indicated earlier, the weights sum to 100%.[4] State results on each indicator were then converted to a scale from 0 to 100, a statistical method that allows for accurate comparisons of different measures. The top five states on each indicator were seen as high, but achievable, measures of performance. In practice, the median of the top five was assigned a score of 100, meaning that potential outliers were eliminated. Finally, each state's score for each category was calculated using the index score on the indicator's and the indicator weights. Once again, the raw category scores are scaled on a 0–100 basis, and grades were assigned to each state in each category using the standard A through F scale common to public schools. When the exercise was finished, each state received five-letter grades (A though F) and an incomplete in the category of learning.

This technique differs from ventures of a similar sort in which absolute standards of performance are determined abstractly, and each entity is measured against

[4] Before the first Report Card was released, the National Center convened a panel chaired by Professor Michael Nettles, to review the weights and suggest changes if necessary. Although a few adjustments were made, the panel argued that the weights were justified based on relevant research.

that standard. In the Report Card, the highest grades are not determined abstractly, but rather represent actual performance delivered by each state. The benchmarking, therefore, is against best practice rather than against a standard that no state may have achieved. The creators of the Report Card believe that this method removes the objection that arbitrary standards have been applied.

Having produced the first Report Card, entitled *Measuring Up 2000*, the next task was to publicize and explain it to the higher education community, the policy community, and the media. Unlike many private research and policy organizations, the National Center has devoted considerable resources to outreach and public relations. The president has met with numerous editorial boards of newspapers, a professional public relations firm worked on the press releases, with one tailored explicitly to each state, and the C-Span television network covered the press release of the first Report Card, an event held at the National Press Club in Washington, D.C. Several members of the board of directors were present, as were key staff and consultants. The effort was rewarded with wide coverage in the press, with the central message being that opportunities for higher education vary widely among the states. In other words, one's chances for higher education depend to a disturbing degree on the state in which one happens to live. This message was portrayed vividly with colored maps identifying the high- and low-scoring states on each measure.

A fascinating aspect of the launch was the reaction of the higher education community, particularly as represented by the national associations of colleges and universities. Before the release, there was considerable nervousness about the Report Card, reflecting a fear that the institutions of higher education would be under attack and poorly graded. It took some time for people to realize that no institution was named, and that indeed, the focus was on state performance, not institutional performance. Once that realization sank in, most college and university presidents simply turned their backs on the report, seeing it as not doing much to help them, as institutions were not identified. In a few instances (Georgia and New York most prominently) there was strong criticism of the grades, particularly the low grades received in the category of affordability. Georgia higher education officials were upset because the method of calculating affordability gave minimal credit to the HOPE scholarship program, a merit-based scholarship that does little to enhance affordability for low-income students. New York officials argued that their state student aid program was not recognized sufficiently by the Report Card methodology; once again, the issue has to do with how those funds are distributed and the impact they make in reducing net cost relative to family income. But beyond some of these debates, the general response of leaders of higher education was to ignore the report, not seeing how they could use it to their advantage.

An important point emerged from reaction to the first Report Card. At least since World War II, most of the policy debate about higher education in the states has focused on the institutions and how well (or how poorly) they are supported, how many new colleges were needed, where they should be built, their missions, and so forth. The Report Card takes a radically different approach, focusing instead on the citizens of the state and the opportunities they have (or do not have) for higher education. This is not the policy discussion with which most higher education leaders are familiar, and the de-centering of the institution from the heart of the conversation was a blow. Indeed, the National Center has taken the unusual stance of focusing its efforts on state policymakers rather than on college and university presidents, and only time will tell whether this effort to change the nature of the state higher education policy debate takes root. Early evidence suggests that this new direction is finding a ready audience in several states.

Having produced a successful first Report Card, the Center began work on the 2002 report and realized as the data were being assembled that now one could not only compare the states to each other in a given year, but one could also determine whether performance in each state in 2002 was improved or declining relative to 2000. This observation posed a communications challenge, as the designers argued that, on the one hand, it was important to replicate the basic design of the first Report Card for comparability purposes, but also that change over time within a state was at least as significant a measure as interstate comparisons. Furthermore, some of the measures (particularly affordability) were objectively getting worse, but the benchmarking approach still required that best practice states be given an A grade. Thus, a state could receive a top grade (even a better grade than in 2000) while actually performing less well. The metaphor that eventually was adopted was that of a race – individual runners might be doing better in the race in 2002 than in 2000, but if other runners outstripped them, their relative standing would decline, and it was the relative standing that determined the letter grade. The second Report Card, *Measuring Up 2002*, made a valiant effort to communicate this complicated situation, but the designers realized that they had a substantive problem to solve. If grades were to be used, then a move by a state from a C to a B should mean objective improvement over time, rather than just relative gain (or even loss). The grades were in danger of not revealing directly and simply the information they were intended to convey. The solution came in the third Report Card, *Measuring Up 2004*.

In the third version, the designers extended the measures back 10 years, to give each state a baseline measure in 1992. That allowed each state to be judged in 2004 not only in comparison to other states in that year, but over a decade against their own earlier performance. The Center published short reports for each state that provided both the current results and the 10-year change. Comments received from the field indicate that this blend of current comparison together with time-trend data provides the most useful information to date. As subsequent editions of the Report are produced, new challenges will undoubtedly arise, but the designers by necessity are becoming adept at finding creative solutions.

Over time other issues have arisen, generally involving data problems, as a survey question may be changed from one year to the next, or a data element may be dropped, or updated surveys not provided. For example, the National Adult Literacy Survey is conducted every 10 years, and while the Center decided it could use older data on adult literacy in *Measuring Up 2000* and *2002*, by *2004* the data were so out of date that the indicator had to be dropped. Similar data problems bedeviled one part of the affordability measure, as availability of data required to determine net cost after financial aid became a problem. In some cases, however, new indicators become available; an example would be the indicator on teacher quality, used for the first time in 2004.[5] On the one hand, it is important to improve the data underlying the graded categories; on the other hand, adding a new indicator means altering the weights and reduces comparability over time. The technical reports accompanying each Report Card give detailed information on how each problem was handled and can be found on the National Center's website. Communicating changes without losing the reader in a shower of technicalities remains a constant challenge for this sort of exercise.

Impact

It is difficult to determine the impact of a project such as this one, as one can never trace precisely the changes in thinking that the Report Cards may have produced. One approach is to examine the media coverage that the Report Cards have achieved, for if the reports failed to gain substantial press coverage, that would be likely to reduce their impact. The National Center has carefully collected information on media coverage; for *Measuring Up 2004*, their statistics show that 2,030 newspaper articles covered the report, including 282 editorials, 38 op-ed pieces, and 34 columns across the country. Coverage was also excellent in the major papers, such as the *New York Times*, *USA Today*, the *Los Angeles Times*, the *Washington Post*, the *Chicago Tribune*, and the *Dallas Morning News*. At least 474 televisions news segments covered the report, including "CNN Headline News" and the "News Hour with Jim Lehrer." Radio and internet coverage were also strong. Most significantly, this third report in the series garnered considerably more coverage than the previous two, indicating a growing familiarity with the report on the part of the media and the sense that the message is important.

Among the states, the response has varied from largely ignoring the reports to making active use of them. Several states, including New Mexico and Oklahoma, have borrowed the format and put out their own state-focused reports, often incorporating state data that are not available in all 50 states, but which give policymakers a better sense of how a given state is doing. One form of follow-up has been creation of the National Collaborative for Postsecondary Education – co-sponsored by the National Center, The National Center for Higher Education Management Systems (NCHEMS), and the Education Commission of the States (ECS). This entity has been the primary vehicle for state follow-up, supported by a grant from the Pew Charitable Trusts to the three organizations. The five states involved are Washington, Virginia, Rhode Island, Missouri, and West Virginia. In each of these states the NCHEMS "drill down" methodology was used, providing results at the county level on each of the *Measuring Up* categories. This effort was followed by

[5] In 2004, the National Center convened an expert panel to review additional indicators that were being considered for inclusion, particularly in the areas of teacher quality and adult learning. The result was inclusion of a new indicator for teacher quality, but no change with the indicators of adult learning.

"policy audit" discussions around the state with local education, business, and political leaders. Feedback from these local meetings was given to a statewide leadership group that each state had to form, including the governor, business, community, public/private 2- and 4-year college university leaders, and K-12 state leaders. This group then determined the priorities, based on the "drill down" exercise and policy audit, for policy focus and change. The Educational Collaborative Program ended in December 2005. Results have varied depending on state leadership – definitely more effective when the governor was involved significantly or chaired the leadership group, as in Virginia and Rhode Island.

In addition, the National Center has worked directly with many more states on one or more of these issues (Kentucky, South Carolina, Arizona, Oregon, Pennsylvania, and Oklahoma, to name a few) and is currently working in-depth in Minnesota through the governor's office to address access and affordability issues.

As a further indication of impact, several national organizations, including the National Conference of State Legislators and the Committee for Economic Development, have devoted resources to producing guides to the Report Card, designed and published for their members. The respective reports are entitled *The Legislator's Guide to the National Report Card on Higher Education* and *Cracks in the Education Pipeline: A Business Leader's Guide to Higher Education Reform.*

A further strategic point about the Report Cards should be noted. No policy recommendations are included in the reports; the data are presented, the state comparisons are made, and the presumption is that the information should start conversations within each state regarding its relative performance and how it might improve. Indeed, the National Center sees its role primarily as moving the policy debates within states forward, while not prescribing any particular set of policies to be adopted. For example, in the first Report Card, two of the states that received A grades in affordability were Illinois and North Carolina. In the case of Illinois, the policy tool was a well-funded program of need-based student financial aid; in North Carolina, the policy tool was relatively low tuition. The Center's report simply demonstrated that there is more than one way to achieve affordability, and circumstances in each state may determine which approach works best in a given state context. In short, the National Center's efforts will be seen as successful if the states investigate and work on the issues measured by the Report Card, and not on the choice of a single set of policy options.

Costs

Over the 2-year development of *Measuring Up*, the National Center spent about \$1.5 million, all costs calculated, including staff and consultant time. After that, for each of the subsequent editions the Center spent about \$1 million. Core funders have included the Pew Charitable Trusts, The Ford Foundation, and Atlantic Philanthropies. In addition, The John S. and James L. Knight Foundation, The

Carnegie Corporation of New York, The John D. and Catherine T. MacArthur Foundation, the William R. Kenan, Jr. Charitable Trust, and the Andrew W. Mellon Foundation provided support for specific aspects of the project.

Comparison

The closest reports to those of the National Center are those prepared by the OECD, entitled *Education at a Glance* (OECD 2004). These periodic publications report on educational trends in the OECD countries, although the apparatus of benchmarking to best practice and providing letter grades is not used in those reports. Individual OECD countries, however, such as Ireland and the Netherlands, have undertaken benchmarking exercises that rely on comparisons with "peer" countries. The National Center has commissioned Dr. Alan Wagner, an economist who worked at OECD for many years, to prepare a paper indicating how the Center's Report Card measures could be integrated with the OECD data to generate comparative international measures of higher education performance.

Wagner notes that several OECD countries are roughly the same size as individual states within the USA, suggesting that comparisons of countries with states may have some value. He points out, however, that countries such as France, Germany, Italy, Poland, and the UK are larger, by an order of magnitude, than California, New York, and Texas (Wagner 2005). Other large OECD countries are Mexico, Canada, Japan, and Korea. As Wagner's work progresses to publication stage, it will be possible to learn more about how the *Measuring Up* reports can be used effectively to enhance international comparisons of educational performance.

A second project worth noting is the publication of the Educational Policy Institute, a private, non-profit organization with offices in the U.SA, Canada, and Australia, entitled *Global Higher Education Rankings: Affordability and Accessibility in Comparative Perspective* (Usher and Cervevan 2005). This report builds on measures of tuition and fees, student maintenance costs, financial aid and public subsidies (including tax expenditures), GDP per capita, and relevant participation and population statistics for several countries. EPI apparently expects to continue this publication, which reflects the growing interest in these types of measures.

References

Committee for Economic Development (2005). *Cracks in the education pipeline: A business leader's guide to higher education reform*. Washington, DC: CED.

OECD (2004). *Education at a glance*. Paris: OECD.

Usher, A., Cervevan, A. (2005). *Global higher education rankings*. Toronto: Educational Policy Institute.

Wagner, A. (2005). *Measuring up, internationally*. Unpublished draft report prepared for the National Center for Public Policy and Higher Education. May 2005.

Part III
State (Direct) Regulation of Academic Quality

David D. Dill and Maarja Beerkens

In addition to instruments of information designed to enhance market competition and novel forms of professional self-regulation, a number of innovative regulatory tools have been implemented by countries around the globe as part of their national strategies to assure academic standards. In this part we will examine the strengths and weaknesses of the most widely adopted and influential state regulatory instruments. We begin with "framework" policies designed to influence the learning outcomes of academic degrees, examine the principal national instruments for external quality assurance (i.e., subject assessment, academic audit, and state-based accreditation), and conclude with analyses of contractual policies and national examinations.

At the outset it is important to admit that government intervention in instances where markets or self-regulation fail may itself prove ineffective. Studies of regulation in other industries have identified a number of potential problems or sources of failure for government regulation (Bishop et al. 1995; Wolf 1993). Regulation may prove ineffective because overzealous or misinformed regulators impose excessively high costs on service providers and discourage process innovations that could increase efficiency. A second reason for regulatory agencies to fail may be that they possess a poor understanding of the "technology" underlying the production process being regulated. For this reason, agencies attempting to regulate academic standards may design an intervention that proves ineffective in improving the quality of education. A third explanation for regulatory failure is "capture theory," in which those being regulated gain control or significant influence over the regulatory agency and alter the regulatory framework to favor their own interests over those of the broader public (Baldwin and Cave 1999). A number of these generic problems of public regulation have also manifested themselves in the new policies designed to assure academic quality (Blackmur 2007).

In response to the changing environment of higher education, a number of countries have adopted new national academic qualifications frameworks (Young 2003). The cross-national Bologna framework of bachelor's, master's, and research doctoral degrees being implemented across the EU, as well as the "Dublin Descriptors" and the UK Graduate Standards Program can also be understood in these terms (Bienefeld et al. 2008). Craig McInnis' analysis of the Australian Qualifications

Framework notes that the major objective of this policy is to provide a coherent national structure for the diverse range of vocational and academic qualifications across the three sectors concerned with post-compulsory education: schools, vocational and training, and higher education. The framework was established in 1995 and serves a number of purposes: facilitating flexible pathways in education and training between the sectors, encouraging cross-sectoral collaboration, and promoting recognition of the Australian higher education courses in the globalized market. The framework has only a broad and indirect impact on the setting and maintenance of academic standards in higher education. By presenting broad descriptors of learning outcomes specific to each level of academic degrees, the framework provides some potential reference points for the National Protocols for Higher Education Approval Processes as well as for the auditing of the standards of awards by the Australian Universities Quality Agency. The Qualifications Framework also has helped encourage a focus on student learning outcomes rather than course content in Australian debates about academic standards.

A more refined example of degree frameworks is the UK Subject Benchmarks program analyzed by Gareth Williams. Subject benchmarking was part of an agglomeration of quality assurance measures that emerged in UK higher education during the 1990s in large part as a reaction to the precipitous transition from "elite" to "mass" higher education in the early years of the decade. Rapid growth of student numbers was accompanied by the development of numerous new academic programs, often in interdisciplinary fields, and academic staff in these emerging subjects frequently lacked clear agreement on academic content and student learning outcomes. This aroused political attention, and the national government, as the main provider of funds, required the higher education sector to take steps to ensure that all its degree programs were fit for the purpose. Over 50 subject-benchmarking committees issued reports between 1998 and 2001, setting out in some detail what degree programs in the specialist subjects might be expected to cover. However, despite the initial intentions the benchmark reports were never used for hard regulatory purposes and instead have become developmental tools. The reports have helped to define and legitimize new academic subjects, generated discussion about appropriate academic standards at the subject and university levels, as well as helped to strengthen internal university processes for new course approvals and academic quality assurance.

One of the most significant changes in national higher education policy at the end of the 20th century was the emergence of what Neave has termed "the evaluative state" (Neave 1988). Many national governments initiated and/or subsidized the creation of new agencies and practices designed to assess quality in existing higher education programs and institutions. These new forms of external quality assessment included subject assessments, academic audits, and innovative approaches to accreditation. Each of these practices adopted a similar sequence of activities – an institutional self-study, an external peer review, and, in sharp contrast to the previously discussed voluntary US accreditation process, a public report of findings – but the primary focus of each of these practices differed.

Subject assessments, as noted in Bjorn Stensaker's analysis of the policy instrument adopted in Denmark, involves systematic evaluations of the quality of delivered performance of study programs with an emphasis on curriculum, teaching, and program relevance to graduates and the economy. With a dual purpose related to accountability and improvement, Denmark initiated this system of external quality monitoring in 1992. Organized through an independent agency (EVA), systematic evaluation of study programs has been the dominant method for quality assurance for a number of years. By emphasizing the use of data, observer triangulation, and stability in procedures, as well as extensive dialogue, the evaluations have resulted in noticeable changes in teaching, learning, and study program objectives. The assessments also triggered dialogue and reflection both within higher education institutions as well as between higher education and its stakeholders. In 1999, the agency responsible for the study program evaluations was made permanent by an act in parliament and has since expanded into evaluating primary and secondary education in addition to higher education. At present, the greatest challenge for the Danish national evaluation system is to adjust a well-functioning domestic system of study program evaluation to the emerging international trends toward convergence in quality assurance processes.

In contrast to subject assessments, academic audit, or what Bill Massy usefully defines as "educational quality audit" in his analysis of the version implemented in Hong Kong, focuses on the processes that institutions use to assure themselves that their chosen standards are being achieved. Academic audit emerged in the UK in the early 1990s and was applied by Hong Kong's University Grants Committee (UGC) in the mid-1990s and again in 2002. The UGC's policy problem was how to discharge its obligation to government and the public to assure the quality of teaching and learning without disempowering the institutions, infringing their autonomy, or spending too much in relation to the results achieved. Its solution was to evaluate the maturity of the universities' "education quality work" (EQW): that is, the organized activities dedicated to improving and assuring educational quality. EQW includes the assessment of student learning and also educational goals, curricula, teaching methods, and quality assurance. Audit differs from subject assessment in that it does not directly evaluate the quality of educational provision. Such evaluations are important, but they are difficult for external bodies to achieve in university education. Audit asks whether the entity itself makes the requisite measurements and what it does with the results. It assumes a delegation of responsibility to the institution and verifies that the delegation is being discharged effectively. The audit mantra is, "Trust but check."

Accreditation as implemented by governments in Europe is most similar to US specialized accreditation in its assessment of a program's capacity for quality and in its binary judgment about the attainment of threshold academic standards (Schwarz and Westerheijden 2004). We examine two innovative approaches to state-based accreditation, the German accreditation system and the government mandated General Medical Council accreditation process in the UK.

As Barbara Kehm's analysis of accreditation in Germany makes clear European public policies on accreditation differ from the US voluntary approach in their focus on study programs rather than institutions, in their comprehensive coverage of subject fields, and in their attention as well to the effectiveness of program quality assurance activities. The German accreditation policy was introduced in 1998 as a response to the Bologna Process and transition to a new degree structure. The accreditation process was expected to ensure minimum quality standards in higher education curricula and to assess labor market relevance of newly established Bachelor and Master programs in Germany. All new study programs are expected to be re-accredited every 5 years.

The structure of the German accreditation system consists of two levels: an Accreditation Council and Accreditation Agencies. The Accreditation Council defines standards, procedures, and criteria for the accreditation. The council also accredits individual accreditation agencies, which perform the accreditation in higher education institutions. There are altogether six accreditation agencies in Germany, specialized either by geographical regions or by disciplines. While the accreditation system has established a comprehensive, external quality assessment system for teaching and learning in Germany, the policy suffers from several problems. For example, the accreditation system cannot keep up with the Bologna reforms, consequently many new programs have had to start without accreditation and the system does not attract sufficient number of respected peer reviewers. Moreover, the accreditation system is strongly influenced by European policies, which further complicates the system. Germany is also a federal system in which responsibility for higher education lies with the 16 German states. This leads to a double layer of decentralized responsibilities (national and European) contributing to the emergence of a super-complex system of quality assurance. Finally, program accreditation is an expensive procedure and the expenses have been absorbed by the institutions without any additional state support.

Lee Harvey examines the accreditation and quality processes of the General Medical Council in the UK. In the UK, most of the professions are controlled to a greater or lesser extent by a professional or regulatory body. In most cases, these bodies, and there are more than 100 of them, have some input into professional education. However, the degree to which control is exercised over the profession and over the training of professionals varies enormously from one body to another. Accreditation of programs in the UK provides approval and recognition of academic and vocational awards. Therefore, accreditation is vital if a higher education institution wishes to run courses that offer awards controlled by professional or regulatory bodies. The approach and extent of accreditation varies from the recognition of courses as representing industry standards of training (such as those accredited by the National Council for the Training of Broadcast Journalists) to the complex and tightly constrained legally binding procedures of the General Medical Council. As Harvey emphasizes there is a considerable difference between voluntary regulation by a professional body, such as the many specialized professional accreditors in the USA, and accreditation by a regulatory body. The former is regulation by a body representing, in the last resort, the interests of the professionals. The latter, as in the case of the General Medical Council, is regulation by a non-membership body, established by law to protect the public.

These analyses suggest that all four of these new external assessment practices had the effect of encouraging dialogue and collaboration among academic staff regarding the improvement of student learning and assurance of academic standards within academic institutions. This is not a negligible impact given the increasing incentives in all of higher education for academic staff to invest time and effort in research. However, if overly focused on external control rather than institutional responsibility for improvement, these assessments can encourage a culture of compliance in which institutions invest time and effort on developing policy documents and erecting quality infrastructures to satisfy external assessors rather than on active efforts to assure and improve academic standards. External subject assessments and program accreditation are also very costly to mount and sustain over time. Because their focus is exclusively on the subject level, these assessments also provide limited incentives for the overall institution to develop an effective internal quality assurance process. Audits by contrast are much less costly, applicable to all types of institutions, and provide some of the same incentives for communication and collaboration on the improvement of teaching and learning. But the potential positive impacts of academic audits may be limited if poorly designed, for example, by too comprehensive an assessment that includes other than core educational processes, or by focusing too much on quality assurance documentation rather than on empirical evidence regarding the validity and reliability of internal processes for assuring academic standards.

In addition to these new external assessment practices an early form of external quality assurance in many countries was the development of performance indicators to help assure academic standards (Cave et al. 1997). Performance indicators are often intended to inform government funding in association with new higher education financial instruments such as performance-based funding or university performance contracts. Outcome measures such as graduate placement and salaries are generally valid and socially relevant quality information and could also be valuable general indicators of effectiveness for academic programs if used by institutions. In contrast readily available output measures such as student marks or graduation rates may be unreliable indicators of academic quality because they can be increased by lowering academic standards. For this reason performance-based funding or contracts, which are usually based upon available input, process, and output measures have proven to be an inadequate instrument for assuring academic standards (Burke and Associates 2002; Jongbloed and Vossensteyn 2001) and usually need to be supplemented by government mandates on quality information provision as well as external quality assessments.

Josep Vilalta and Joaquim Brugue analyze the contract-based strategy used in the Catalan university system (Spain) for improving academic quality and funding. Since 1997, the autonomous government of Catalonia (Spain) and the public universities have developed a new tool for university management: The 4-year *programme-contract*. This tool represents a pioneering initiative in Spain. An initiative based on the formulation of objectives for improving universities' institutional and academic quality, reflecting both the priorities of government higher education policy and the individual strategies of each university. The chapter describes in detail the characteristics of the initiative and the principal repercussions and results, as well as giving a critical evaluation from the perspective of university policy and public management.

While common graduate exams exist in certain professional fields such as medicine or teaching and have been utilized in association with external assessments by professional accrediting agencies in the UK and the USA to improve academic standards, common exams do not exist in all subjects. Simon Schwartzman analyzes Brazil's ambitious experiment to nationally assess higher education courses. The assessment consisted of a mandatory test applied to all students graduating from specific course programs in the country. Initially tests were applied the programs with the largest attendance: law, administration, and civil engineering. Ultimately the exams were to be applied to 470,000 students graduating in 26 different fields in 6,500 course programs across the country. The objective of the tests was to provide information to the public on the quality of higher education courses, helping the students and their families to choose where to study, and to provide the Ministry of Education with information that could be used in the accreditation and re-accreditation of higher education institutions. The exam was introduced without previous consultation and was received with strong opposition from student associations, teachers' unions, and many higher education institutions. However, from the beginning, it received strong support in public opinion and in the press. The criticisms ranged from objections to the way the tests were conceived and the results presented, to the failure of one-time exams to measure the actual value-added by an academic program, to broad objections to any kind of measurement of education outcomes. Nonetheless, once in place the results became widely used by students in their choice of institutions and by institutions themselves, particularly in the private sector, to publicize their results or to try to improve them. Poor results, when persistent and associated with other indications of low quality, were supposed to lead to the closing down of the course programs by the education authorities, but, in practice, this seldom happened. In addition the exams generated intensive discussion and consultations among academics about the contents and standards of the different careers. However, when the opposition Laborer's Party won back the presidency, the exams were modified to a voluntary survey of a sample of graduates, thereby diminishing the validity and reliability of the performance information provided.

Taken together the experiences with these new regulatory instruments, the new forms of professional self-regulation, and the innovative instruments of information begin to suggest the outlines of a national policy appropriate to assuring academic standards in the new environment of global economic forces and mass higher education. In our concluding chapter we will synthesize the findings from the PPAQ project and suggest the key components of such a framework.

References

Baldwin, R., Cave, M. (1999). *Understanding regulation: Theory, strategy, and practice*. Oxford: Oxford University Press.

Bienefeld, S., Harris, N., Helle, E., Hopbach, A., Maguire, B., Michalk, B., Rozsnyai, C., Sohm, K., Zaharia, S. E. (2008). *Quality assurance and qualifications frameworks*. Helsinki: European Association for Quality Assurance in Higher Education.

Bishop, M., Kay, J., Mayer, C. (1995). *The regulatory challenge*. New York: Oxford University Press.

Blackmur, D. (2007). The public regulation of higher education qualities: Rationale, processes, and outcomes. In D. F. Westerheijden, B. Stensaker, M. J. Rosa (Eds.). *Quality assurance in higher education: Trends in regulation, translation and transformation* (pp. 15–46). Dordrecht: Springer.

Burke, J. C., Associates (2002). *Funding public colleges and universities for performance: Popularity, problems, and prospects*. Albany, NY: Rockefeller Institute of Government, State University of New York.

Cave, M., Hanney, S., Henkel, M., Kogan, M. (1997). *The use of performance indicators in higher education: The challenge of the quality movement* (3rd ed.). London: Jessica Kingsley.

Jongbloed, B., Vossensteyn, H. (2001). Keeping up performances: An international survey of performance-based funding in higher education. *Journal of Higher Education Policy and Management*, 23(2), 127–145.

Neave, G. (1988). On the cultivation of quality, efficiency and enterprise: An overview of recent trends in higher education in Western Europe, 1986–1988. *European Journal of Education*, 23(1/2), 7–23.

Schwarz, S., Westerheijden, D. (2004). *Accreditation and evaluation in the European Higher Education Area*. Dordrecht: Kluwer.

Wolf, C. (1993). *Markets or governments: Choosing between imperfect alternatives*. Cambridge, MA: MIT Press.

Young, M. F. D. (2003). National qualifications frameworks as a global phenomenon: A comparative perspective. *Journal of Education and Work*, 16(3), 223–237.

Chapter 8
The Australian Qualifications Framework

Craig McInnis

Introduction

Australia is a federation of state and territory governments. The relationship of the Commonwealth and States with respect to the Australian higher education system is historically complex and has ambiguities that make a national policy on qualifications and standards problematic. Responsibilities for the universities are shared between the Commonwealth and eight States and Territories. The States and Territories are responsible for legislation establishing universities and for accreditation approval processes. The Commonwealth provides most of the funds for the universities and is able to steer the policies of the universities by way of financial incentives and penalties related to a range of compliance measures.

All but three of the 39 universities in Australia are public, but the level of Commonwealth funding has declined to a point where it is only a small proportion of the funds in the major institutions. The public universities in Australia have more autonomy than their counterparts in most other countries. Importantly, Australian universities are 'self-accrediting' institutions. Neither the State nor Commonwealth governments directly control or manage what is taught, or how it is taught, or by whom. Universities set their own entry standards as well as the academic standards of their courses. As long as the universities have in place internal mechanisms to assess new course proposals and accredit courses, they are generally free to design and deliver programs and to set standards of student achievement, without the scrutiny of external bodies.

Traditionally, there have been few objective reference points for the standards of Australian awards and there are no significant external moderation processes. From time to time there have been national discipline reviews and public enquiries to examine academic standards. The discipline reviews typically focused on the performance of students in the honours programs (the 4-year undergraduate degree) that had a common currency nationally as a preparation for entry into research

C. McInnis (✉)
University of Melbourne, VIC, Australia

D.D. Dill, M. Beerkens (eds.), *Public Policy for Academic Quality*, Higher Education Dynamics 30, DOI 10.1007/978-90-481-3754-1_8,

higher degrees. Benchmarking across universities, national and international, is increasingly being used, but on the whole the approaches have not been particularly systematic or effective (Stella and Woodhouse 2007, p. 22).

From a series of Australian case studies of universities, Anderson (2001, p. 1) observed that none of the vice chancellors and deans interviewed on the subject of quality assurance 'had any reliable or valid means of knowing how good their degrees were, for example, how intellectual standards might change over time, vary between fields or compare with other institutions.' Anderson found the methods typically used by universities to check on the academic standards of their degrees included:

- Self-reporting of the standards of knowledge and the intellectual characteristics which distinguish between important levels of achievement;
- Graduate destinations, particularly admission to selective higher degree programs;
- Systems of examining that includes external examiners;
- Academic Standards Panels working from a 'Code of Practice for monitoring academic quality and standards' as operated by the Australian Vice-Chancellor's Committee (now Universities Australia) a decade ago;
- University-initiated rolling reviews of departments (or programs) that include scrutiny of the assessed work of students; and
- Scrutiny by professional organizations operating on the assumption that a minimum standard for entry into the relevant profession is identified.

The lack of clear national reference points compounds the problems associated with a lack of consistency or transparency as to how academic standards are determined, applied, monitored, and maintained in Australian universities.

Despite the Commonwealth's interest in quality assurance, it has not pursued any direct measures or processes to set or monitor the academic standards of courses. However, a Review of Australian Higher Education (DEEWR 2008) has proposed a raft of major changes, among them, a new accreditation, quality assurance, and regulatory framework for tertiary education that fully embraces higher education and the vocational and training sector. These proposals were informed by an inquiry into the desirability of a national higher education accreditation body (PhillipsKPA 2008). The review gives particular attention to the need for universities to demonstrate learning outcomes and academic standards.

The Policy Problem

The Ministerial Council on Education, Employment, Training and Youth Affairs (MCEETYA) brings together the various State and Commonwealth ministries and agencies to coordinate policies and the development of national agreements. The

Australian Qualifications Framework (AQF) was introduced in 1995 and phased in over 5 years. It was initiated primarily to bring a diverse array of qualifications and titles together as one nationally recognized scheme. It specifies qualifications titles, their characteristic learning outcomes, and pathways to them (DEEWR 2008, p. 129). This is the critical reference document for universities and other providers as they develop new courses or amend courses to change their standing.

The main focus in the initial formulation of the AQF was on supporting reforms in the vocational and training sectors across the States. In the vocational and training sectors what employers and institutions needed to know, from an increasingly complex array of award-bearing courses across the states, was what students had learned. This was most obviously critical in terms of credit transfer within the sector, but as universities made their existing course more accessible and created more vocationally oriented courses, the credibility of their programs was at risk at the entry point where judgments are made about the relative standing of prior qualifications.

The policy context has changed dramatically since the AQF was first devised and more so over the past 5 years. There has been ongoing concern about the international recognition of Australian qualifications and its impact on both the recruitment of international students and the general reputation of Australian higher education. International recognition of the quality of its qualifications is of critical concern to the Australian tertiary education sector. In particular, Australian universities are heavily dependent on their capacity to attract international fee-paying students and can ill-afford to allow any uncertainty or ambiguity about course structures, pathways, and standards.

The Commonwealth and State governments are highly sensitive to the economic and social significance of international students. The assessment of quality by overseas students, institutions, and governments starts with their recognition of the nature, currency, and transferability of qualifications offered by Australian institutions and where they stand in the international market. Of growing importance is the recognition factor for domestic students, many of whom expect to work or study overseas and who increasingly will be competing for places in overseas universities. Credit transfer arrangements are motivated by the need to build a national skill base, to enable the system to respond to rapid changes in workplace requirements for skilled workers, and to promote lifelong learning. From the Commonwealth perspective Australia has been relatively weak on the clarity of its credit transfer between institutions and therefore exposed to the emerging competition from universities developing the Bologna model.

The AQF provides a means of resolving or at least managing these policy problems and minimizing three levels of tension, that is, State and Commonwealth priorities, the different missions of the higher education and vocational sectors, and the relationship between industry and institutional qualifications. By providing a systematically formulated frame of reference for the assessment of the expected broad learning outcomes of higher education qualifications, the Framework fills a policy gap that gives other quality assurance mechanisms some reference points for the accreditation of providers and courses and for auditing the performance of institutions.

The move toward a set of formal protocols for university accreditation was prompted by an application in 1999 by Greenwich private university for listing on the National Register. It exposed the lack of a national approach to the accreditation of higher education. Since the AQF was not designed with quality assurance and academic standards as a central issue, its contribution to resolving the policy problem of academic standards relates to its inclusion as part of a national quality assurance framework introduced in 2000 following the Greenwich issue. The main policy concern in this respect was the potential threat to Australia's reputation for high academic standards in the face of new private for-profit providers and a more general public concern about lower standards or 'soft-marking' for international students.

The sharper focus of the AQF on the higher education sector was also influenced by the rapid expansion of student numbers in the late 1990s and the dramatic increase in Australia's share of the international student market. At that time, the forms of quality assurance for higher education at a national level were limited, and the sophistication of processes at the level of the institution varied considerably. As the imperative for a 'truly national system, recognized for its high quality' became more pressing, the Commonwealth and States agreed in 2000 to develop the Australian Higher Education Quality Assurance Framework (AHEQAF). At this point the AQF took on new significance with its broad descriptors of levels of knowledge, skills, and abilities characterized as learning outcomes specific to each level of academic qualification.

At another level of policy problem, the arrangements between the Commonwealth and the States have been the subject of ongoing debate as the Commonwealth argued for a rationalizing of responsibility. The AQF is one example of an attempt to deal with this policy problem of:

> . . . complexities in the shared arrangements, a lack of consistency to the largely historically determined nature of them, and limitations and constraints that result from having nine jurisdictions involved. (DEST 2005, p. 1)

The Commonwealth pointed to the potential confusion created by inconsistencies in the implementation of National Protocols for Higher Education Processes by the States (DEST 2005) and the risks involved for consumers who need to be assured that providers of higher education in Australia have met certain criteria and standards.

Main Elements of the Australian Qualifications Framework

The AQF is broadly characterized as a national policy instrument within the AHEQAF to protect the quality of Australian education and training. It comprises a unified system of national qualifications for all education sectors to ensure that there is nationally consistent recognition of the outcomes of qualifications awarded by Australian schools, colleges, and universities. The Framework links together 15 qualifications (Table 8.1) as a highly visible quality-assured national system.

Table 8.1 Qualifications by the educational sector in which they are most commonly used

Vocational Education and Training Sector	Higher Education Sector
	Doctoral degree
	Master's degree
Vocational graduate diploma	Graduate diploma
Vocational graduate certificate	Graduate certificate
	Bachelor's degree
Advanced diploma	Associate degree, advanced diploma
Diploma	
Senior secondary certificates:	
Certificate IV	
Certificate III	
Certificate II	
Certificate I	

The conceptual approach embedded in the operational objectives of the AQF is essentially one of providing a formal classification process that regulates by identifying and authorizing the various agencies responsible for accrediting the qualifications and maintaining a public register of those authorized. It defines the differences between the qualifications in terms of the levels, expectations, and learning outcomes (broadly conceived).

The Framework has multiple purposes and is expected to meet the needs of a diverse range of stakeholders. It embraces, for example, the lifelong learning agenda of particular interest to the vocational and training sector, the promotion of access and equity policy, the provision of more and higher quality vocational training, and the recognition of prior learning.

The key elements of the Framework are

- A set of national guidelines and descriptors of all awards for each of the current qualifications issued in Australian schools, vocational education and training, and higher education sectors;
- A set of principles for articulation and credit transfer;
- A register of authorities at the State levels empowered by the Australian Government to accredit post-compulsory education and training to accredit qualifications and to issue qualifications; and
- A series of protocols for issuing qualifications and a structure for monitoring implementation of the AQF and advising ministers, including recommending any changes.

Table 8.1 perhaps gives the impression that there are tight boundaries between the qualifications. However, the qualifications are actually grouped according to the sector in which they are most commonly used. Where the vocational and higher education sectors have qualification titles in common, the AQF Guidelines suggest they are ‘equivalent although sector-differentiated’; that is, they have parity of esteem: ‘there are no standardized rankings or equivalences between different qualifications issued in different sectors, as these qualifications recognize different types of learning reflecting the distinctive educational responsibilities of each sector.’

The Framework guidelines (AQF Implementation Handbook 2005) spell out the main criteria for defining the qualifications listed in Table 8.1. The characteristics of each qualification are expressed in terms of learning outcomes in an attempt to provide common ground for qualifications across the sectors. The table suggests a hierarchy of qualifications and a vertical pathway to the top. The pattern of progression for individual students is not necessarily along that line and there is blurring of boundaries between the sectors, which makes the setting, monitoring, and consistency of standards more difficult to manage at the system level.

The guidelines for the bachelor’s degree illustrate the detail of the Framework and the extent of its influence. First, concerning who has the authority for the learning outcomes of the degree, there is a generic statement to the effect that the universities have autonomy on these matters. Second, with respect to standards in the higher education sector, the guidelines refer to the responsibility for the assessment of individuals as resting ultimately with the provider institution or organization and not those who actually conduct the testing of achievement.

The third element is the ‘Characteristics of the Learning Outcomes’ provided for each level of qualification. For the bachelor’s degree, they include, for example,

- The acquisition of a systematic and coherent body of knowledge;
- The development of academic skills and attributes necessary to undertake research and to comprehend and evaluate new information;
- A foundation for self-directed and lifelong learning; and
- The acquisition of interpersonal and teamwork skills appropriate to employment and/or further study.

Reference is also made to the ‘significant depth and progressive development of the course content’ in the bachelor’s degree as the basis for postgraduate study and professional careers.

The introduction of the associate degree provides an example of the limited extent to which the AQF can contribute to the assurance of quality. The associate degree is of 2 years’ duration following the end of secondary school, that is, post–year 12. Again, it is the universities that have the authority to set the objectives and academic requirements of the courses. There is potentially a high level of permeability in the boundaries between the associate and bachelor’s degrees and between associate degrees and advanced diplomas shown in Table 8.1.

The characteristics of the learning outcomes for the associate degree specified by the AQF include the ‘acquisition of the foundational underpinnings of one or more disciplines’ to emphasize that the degree is ‘generally but not exclusively articulated with relevant bachelor’s degree programs’. The degree is also intended to provide a broad-based point of entry to employment especially in the associate professional occupations.

The AQF suggests that the distinctive features of the associate degree include the foundational research-based knowledge of an academic discipline and the broad, often multidisciplinary content. The critical point with respect to academic standards, from the higher education perspective, is that the associate degree may also be offered by technical and further education colleges and by private training organizations, although it does not appear in the vocational section of Table 8.1. The nature of the research base is open to interpretation. However, the 2008 Review gives considerable attention to the significance of research. The relationship between the AQF and the National Protocols is not helpful on this point. As the Guthrie Review (2005) – charged with further development of the approvals process – pointed out, the Protocols were 'silent on the requirements for scholarship and research to underpin the course approved as is the expectation at Australian universities'. It noted that the States and Territories have taken different stances on this issue with some assessing the qualifications of staff as a means of meeting the requirements, while others expecting evidence of staff research activity. This then raises questions as to the absence of a clear distinction between university and other higher education providers.

The Guthrie Review (2005, p. 22) also drew attention to the potential for undermining both the AQF and the awards themselves, as a result of the different historical arrangements across the States and Territories. It identified a great deal of variation across the States and Territories in the approvals process and criteria and noted that 'there is a long way to go to achieve consistency' (Guthrie et al. 2005, p. 9). One of the enduring problems embedded in such an arrangement is that the ability of the universities to maintain standards is not really tested closely by the AQF or by the protocols. The Guthrie Review (2005, p. 6) appeared to accept the assumption that the universities have 'long standing traditions about required standards for awards and established academic processes to monitor those standards'. It argued that the task of the third 'leg' of the quality assurance framework, AUQA, is to audit the extent to which this occurs, but not the actual standards of student learning outcomes.

The National Register

The National Register in the Qualifications Framework has five sub-categories:

1. Government Accreditation Authorities;
2. Universities and Other Self-Accrediting Higher Education Institutions;
3. Non-Self-Accrediting Higher Education Institutions and their AQF-approved qualifications;
4. Registered Training Organizations and their AQF-approved qualifications (VET sector); and
5. Overseas Higher Education Institutions and their AQF-comparable approved qualifications.

To be listed on the National Register a university must have been recognized or established by a State or Territory Minister. Again, it needs to be noted that the Commonwealth does not have the authority to approve the establishment of providers. All Australian universities that have been recognized or established by a MCEETYA Minister are listed on the Register. Higher education courses delivered by approved non-self-accrediting providers are also listed. Only government accreditation authorities listed on the AQF Register are able to accredit courses. The effect of this is that by approving a course for listing, the State and Territory ministers are deemed to vouch for the quality of the higher education provider, the course, or the accreditation authority.

Advisory Board

An Advisory Board manages the AQF. The Advisory Board has a modest secretariat, and most of the cost of accrediting new non-self-accrediting providers is borne by the State agencies. This process does not occur very often, and the cost of each exercise is therefore not readily available.

The Advisory Board has an essentially custodial rather than operational role in the implementation of the AQF. It manages a fairly straightforward bureaucratic process of inventory keeping and procedural measures that provide national order and consensus to deal with the diversity of qualifications across sectors. It is not empowered to take initiatives to shape the standards of the qualifications it registers, although the members, representing the diverse cross-sectoral interest groups, bring their expertise to bear by informing and shaping the national agenda. Similarly, the extensive quality assurance processes that underpin the Framework qualifications are the responsibility of each of the sectors.

Implementation of the Framework

It was not until 2000 that the AQF took a more central place in the overall scheme of quality assurance for higher education. In 2000, MCEETYA agreed on an Australian Higher Education Quality Assurance Framework with five elements (Fig. 8.1) including the Qualifications Framework. The two significant new elements were the establishment of the Australian Universities Quality Agency (AUQA) and formulation of the National Protocols for Higher Education Approval Processes.

AUQA and the Qualifications Framework

AUQA conducts quality assurance audits of higher education institutions on a cyclical basis, and, importantly, it also audits and reports on the accreditation bodies of the states and territories. AUQA is required to report on the 'relative standards of the Australian higher education system ... including their international standing' and

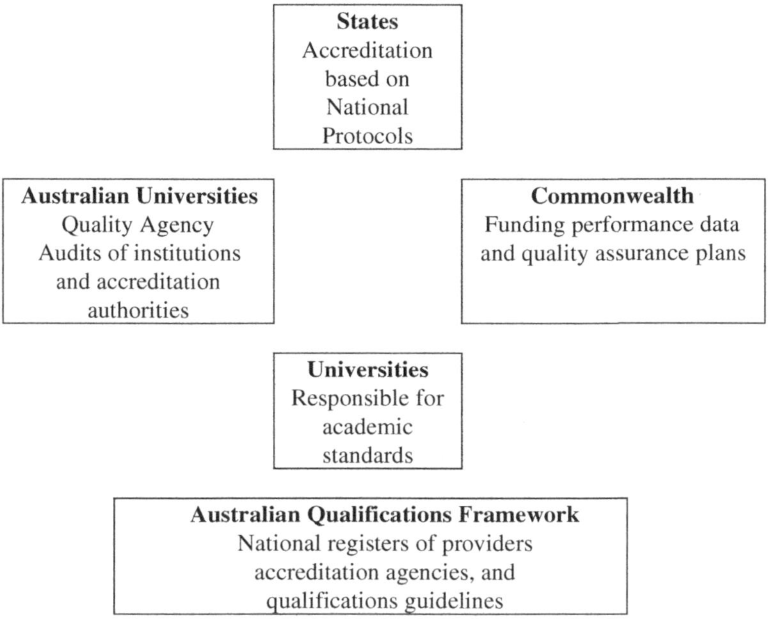

Fig. 8.1 The Australian higher education quality assurance framework

to address the issue of the maintenance, deterioration or improvement of academic standards. To do this, AUQA looks at the ways in which institutions set and access standards including moderation methods, formal benchmarking (including international benchmarking), and less structured inter-institutional comparisons (including international comparisons).

The Qualifications Framework provides AUQA with a reference point for national standards to the extent that AUQA audits cover the basic expectations of awards in terms of learning outcomes, as well as accreditation processes and the operations of the agencies that conduct the accreditations. AUQA's connection to the National Protocols and academic standards comes through its role in the audit process of assessing whether a university's objectives are consistent with the established criteria for a university. AUQA also checks to verify that agencies involved in the accreditation process of non self-accrediting institutions are applying the protocols correctly.

MCEETYA advises AUQA on new and emerging issues related to the Qualifications Framework. It has, for example, suggested to AUQA that it might take a more active role in auditing universities against a set of 'Good Practice Principles for Credit Transfer and Articulation' it developed and adopted. This does not enable MCEETYA, AUQA, or AQF to set standards, but it does in principle provide a check against university practices that might inflate the value of one qualification or dilute the standards of another in the process of determining their relative merit for selection purposes.

The National Protocols for Higher Education Approval Processes

The National Protocols for Higher Education Approval Processes are the key regulatory mechanism for defining and accrediting universities in Australia. The National Protocols were revised in July 2006 for implementation in December 2007. The States and Commonwealth agreed on significant revisions to clarify approval processes and criteria to address the increasing diversity of higher education providers. They have the express purpose of ensuring that consistent criteria and standards are used in the approval process for new institutions and courses.

> The five separate Protocols set out criteria and processes for approving universities and other types of higher education institutions. State and territory governments accredit courses where the institution is not authorized to do so (PhillipsKPA, 2008, p. 8). The protocols cover the following:

- Protocol A relates to all higher education institutions;
- Protocol B relates to the registration of non-self-accrediting higher education institutions and the accreditation of their higher education courses;
- Protocol C relates to awarding self-accrediting authority to higher education institutions other than universities;
- Protocol D relates to establishing Australian universities; and
- Protocol E relates to overseas higher education institutions seeking to operate in Australia.

The revised protocols specify standards for the registration and/or accreditation processes that enable certification that an institution, or a course, meets appropriate standards. They were accompanied by a set of National Guidelines for each category of higher education institutions. The guidelines provide a detailed specification of the requirements outlined in the protocols with the aim of improving the level of national consistency in the application of the revised protocols. The intent of the governments to date has been to work toward harmonization in arrangements rather than absolute uniformity (PhillipsKPA 2008, p. 12).

In the previous version the protocol spelling out the criteria for recognition as a university was the outcome of a major debate. This included offering a broad range of disciplines, engaging in research, and having a culture of sustained scholarship. The term university is now protected under a Commonwealth Corporations Act (2001), but it is the States that can take action to prevent and penalize 'degree mills', which may be falsely presenting themselves as Australian universities. The Australian strategy has been to define what makes a university and then to grant institutions the right to be self-accrediting. There is generally no regular re-accreditation process. However, the 2008 Review of Higher Education recommends that all universities should be re-accredited and suggests a 10-year cycle.

It is important to note that existing higher education institutions (including universities) are to be assessed regularly through the standard quality assurance processes, including external quality audits that apply to each institution. For example, the universities would satisfy this requirement through the regular external audit by

AUQA whereas overseas higher education institutions operating in Australia would be subject to the quality assurance requirements of their overseas accrediting authority and also any other requirements specified as conditions of the approval to operate in Australia. The growth of private providers has been a key factor in testing the value and effectiveness of the frameworks. There are currently around 150 non-university providers in Australia. The number of non-self-accrediting institutions grew by 35% in the 3 years from 2005 to 2007 and is anticipated to grow at an increasing rate (PhillipsKPA 2008, p. 25).

Mutual recognition under the National Protocols and Guidelines is a major point of concern for governments, providers and students. This involves a standard being recognized as equivalent between jurisdictions. That is, an organization registered in one State should be able to operate in another since the standards underpinning registration should be the same.

The arguments for maintaining the current arrangements are largely centered on the belief that local variations are appropriate and respond to jurisdiction-specific policies and priorities. The arguments against the current arrangements include the views that the integrity of Australian qualifications in off shore markets is potentially at risk, and that there are considerable inefficiencies associated with eight different regulatory bodies. More particularly, it is argued that

> ... higher education policy is effectively a national responsibility and higher education provision is increasingly taking place in the context of national and international markets for education, requiring the maximum level of national consistency in costs, processes and decision-making for recognition and accreditation.... (PhillipsKPA 2008, p. 48)

Two models of accreditation were proposed by this inquiry: a harmonized model that would coordinate current arrangements to maximize national consistency while allowing for local conditions at the State levels and variations on a uniform model that would bring together all accreditation processes and standard setting activities under a central body.

The revised National Protocols also outline obligations on government accreditation authorities, including the requirement for all jurisdictions to undergo regular external quality audits by AUQA. The revised National Protocols will apply to both new and existing institutions. Compliance will be regularly assessed through the standard quality assurance processes that apply to each institution (MCEETYA 2006). To give effect to these revised protocols, National Guidelines for Higher Education Approval Processes were developed. However, the implementation has been overtaken by the MCEETYA inquiry, and the 2008 Review of Higher Education has effectively put these developments on hold.

Regardless of the specific arrangements that result from these initiatives, this represents the start of a new era in national consistency across the states and territories. The guidelines were aimed at achieving greater national consistency in higher education approvals to allow for the proposed introduction of new types of higher education institutions in Australia, including

- *Specialist universities:* High-quality higher education institutions meeting the same requirements as other universities, with the exception of breadth of fields of study. They will be required to offer courses including research masters and doctorates, and undertake research activity, in one or two fields of study only.
- *Self-accrediting institutions other than universities:* Selected non-self-accrediting providers, usually with a strong track record in re-accreditation, will be able to seek authority to accredit their own courses.
- *University colleges:* This title will be protected under the revised National Protocols, reserved for use by new universities, which at point of establishment need only undertake research and research training in one field. It may also be used by provisionally approved 'greenfield' institutions, based on a plan, which would normally be mentored by an existing university.
- *Overseas institutions:* Clearer rules around entry and their use of university title will assist more overseas institutions to establish a presence in Australia and offer their own qualifications, thereby increasing choice for students.

Again, these have been developed further by the 2008 Review proposals that give particular attention to research and the nexus between research and teaching, as a distinctive feature of universities. The review proposes modifications to the 2007 National Protocols with three major types of institutions:

- *Comprehensive universities:* These should provide research higher degrees in at least three broad fields and undertake research in all fields in which research degrees are offered. They should also conduct research in the fields in which coursework degrees are offered;
- *Specialist universities:* These should provide research higher degrees in one or two broad fields and research in those fields. They should also conduct research in the fields in which coursework degrees are offered;
- *Other higher education institutions:* Unlike the first two types, these will not be required to conduct research. They will focus on delivering degrees at levels and in any number of fields as they are accredited.

Under these proposals university colleges could still be established on a pathway to full university status by initially delivering qualifications up to master's coursework degrees in at least three broad fields of study and research master's and PhDs in at least one field.

Impact of the Australian Qualifications Framework

The AQF was first developed with the brief to 'protect' the qualifications guidelines and to 'promote and monitor' national implementation of the Framework. The passive nature of the language is telling. While the Advisory Board can advise, protect, guide, register, inform, promote, and monitor, it is not in a position to directly initiate new qualifications or to set academic standards. It has a role in responding to initiatives from providers or from governments.

There has been no serious opposition to the Framework from the higher education sector, and it has been uncontroversial, perhaps because it is relatively weak in character (Young 2003, p. 226). It serves the purpose of informing students as to what different qualifications mean relative to others, and it provides something of a gatekeeper role for institutions introducing new courses. Like the national qualifications frameworks of Ireland and Scotland, it has, as Young suggests, limited and indirect power. Clearly, if the 2008 Review proposals to 'modernize' the AQF are adopted, this will change (DEEWR 2008, p. 193).

It is noteworthy that while the peak body of the universities, Universities Australia, does not appear in the AQF it is typically referred to as having a long-standing role in developing guidelines relevant to quality assurance, for example, a 'Code of Practice' for maintaining and monitoring academic quality and standards in higher degrees. In practice, it appears there is tacit agreement at the level of the AQF Advisory Board that the qualifications descriptors are effectively 'owned' by the universities through the membership of Universities Australia on the Advisory Board. Universities Australia provides advice and comment on the Framework descriptors and proposals for new qualifications.

Lifelong Learning

As part of the national strategy for lifelong learning, another test of the impact of the AQF arose from a suggestion that it could be used as the basis for a customized portfolio approach that recognizes modules of learning completed, through different providers at different times. It was argued, hypothetically, that a student might take a mix of subjects from diverse providers and the 'total package' could be recognized as a credential such as a bachelor's degree. This fairly unlikely scenario provided a fundamental test of the role of the AQF as a policy instrument to assure quality. A key element of the descriptor for the bachelor's degree is 'the acquisition of a systematic and coherent body of knowledge'. The portfolio proposal raised the question of how significant the coherence of a degree program needs to be and how that is addressed in quality assurance processes.

The Advisory Board responded that while flexibility in pathways is central to the recognition of prior learning in all sectors, the portfolio notion was likely to devalue the degree awarded. It points to the importance of establishing appropriate assessment guidelines based on agreed national standards for the qualification level. All that the Framework can do is to verify indirectly that the qualifications awarded in higher education have taken cognizance of the AQF descriptors and learning outcomes appropriate to the level of the award. To close this gap, the Advisory Committee argued that the only way to deal with a pathways or portfolio approach is to establish an assessment authority that would be responsible for assessment-only pathways or portfolios.

Conclusions

The broad educational goals of the Australian Qualifications Framework are common to most national qualifications systems. That is, they create transparency for the users, minimize barriers to progression, and maximize access, flexibility, and portability between different sectors (Young 2003, p. 224). There are, however, deficiencies and limits to the impact of national qualifications frameworks generally as quality assurance instruments that promote and enhance academic standards (Blackmur 2004; McInnis 2003; Young 2003), and these are being addressed by the current policy developments in Australia.

The proposed reforms from the 2008 Review of Australian Higher Education in relation to accreditation, quality assurance, and the regulatory framework focus directly on the need to demonstrate outcomes and standards. At the time of writing it is not possible to predict just how the Commonwealth will respond to the recommendations of the review. Nevertheless, the outcomes of the review provide a picture of some urgency to meet the significant and growing concerns about the complex, fragmented, and inefficient arrangements outlined above. Among other things the review concluded that:

- The quality assurance framework is too focused on inputs and processes and does not give sufficient weight to assuring and demonstrating outcomes and standards;
- Arrangements for mutual recognition of providers and courses operating across state and territory boundaries are inefficient and do not operate effectively; and
- Within higher education the framework is applied unevenly so that not all providers are reaccredited on a regular basis.

The primary motive of the proposed changes is to ensure that Australia must maintain confidence in accreditation and quality assurance to enhance the international position of its higher education system. The key recommendation is for the adoption of a national framework for higher education accreditation, quality assurance, and regulation. The features of this model include an AQF 'with enhanced architecture and updated and more coherent descriptors of learning outcomes.' (DEEWR 2008, p. 116). The review points out that, with the exception of the associate degree, there has been little change to the AQF since it was introduced. The review argues that since the AQF is a key element of the quality assurance framework for both the vocational and higher education, responsibility for a revised qualifications framework should rest with a national regulatory body.

The proposal for an independent national regulatory body is a major initiative that if adopted will, in addition to taking responsibility for the AQF, accredit new providers including new universities, re-accredit existing universities over a 10-year cycle, and carry out quality audits of all providers focused on the institution's academic standards and the processes for setting, monitoring, and maintaining them. It is proposed that the regulatory body will have the power to remove the right of institutions to operate if necessary.

Regardless of the extent to which the AQF is currently limited in its capacity to directly ensure academic standards, it is now firmly embedded as part of a suite of measures and organizations that MCEETYA and the Commonwealth can call into play at a national level. Unlike some models elsewhere, the profile of the AQF has been enhanced beyond its initial goals by the quality assurance function. The potential of the Framework as a policy instrument to monitor and improve academic standards is likely to be strengthened if assessment processes and standards at discipline level are clearly linked to learning outcomes at the appropriate level of the AQF (DEEWR 2008, p. 136).

References

Anderson, D. (2001). When somebody's watching: Perspectives on academic audit. Victoria University Public Seminar, 6 July 2001, Melbourne, VIC.

Australian Qualifications Framework Implementation Handbook (AQF) (2005). http://www.aqf.edu.au/implem.htm. Accessed 12 June 2009.

Blackmur, D. (2004). Issues in higher education quality assurance. *Australian Journal of Public Administration*, 63(2), 105–116.

Department of Education, Employment and Workplace Relations (DEEWR) (2008). Review of Australian Higher Education, Final Report. Canberra: Australian Government. http://www.deewr.gov.au/HigherEducation/Review/Pages/ReviewofAustralianHigherEducationReport.aspx. Accessed 12 June 2009.

Department of Education, Science and Training (DEST) (2005). Building better foundations for higher education in Australia: A discussion about re-aligning Commonwealth-State responsibilities. http://www.dest.gov.au/sectors/higher_education/policy_issues_reviews/key_issues/commonwealth_state_responsibility_higher_ed.htm. Accessed 12 June 2009.

Guthrie, G., Johnston, S., King, R. (2005). *Further development of the national protocols for higher education approval processes*. Canberra, ACT: DEST. http://www.dest.gov.au/sectors/higher_education/policy_issues_reviews/reviews/guthrie_review/www. Accessed 12 June 2009.

McInnis, C. (2003). Touchstones for excellence: assessing institutional integrity in diverse settings and systems. In E. De Corte (Ed.), *Excellence in higher education*. London: Portland Press.

Ministerial Council on Education, Employment, Training and Youth Affairs (MCEETYA) (2006). National protocols for higher education approval processes. http://www.mceetya.edu.au/verve/_resources/RevisedNationalProtocols20081.pdf. Accessed 14 March 2007.

PhillipsKPA (2008). Inquiry into the desirability of a national higher education accreditation body. Final report to the Joint Committee on Higher Education. http://www.mceetya.edu.au/verve/_resources/JCHE_National_HE_Accreditation_Agency_June_08_Final_Report_.pdf.

Stella, A., & Woodhouse, D. (2007). *Benchmarking in Australian higher education: A thematic analysis of AUQA audit reports*. Melbourne, VIC: Australian Universities Quality Agency.

Young, M. F. D. (2003). National qualifications frameworks as a global phenomenon: a comparative perspective. *Journal of Education and Work*, 16(3), 223–237.

Chapter 9
Subject Benchmarking in the UK

Gareth Williams

Background

Before 1989 the UK higher education system consisted of two sectors, an autonomous university sector and a public sector under the control of local education (county) authorities. The universities, though publicly funded up to about three-quarters of their income, had almost complete financial and academic independence. Public funds were unconditional provided they were spent in accordance with the universities' charters, which were couched in very broad terms. There were only two external brakes on their freedom to teach their students what they wanted and how they wanted. One was the external examiner system, whereby all award-bearing programmes of bachelor's degree level and above had at least one examiner from another UK university or, very occasionally, a university considered to be of equivalent standard in another country. The other was the professional and statutory bodies (PSB) such as those for medical doctors, law and various branches of engineering. Their interest was based on the fact that recognised university qualifications gave certain exemptions to candidates for professional qualifications. However, the real guarantor of quality and standards was the fact that it was a meritocratic (see, e.g., Young 1961), or elite, scholarly system in which most of the students and the staff who taught them were from the top levels of the ability range and universities were jealous of their reputations.

'Public sector higher education' institutions were under tighter regulation. Their finances were controlled by the local education authorities which owned them, and their degree level teaching was regulated by the Council for National Academic Awards (CNAA) created in 1965, whose essential function was to ensure that the qualifications awarded by the polytechnics and other colleges were equivalent in standard to those of the universities. By the 1980s the polytechnics were chafing under what they considered a restrictive regime compared with the universities and the government was becoming dissatisfied with the control exercised by the local

G. Williams (✉)
Institute of Education, University of London, London, UK

D.D. Dill, M. Beerkens (eds.), *Public Policy for Academic Quality*, Higher Education Dynamics 30, DOI 10.1007/978-90-481-3754-1_9,

education authorities. The 1988 Education Reform Act gave the major public higher education institutions almost as much autonomy as the universities and the process was completed in the 1992 Higher and Further Education Act which abolished the CNAA and enabled all but a rump of smaller institutions to be transformed into autonomous universities.[1]

At the same time as these reforms were occurring, the government stimulated, through a new formula funding methodology, an explosive expansion of student numbers (see Bekhradnia 2003). Student numbers increased by 75 per cent between 1989 and 1994 and UK higher education was transformed from an elite to a mass system within half a decade. All universities, including the new ones, were allowed to start new degree courses with no need for authorisation from outside the university. Many new degree courses were established in response to student demand. Public funding grew much less rapidly than student numbers following the introduction of student number based formula funding, which encouraged universities and colleges to expand student numbers at marginal costs (see Williams 2004). Huge increases in university and college admissions and declining income per student from public funds led to growing concerns in government both about the academic potential of some of the students embarking on higher education programmes and on the capacity of the institutions to provide them with teaching of satisfactory quality.

Subject benchmarking was one of the many quality-related innovations in the national regulation of higher education that emerged in the 1990s in response to these concerns. It reflected a gradual realisation that the assurance and enhancement of the quality of learning and teaching in higher education was not solely an issue of institutional resources and management arrangements, nor of the processes of teaching and learning, though both have been the subject of considerable policy concern over the past two decades, but also of the content of academic programmes.

The Policy Issue: 'Graduateness'

One debate of the mid-1990s was the relationship between 'fitness for purpose' and 'fitness of purpose'. The early work of the Academic Audit Unit (AAU) set up at the end of the 1980s by the academically elite pre-1992 universities was concerned primarily to ensure that the universities had teaching and examining procedures that underpinned their own criteria for the award of degrees. Were the procedures fit for the purpose for which they were designed? The underlying ideology was that the intentions and capabilities of these institutions were good but they may have not been careful about the administrative details.

---

[1] This process came to fruition in 2004/2005 when most of the remaining higher education institutions were transformed into teaching universities.

However, this approach was deemed to be inadequate by the government in the early 1990s, and it established its own Higher Education Quality Council (HEQC), in part as a replacement for the CNAA which had been responsible for standards in the non-university sector. The immediate origins of the Graduate Standards Programme of the HEQC lay in the concerns about degree standards first voiced publicly by the Secretary of State for Education and Science, John Patten, in April 1994. With the rapid expansion of the early 1990s, the transformation of a large number of previously publicly regulated polytechnics and other higher education institutions into autonomous universities and the establishment of a very much wider range of degree course subjects, concern began to shift from ensuring that universities were meeting their own quality criteria (fitness *for* purpose) towards a concern that the degree programmes being offered throughout the system were appropriate for bachelor's and higher degree programmes (fitness *of* purpose).

Patten asked the Higher Education Quality Council (HEQC) to investigate the broad comparability in the standards of degrees offered by different institutions. In response, the Council embarked on a 2-year programme to investigate the issue of 'graduateness' – what attributes someone who had qualified as 'a graduate' might be expected to possess. The Council's own rationale for this programme included the following:

> UK higher education has vastly increased in size and heterogeneity over the last generation and especially during the 1990. The informal mechanisms that were believed to ensure comparability in a small, homogeneous system seem increasingly unlikely to be effective in the present, greatly diversified one. What is more, the rapid increase in the scale and cost of higher education continues to fuel demands for its activities to be more transparent and more publicly accountable; the large-scale introduction of modular programmes has necessitated greater explicitness of purpose and has focused attention on key issues relating to assessment, comparability and related matters that were less visible before;
>
> many new subjects have entered HE, or been developed within it, in which degree qualifications had not previously been awarded: these are now confronted with defining their own understandings of graduateness;
>
> the diversity of types of programme of study now available has increased. This enables students to attain a degree by many different kinds of learning experience but raises the issue of how to establish the comparability of outcomes;
>
> the growth of collaborative work of various kinds (including franchising [sometimes overseas], the validation of the awards of one institution by another or partnerships between higher education and further education institutions) reinforces the need for clarity about the concept of 'graduateness' that is being shared;
>
> growth in the number of students at a time of declining per capita resources makes it necessary to establish clearly and publicly what is represented by a degree;
>
> increasing diversity in the qualifications of those entering HE programmes, a greater variety of modes of study (including innovations such as work-based learning or the accreditation of prior learning) tend to make insufficient the conventional assumption that a degree represents the successful completion of 3 or 4 years of full-time study following the award of A-levels;
>
> the growing internationalisation of higher education has made it more important to clarify the standards of UK degrees in relation to those in other countries. (HEQC 1995, p. 2)

The 'Graduateness' study set out to determine the following:

- whether it is possible to identify shared attributes (that is to say, attributes that go beyond the knowledge, understanding, skill and other qualities that are specific to their field[s] of study) that graduates are expected to possess;
- the extent to which such attributes are common to all programmes of study or to particular clusters of programmes;
- whether any particular attributes may be identified that are specific to a given subject yet would appear to be applicable beyond that subject;
- whether generic attributes could be useful in helping to define and establish threshold standards for all degrees, clusters of degrees, or degrees in certain subjects, fields, or sub-fields;
- if it were judged possible to define generic attributes that might play a part in the definition of threshold standards for degrees, how the student's possession of these attributes might best be assessed. (HEQC 1995, p. 3)

There were two main lines of development work: in the first, a pilot project on benchmarking assessment practice, the Council worked with five subject communities[2] in order to establish the feasibility of defining threshold standards. The second stage of HEQC's work encompassed fourteen subject communities[3] in a bid to examine the feasibility of using the concept of 'graduateness' – the attributes that a person graduating with a degree might be expected to possess. This project developed a profile of graduate qualities on to which subject groups could map their disciplines and identify the qualities of their graduates (Wisby 2002).

The outcome of the programme was inconclusive and it became clear that apart from some generalisations about 'communication skills' and 'critical thinking' the concept of a graduate, at least in the UK context was very subject specific.

Further development work by the HEQC was overtaken in 1997 by the abolition of the AAU and the HEQC and the establishment of the Quality Assurance Agency (QAA). At the same time a National Committee of Enquiry into Higher Education – the Dearing Committee – was established. The development of subject-based benchmark standards was a key recommendation of the Dearing report (Dearing 1997). The Committee recognised that the massive expansion of the early 1990s had led to much greater diversity in higher education provision but considered that 'the task facing higher education is to reconcile that desirable diversity with achievement of reasonable consistency in standards of awards' (para 10.3).

As one of a number of recommendations covering 'a national framework of qualifications', 'standards of awards' and 'quality assurance of the students' learning experiences', recommendation 21 proposed that institutions of higher education should develop, for each programme they offer, a programme specification which...

---

[2] Art and design, biology, business and management, English, and music and drama.

[3] Accountancy, art and design, biological sciences, classics, communication and media studies, economics, English, European studies, French, geography, history, hospitality management, law, and philosophy.

gives the intended outcomes of the programme in terms of

(1) the knowledge and understanding that a student will be expected to have upon completion;
(2) key skills: communication, numeracy, the use of information technology and learning how to learn;
(3) cognitive skills, such as an understanding of methodologies or ability in critical analysis;
(4) subject specific skills, such as laboratory skills.

The report recommended the establishment of small, expert teams to provide benchmark information on standards, in particular threshold standards, operating within a framework of qualifications that was also proposed by the Committee.

The government accepted this proposal, and the task of formulating these benchmark standards was assigned to the newly established Quality Assurance Agency.[4]

The Policy Instrument: Subject Benchmarks

Pilot Benchmarking Studies

The purposes of subject benchmarking were to assist higher education institutions in planning programmes of study, to provide baseline information for quality assessors working for the QAA and to inform potential students, professional bodies and employers about the knowledge and competences that can be assumed to be possessed by individuals with particular specialist first-degree qualifications.

Pilot subject benchmarking groups for chemistry, history and law were appointed before the end of 1997 shortly after the publication of the Dearing Report and the three pilot subject benchmark reports appeared in 1998. These three subjects were chosen because they represented different academic traditions: in the words of the QAA in its report on the pilot study 'each statement is different, reflecting the differing traditions and cultures of individual academic disciplines' (QAA 1998, p. 1). They represented three different degrees of involvement with employment interests. Each pilot group was given open terms of reference to consider in setting out benchmarks for first degrees in the subject.

[4] In view of its central role in the development and use of subject benchmarks it is important to realise that the QAA is an independent body funded by subscriptions from UK higher education institutions, and through contracts with the main UK higher education funding bodies. It is owned by the higher education institutions through Universities UK (UUK) and the Standing College of Principals (SCOP), membership organisations whose members are heads of all the higher education institutions in the UK. However, legally and contractually it does have obligations to perform certain tasks on behalf of the government owned Higher Education Funding Councils.

In developing their policy on subject benchmarking the QAA claimed that it sought to work not only with subject providers but also with appropriate employment interests and relevant professional and statutory bodies. However, the membership of each of the groups consisted entirely of practising members of academic staff of UK higher education institutions with a slight exception in chemistry where one member was a representative of the Royal Society of Chemistry, the pre-eminent professional body in the subject. The chair of each group was a prominent member of the relevant higher education subject association, and other members were selected by the Quality Assurance Agency after consultation with the Chair and with subject associations and other bodies with an academic interest in the subject. There were no representatives from universities or academic associations in other countries and no representatives of employers or government agencies. The history panel had 16 members, chemistry 14 and law 13.

The QAA, in its introduction to each of the pilot reports, claims that they represent the first attempt to make explicit the general academic characteristics and standards of an honours degree in this subject area in the UK. Certainly they provide the first formal documented evidence (as opposed to informal evidence in novels, autobiographies, etc.) of the very different academic experiences provided for bachelor degree students in different subjects that result from the extreme specialisation of British first-degree courses.

In the UK with its strong tradition of subject specialisation, chemistry is a discipline in which the university curriculum is more or less linearly sequential with what students have learned in secondary school. It is a long-established university subject, and there is fairly wide agreement amongst its professionals about what a new graduate should know. The assumptions of those who prepared the benchmark report are clearly that many graduates in chemistry will have entered higher education with a good knowledge of the subject and are likely to use the specialist knowledge they have acquired after graduation.

This contrasts with the claims made by the pilot benchmarking group in history, even though this too is a subject that is taught as a specialism in secondary schools. However, the history group was very concerned that the subject benchmark should not ossify the teaching of history in higher education. It aimed 'to lay out criteria for judging the suitability and adequacy of single-honours degree courses in History; to do this in a way that is as specific as possible without undermining the principle that there are many different suitable and adequate ways of constructing and making available the great richness and diversity of History; to do it in a way that recognises also the need for adaptability to new academic developments in the field, and innovations in course structures and teaching methods. We insist that teaching and learning are evolving processes and that it not our intention to freeze the teaching of History in a particular model. Our benchmarking statement should be seen as a starting point ... We accept variation in how the vast body of knowledge which constitutes the subject is tackled at undergraduate degree level.' (History Benchmark statement, para 4)

Law is a subject that most students study for the first time at university, and indeed they often graduate in other subjects before they begin to take specialist law courses. This benchmarking group did not make a statement of the aims of a law degree: it appears to have been taken for granted that the purpose of a law degree is to acquire at least some of the knowledge that enables the graduate to proceed to specialist training to prepare them for practice in some part of the legal profession. Their report concentrates on setting out the 'minimum achievement which a student should demonstrate before s/he is awarded an honours degree in Law'.

The long recognised differences between studying for a degree in the physical sciences and in the humanities and social sciences come across clearly in these pilot subject benchmark reports. Both law and history made strong recommendations concerned with 'analysis, synthesis, critical judgement and evaluation'. They refer to 'autonomy' and 'ability to learn' in some form or other. Law, for example, considered that A student should demonstrate a basic ability

- to recognise and rank items and issues in terms of relevance and importance;
- to bring together information and materials from a variety of different sources;
- to produce a synthesis of relevant doctrinal and policy issues in relation to a topic;
- to make a critical judgement of the merits of particular arguments;
- to present and make a reasoned choice between alternative solutions.
- to act independently in planning and undertaking tasks in areas of law which she or he has already studied;
- to be able to undertake independent research in areas of law which he or she has not (studied);
- to reflect on his or her own learning, and to seek and make use of feedback. (Law benchmark statement, p. 3)

The historians consider that their graduates should have 'basic critical skills: a recognition that statements are not all of equal validity, that there are ways of testing them . . ..' and 'intellectual independence'.

In contrast the chemists laid particular emphasis on familiarity of 'chemistry related cognitive abilities and skills'. According to their benchmarking report, the main aims of bachelor's honours degree programmes in chemistry should be:

- To instil in students a sense of enthusiasm for chemistry, an appreciation of its application in different contexts and to involve them in an intellectually stimulating and satisfying experience of learning and studying.
- To provide students with a broad and balanced foundation of chemical knowledge and practical skills.
- To develop in students the ability to apply their chemical knowledge and skills to the solution of theoretical and practical problems in chemistry.
- To develop in students, through an education in chemistry, a range of transferable skills, of value in chemical and non-chemical employment.
- To provide students with a knowledge and skills base from which they can proceed to further studies in specialised areas of chemistry or multi-disciplinary areas involving chemistry.
- To generate in students an appreciation of the importance of chemistry in an industrial, economic, environmental and social context. (Chemistry benchmarking statement, p. 2)

The pilot benchmarks and the processes by which they were arrived at were the subject of a rather cursory small scale evaluation by the QAA consisting of a review of documentary material, observation of the benchmarking meetings and interviews with about half the members of the pilot benchmarking groups (BMGs) (QAA 1998).

The review dealt with the issue of membership of the BMG and concluded that the process of identifying members for the BMGs works well where, as in chemistry and law, there is an accepted 'lead body', a degree of commonality of provision across institutions and limited fragmentation or factions within a discipline. The QAA concluded that where, as in history this is not the case, or where there are competing bodies, they should establish a formal nominating committee drawn from different groups. In all cases higher education institutions and relevant subject departments should be consulted. However, the QAA concluded clearly that

> The criteria for selection of BMG members should be clear and might include: a range of experience of different forms of provision in the subject; representation from the range of higher education institutions; a balance of gender and age; an appropriate spread of knowledge of the main elements of the subject. Experience of external examining, accreditation or other QA processes is also useful. (QAA 1998, para 8.14)

The evaluation concluded that each benchmark report could be expected to take about a year to produce with each member of the (very part time) panel contributing up to 3 weeks work. In its substantive evaluation the QAA concluded that

> overall, there was support from all members of BMGs for the process of Benchmarking and broad satisfaction with the information produced, in several cases despite initial scepticism .... The views expressed by BMG members suggest that the Benchmarking process and its outcomes represent an advance on current practice in the articulation and judgement of standards within subjects. The majority saw the Benchmarking information as providing a national framework or 'meta-level' guide to the subject and for the subject as well as for other interested parties, including students. The frameworks produced were seen as useful for a variety of purposes including design and validation of programmes, examination and review. (QAA 1998, para 6)

However, a number of technical and practical issues arose. All the groups reported considerable difficulties in agreeing on threshold standards caused by:

> difficulties in identifying a single acceptable threshold of attainment within a classification system that has 5 thresholds (fail/pass, pass/third, third/lower second, lower/upper second, upper second/first);
>
> grading conventions and performance criteria that appear to identify an ideal performance level and relate other levels to this;
>
> the widespread use of norm rather than criterion-referenced assessment;
>
> differences in grade points, grading conventions, and classification criteria across the UK higher education system;
>
> a culture within subjects that regards a minimum threshold of attainment (at the pass/fail or pass/third boundary) as 'unsatisfactory', 'not worthy', 'negative in terms of attainment and public acceptability', 'not representative of the majority of students', 'certifying attendance rather than attainment'. Each group came to a different solution to this problem so that there is a lack of consistency across the documents. (QAA 1998, para 8.1)

There was also concern about whether the benchmarks were intended to represent what all successful graduates of the programme should be expected to have achieved or whether the concern was with what the course offered them. It was felt that this distinction, although in some senses pedantic does have implications for the work of institutions, external examiners and reviewers as well as the perceptions of stakeholders and possibly as evidence in legal claims against higher education institutions.

Subject Benchmarking Across the Whole of Higher Education

Despite these reservations by members of the pilot BMG the three pilot reports were quickly followed by the establishment of another 19 subject benchmarking groups. This second group included not only well established university subjects such as economics, engineering and English, but also a number that were, in the UK at least, much more recently established as degree courses, such as 'Hospitality, Leisure, Sport and Tourism'. The benchmarking groups still had between ten and twenty members drawn from senior teaching staff from across the whole higher education sector supplemented, where appropriate, by members of relevant professional bodies and very occasionally employing organisations. Although there was no mandatory template all the new reports included the following sub-headings: Defining principles; Nature and extent of the subject; Subject knowledge and understanding; Teaching, learning and assessment; Standards and levels of achievement – or something very near to them. There were however fairly wide variations in the way the headings were interpreted, not least in the length of the reports which varied from about 2,500 words to over 10,000.

The Learning and Teaching Support Network evaluated the first 22 subject benchmark reports (Yorke 2001). In Yorke's opinion

> Whilst benchmarking can relate both to developmental work and to regulation, the subject benchmarking exercise sponsored by the QAA leans towards the latter. ...... Regulation has come to the fore, with the intention being to use benchmarking to provide explicit standards against which institutions' performances can be measured. Jackson (1998, p. 5) observed that, in the context of the assurance of quality and standards in the UK, benchmarking might more appropriately be defined as a learning process to facilitate the systematic comparison and evaluation of practice, process and performance to aid improvement and regulation.... The benchmark statements are broad in character since they have to cater for variety in the approach to subject disciplines and, in some cases, transdisciplinary spread. As a result, their relationship with standards is loosely-coupled and open to interpretation. It is argued that attempts to achieve a high degree of precision in specification are likely to prove counter-productive. (Yorke 2001, p. 1)

By the end of 2004 another 25 subject benchmark reports for bachelor's 'honours' degrees had been published so by 2005 the courses taken by the great majority of first-degree students were covered by a subject benchmark. This final group of benchmark reports is very similar in structure and approach to the previous block but it may be significant that the title of this final group in all cases is 'Subject

Benchmark Statement: Academic Standards: (name of subject)', whereas in the earlier group of subjects the title of the report was 'Subject Benchmark Statement (name of subject)' and Standards was not always mentioned even as one of the subheadings. The structure of the reports was not fundamentally different but it is of some interest that the benchmarking panels had been reminded that their primary purpose was to establish threshold 'standards' for an honours degree.[5] Several also indicated what some described as modal standards, i.e. levels of achievement that half the graduates could be expected to have met.

Overview of the Benchmarks

All the benchmark reports open with a statement from the Quality Assurance Agency.

> Subject benchmark statements provide a means for the academic community to describe the nature and characteristics of programmes in a specific subject. They also represent general expectations about the standards for the award of qualifications at a given level and articulate the attributes and capabilities that those possessing such qualifications should be able to demonstrate.
>
> Subject benchmark statements are used for a variety of purposes. Primarily, they are an important external source of reference for higher education institutions when new programmes are being designed and developed in a subject area. They provide general guidance for articulating the learning outcomes associated with the programme but are not a specification of a detailed curriculum in the subject.
> Subject benchmark statements also provide support to institutions in pursuit of internal quality assurance. They enable the learning outcomes specified for a particular programme to be reviewed and evaluated against agreed general expectations about standards.

However, it is in the final section of this paragraph that offers a glimpse of the intended regulatory teeth.

> Finally, subject benchmark statements are one of a number of external sources of information that are drawn upon for the purposes of academic review and for making judgements about threshold standards being met.

However, the detailed benchmarks are drawn up by broad cross sections of the academic teaching profession and not surprisingly most of the specific proposals are often bland and couched in rather general terms. The teeth are filed down.

In a subject known to the present author the economics report starts with a general statement about the areas of knowledge covered by the subject.

---

[5] An honours degree in UK higher education is essentially a basic first degree. There are many local variations and in a few cases 'pass' degrees are still awarded. Most degrees are classified into first class, upper second class, lower second class, and third class. (This is currently under discussion and there are moves to substitute a Grade point average system). More than half of today's graduates obtain an upper second class degree so the modal threshold is somewhere in this range.

> Economics is the study of the factors that influence income, wealth and well-being. From this it seeks to inform the design and implementation of economic policy. Its aim is to analyse and understand the allocation, distribution and utilisation of scarce resources. Economics is concerned both with how present allocations arise and how they may change in the future. Study of Economics requires us to understand how resources are used and how households and firms behave and interact. This understanding is required at both the individual (micro) and the aggregate (macro) level. The analysis is both static (dealing with output, employment, income, trade and finance) and dynamic (dealing with innovation, technical progress, economic growth and business cycles). The study of Economics requires an understanding of resources, agents, institutions and mechanisms. Moreover, since virtually no economy operates in isolation, it is important that these phenomena are studied in an international context. (Economics benchmarking report, para 1.1)

The report goes on to identify

> the study of the factors that characterise the economist's approach. First there is the ability to abstract and simplify in order to identify and model the essence of a problem. Second is the ability to analyse and reason – both deductively and inductively. Third is the ability to marshal evidence and to assimilate, structure, and analyse qualitative and quantitative data. Fourth is the ability to communicate concisely results to a wide audience, including those with no training in Economics. Fifth is the ability to think critically about the limits of one's analysis in a broader socio-economic context. Sixth is the ability to draw economic policy inferences and to recognise the potential constraints in their implementation. (ibid., para 1.3)

The report covers the normal content of the subject in most universities and colleges but makes little attempt to be prescriptive. The stressed passages in the extract below are mine.

> any single honours degree in Economics *normally* comprises the following elements.
>
> A coherent core of economic principles. The understanding of these might be *verbal*, graphical or *mathematical*. These principles should cover the microeconomic issues of decision and choice, the production and exchange of goods, the interdependency of markets, and economic welfare. They should also include macroeconomic issues, such as employment, national income, the balance of payments and the distribution of income, inflation, growth and business cycles, money and finance. The understanding should extend to economic policy at both the microeconomic and macroeconomic levels. In all these, students should show an understanding of analytical methods and model-based argument and should appreciate the existence of different methodological approaches. (ibid., para 3.1)

There are let-out clauses for individual institutions and for courses that may include some economics.

> It is recognised that, in both single honours degrees and in many degrees that involve a substantial amount of Economics, content will be adapted to suit the nature and objectives of the degree programme ....
>
> In degrees that are not single honours Economics, not all the core elements ... may be covered. It is also recognised that the forms of analysis chosen may differ and may be tailored to best serve the skills that students bring with them into their degree programme. It is neither the function nor the objective of this benchmarking document to prescribe what these forms of analysis might be; this is a matter for institutional choice and decision. (ibid., para 3.2)

The passages quoted above illustrate how, even in a fairly well defined subject such as economics the subject benchmarking report allows course teams very wide latitude in deciding the actual content of their courses.

It is clear that one implicit aims of the benchmark report is to 'sell' the subject to employers and to future students and their advisers. Along with all the other subject reports, the economics benchmarking statement is at pains to point out that:

> Some of the attributes that a graduate in Economics possesses are generic and not specific to the study of the subject. Their enhancement would be part of any degree programme. These would include general intellectual skills such as literary and information-processing skills, as well as interpersonal skills, such as communication. Economics degree programmes, therefore, provide a learning environment that facilitates and encourages the development and use of such skills. (ibid., para 4.1)

Some brief comparisons with other subjects highlight variations in content and treatment. The philosophy panel claims that its graduates will be expected to have acquired intellectual abilities which are readily transferable to other contexts including (inter alia) articulacy in identifying underlying issues in all kinds of debate, precision of thought and expression in the analysis and formulation of complex and controversial problems, clarity and rigour in the critical assessment of arguments, ability to abstract, analyse and construct sound arguments and to identify logical fallacies, ability to recognise methodological errors, rhetorical devices, unexamined conventional wisdom, unnoticed assumptions, vagueness and superficiality. Sociology graduates will have been 'enabled to develop competence in' judging and evaluating evidence, appreciating the complexity and diversity of social situations, assessing the merits of competing theories and explanations, gathering, retrieving, and synthesising information, making reasoned arguments, interpreting evidence and texts, developing the ability to reflect on their own accumulation of knowledge. Apart from a wide range of subject specific skills the music panel lists 'a wide range of transferable skills, ... many of which are applicable to issues of musical and non-musical origin'. There are ten 'intellectual skills', nine 'communication and interaction skills', eleven 'skills of personal management' and four involving 'enhanced powers of imagination/creativity'. Physics is more succinct and more self-confident about the transferable skills. A physics graduate may be expected to have acquired: problem-solving skills, investigative skills, communication skills, (physics and the mathematics used in physics deal with surprising ideas and difficult concepts; good communication is essential.), analytical skills, IT skills and personal skills (ability to work independently, to use their initiative, to organise themselves to meet deadlines, and to interact constructively with other people).

A simple word count of the 47 reports overall (see Appendix) makes it clear that 'knowledge', 'understanding' and 'skills' are thought to be the most pervasive aims of most UK first-degree programmes. There are, however, some interesting variations between subjects. 'Knowledge' is mentioned ten times as frequently in accountancy and biomedical studies as it is in 'religious education'. 'Skills' are particularly likely to be mentioned in the economics, business and management, architecture, accountancy, dance and drama, music and health studies reports, but considerably less likely to be mentioned in archaeology, educational studies and religious studies. A second batch of key words are critical and analysis, which appear about half as often as knowledge, understanding and skills. 'Critical' is used at least twice as often in English as in most other subjects. 'Analysis' is particularly favoured in economics, health studies and linguistics where it is used twice as often as in other subjects, but 'analysis' is relatively rare in librarianship, medicine, dentistry and veterinary science. Dance and drama, psychology, and health studies appear particularly likely to favour students doing independent work, while computing, economics, educational studies, materials and town and country planning do not mention this at all. Employers or employment are mentioned rarely as are words that Yorke thinks ought to be included such as 'original', 'creative/creativity' and 'synthesise'. As might be expected, creative or creativity appears frequently in art, music and communication studies but not at all in 12 other subjects, including accountancy. Students of education and those of health studies are particularly likely to be expected to 'reflect' on what they are learning while the word is not mentioned by the accountants, the economists, the engineers, the physicists and six other subject areas.

One of the entrenched features of first degrees in all UK universities is that degrees are classified – first class, upper second class, lower second class and third class are the usual categories used. In recent years upper second class honours has become the modal category and third class the threshold. In its pilot study the History Panel confronted this issue directly and gave some indications of what it considered appropriate for first, second and third class honours degrees. Most of the reports in the main body were content with a distinction between minimum or 'threshold' standards and the standards that the 'modal' or 'typical' or 'focal' graduate may be expected to achieve.

The economics panel distinguishes between 'modal' and 'threshold' attainment.

A graduate in economics who has attained the threshold level should:

- Demonstrate *knowledge* of economic concepts and principles.
- Demonstrate *knowledge* of economic theory and modelling approaches.
- Demonstrate *awareness* of quantitative methods and computing techniques appropriate to their programme of study, and show an appreciation of the contexts in which these techniques and methods are relevant.
- Display *knowledge* of the sources and content of economic data and evidence and appreciate what methods might be appropriately applied to the analysis of such data.
- *Know* how to apply economic reasoning to policy issues
- Demonstrate *knowledge* in an appropriate number of specialised areas in Economics.
- Display *awareness* of the possibility that many economic problems may admit of more than one approach and may have more than one solution. (Economics benchmark report, para 6.2)

For the 'modal' level of achievement 'understanding' replaces 'knowledge' and 'proficiency' replaces 'awareness' and the modal graduate will 'know how to apply economic reasoning to policy issues in a *critical* manner'. 'Knowledge' and 'understanding' are given the following meanings.

> Knowledge is the ability to reproduce theory and evidence as taught, understanding is a term applied to constructive and critical use and analysis of that material. (ibid., p. 5 footnote 4)

In the most rigorous analysis of the subject benchmarks so far available Wisby (2002) made a detailed observation and interview based case study of the Sociology subject benchmarking group, and conducted in depth interviews with the chairs of twelve other benchmarking groups, selected to be indicative of the type of subject: humanities, social sciences and physical sciences and of the 'market position' of subjects in terms of the pressure of applications to study in those areas and the perceived employability of their graduates. Arising out of these interviews she distinguishes a difference in attitudes toward benchmarking by three broad categories of subject: 'established' subject areas that have a relatively long history in higher education; 'accredited' subjects, those disciplines that are subject to accreditation by one or a number of professional and statutory bodies (PSBs) and 'new' subject areas that have arrived only relatively recently in the UK higher education sector and usually have a strong vocational emphasis. This last group tends 'to be located predominantly in the post-1992 university sector', which may be significant because this sector has a much longer experience of external involvement in the content of its teaching.

Wisby found that the established subject providers could be ambivalent towards the external regulation of provision, and that they were also likely to be attached to a 'liberal' model of teaching and learning. These subject areas and their benchmarking groups tended to be suspicious of the benchmarking exercise.

In the case of the accredited subjects the PSB

> play an important role in the current regulatory regime in protecting standards in professional and vocational education. Their remit typically incorporates all aspects of teaching provision – from entry standards and curriculum content to methods of teaching, learning and assessment, as well as resources to support learning. Notably, this activity often resembles the frameworks now being established through standards-based quality assurance – not least the provision of threshold standards. These disciplines, then, are obviously used to the external monitoring of provision and to working to external requirements. (Wisby 2002, p. 139)

The most interesting category for Wisby were the 'new' subject areas. They appeared to have more to gain from the benchmarking exercise than the others. For them

> the main priorities in the benchmarking exercise were using the benchmark statements to establish subjects as valid areas of study; to demarcate subject areas; and to address poorer provision at the margins in order to protect subjects' reputation or academic standing. This entailed a more strategic use of the benchmarking exercise. (ibid., p. 140)

One of Wisby's respondents from a new subject area claimed that benchmarking gave them an 'opportunity to strengthen our subject' and 'has given us a little bit of credibility, it's given us some ammunition' (ibid., p. 159).

Impact of the Benchmarks

Subject benchmarks cannot be viewed in isolation. They need to be seen in the context of introspection about the aims and purposes of higher education and a significant cultural shift, inspired in large part by concerns about the implications of the sudden chaotic shift to mass higher education in the early 1990s. They are part of an interlocking network of quality assurance and enhancement measures imposed on UK higher education since 1990. A related contextual factor is pressures on academic departments to concentrate effort on research at the expense of undergraduate teaching in order to take advantage of the substantial financial rewards available to those university departments that are successful in the Research Assessment Exercises. A detailed blow-by-blow analysis of the 'quality wars' of the 1990s is provided in Brown (2004).

Among other developments is a national qualifications framework, recommended by the 1997 Dearing Committee, which attempts to ensure for the first time that all UK higher education institutions have similar structures of sub-degree programmes, first degree, taught postgraduate and research degree programmes. There are 'institutional audits' of all universities and other degree awarding institutions, which monitor and assess the quality of the learning programmes and the standards of the awards in teaching departments and audit the institutions' ultimate responsibility for what is done in their names and through the exercise of their formal powers. There are specific subject-based 'academic reviews' in all higher education courses provided in the lower level further education institutions every 6 years. Reviewers test, by means of their own observations and analyses of the evidence provided by the college, the statements made in a self-evaluation. There is a 'Higher Education Academy', which is concerned with improving teaching procedures at individual and departmental levels with a membership of individuals who have attended a recognised course on teaching in higher education.

The original expectation was that the subject benchmarks would be an important input into Discipline Audit Trails (DAT) which were intended to be a significant component of the new round of institutional quality assurance reviews established by the QAA in 1999. Programmes of study were expected to be consistent with the benchmarks or the university would be required to provide a convincing explanation of why alternative content or approaches to teaching the subject were being adopted. In practice this intention was not achieved for two main reasons. One was that as a result of pressure from the higher education institutions, and in particular the powerful research led universities, but also because of changes in the leadership of the QAA, the external quality assurance reviews were modified and a so-called light touch procedure adopted whereby established universities needed merely to demonstrate that their own quality assurance procedures were effective: the DAT were effectively discontinued for these institutions. The other was that the benchmarking groups framed the subject benchmarks in ways that allowed course teams considerable leeway in interpreting them. There were also issues of the personalities and changing aims of senior politicians and the heads of the various agencies responsible for quality assurance in higher education.

Such obstacles to the achievement of hard line regulatory interpretations of the subject benchmarks illustrate the point made by Adam Smith over two centuries ago: 'All that such superiors can, however, force him (the university teacher) to do is to attend upon his pupils a certain number of hours, that is, to give a certain number of lectures in the week or in the year. What these lectures shall be must still depend on the diligence of the teacher ....' Rae (1895, p. 248)

Wisby (op cit) makes frequent mention of the 'developmental "offshoots"' of what had originally been conceived of as a regulatory exercise. These benefits were understood in terms of encouraging reflection on teaching practice, and generating discussion about teaching and learning issues – whether at the level of the individual academic, the department, or across institutions.

The situation has changed somewhat since the 1770s. The individual university teacher is not quite so autonomous as he was then. But it remains the case that it is the academic profession that must, in the last analysis decide what is taught in universities, no other group has the knowledge or expertise and the subject benchmarking episode in British higher education makes this point quite clearly. One finding to emerge from Wisby's (op cit) fieldwork, 'is the enduring collegialism within subject communities and the way in which this provides some protection from unwelcome external pressures'. She also notes a remark from one of the subject committee chairs that she interviewed:

> The fact that the exercise to a large extent has been taken over by the academic community, means that academics themselves have, certainly in the Group that I chaired,. . .insisted that this should not be a. . .regulatory document, but that it should allow for academic freedom within it. The whole document was framed. . .in a way that really shifted the boundaries of the benchmarking exercise. When benchmarking started I think QAA had seen it as more prescriptive than in fact it has turned out to be. (op cit, p. 197)

However, the subject benchmarks are having some impact on course content in many higher education institutions. In one college of London University the teaching of geography was discontinued and the geographers transferred to a neighbouring university in order to meet the benchmark recommendations that a geography degree programme should include both physical and human geography. More generally, it is usual for new course proposals to be required by their institutions' course approvals committees to state what account they have taken of the relevant subject benchmarks. There are many cases of course documentations and assessment criteria being rewritten to take account of subject benchmark recommendations. Most departments appear to be aware of the need to ensure that either all their programme specifications are consistent with the relevant benchmark statement or statements, or to be able to provide a convincing explanation of why they do not. In future academic and institutional reviews by the Quality Assurance Agency, universities and colleges will be expected to show that they have conducted regular internal quality assurance reviews of each programme, and that a programme specification has been published and used for each programme. These will not be required to follow subject benchmarks but it is likely that most will make some reference to them. In further education colleges, which are subject to more detailed appraisal by the QAA, there will need to explicit evidence of having taken account of the benchmarks.

Subject benchmarks have not engendered serious hostility from within the academic profession, at least compared with many of the other activities of the Quality Assurance Agency. One reason for this is that the benchmarks have proved not to be external regulatory impositions on academic staff. They are seen as formative and developmental, representing the considered views of senior academics across the higher education sector about the aims and content of first-degree courses in each subject. Related to this is the crucial fact that unlike many of the quality assurance innovations of the past 20 years the benchmarks are subject based and, therefore are accepted as being at the heart of academic identity. Despite the modularisation and hybridisation of courses that accompanied and followed the expansion of the early 1990s most British academics still see themselves as being subject specialists (Becher and Trowler 2001; Henkel 2000) and subject benchmarks are seen as having more direct relevance to their daily work than most of the other institutional management and process based evaluations of the Quality Assurance Agency. In the last analysis, however, developments since 2001 have, in effect enabled course teams in universities to disregard the subject benchmarks if they wish to do so.

Concluding Comments

In the UK each university is responsible for the quality of the degrees and diplomas it awards to students and it is a basic belief that any external system of quality evaluation should not undermine this responsibility. Nevertheless a fundamental change since 1990 is that it is now generally accepted that students and employer have a right to be assured that the university is fulfilling its responsibility to offer worthwhile qualifications. Worthwhile in this context is recognised as having two dimensions – fitness for purpose and fitness of purpose. The first allows each university to set the standards for its own degrees and leave it to the students and employment markets to differentiate between the programmes. The second involves some way of ensuring that all stakeholders, and especially students and employers of graduates, are not misled about the signals that possession of a degree sends out.

At least until the late 1990s it was a firm belief in UK universities that all degrees should be at least broadly equivalent. A degree from one university should be equivalent to a degree in the same subject from another university. This has been most explicit in the postgraduate fellowships awarded by the research councils. The prime consideration in the award of a postgraduate fellowship has been the class and subject of the first degree: the institution where it was awarded was less explicitly considered. This requires that the content of a first-degree programme and the levels of attainment expected are at least broadly comparable with those of other universities awarding the same qualification. Before the 1990s, when British higher education was by most international standards an 'elite' system with a relatively small proportion of the population obtaining degrees, it was possible to maintain this belief through the informal system of external examiners which in broad terms ensured that standards of performance in all universities were comparable.

Subject benchmarking can be seen as an attempt to maintain this comparability over a very much larger and more diversified system of higher education. However, any endeavour to create benchmarks to facilitate comparability, while at the same time not imposing rigid straitjackets, is beset by obvious problems. Wisby's distinction between 'established', 'accredited' and 'new' subjects provides a useful analytical tool. The 'established' subjects are, for the most part taught in established universities, and even where they are taught in less august institutions, the subject associations and other forms of informal authority influence the ways in which a subject is treated. Subject benchmarking for them was an interesting exercise in collectively exploring the nature of their subjects. None of the benchmarks attempts to impose specific content on individual degree programmes and within very broad limits universities can accept or ignore them according to the professional judgements of their own academic staff in the relevant areas.

In most accredited subjects such as law and medicine, there is a distinction between acquisition of a degree and obtaining a licence to practise. Where the degree gives exemption from all or some of the courses needed to obtain professional qualifications it is the relevant professional body that validates the course and it is this professional validation that is the most important consideration for the course providers. This is made clear in the subject benchmark for medicine:

> The benchmarks for medicine are but one of the external reference points for the undergraduate medical curriculum and must be considered together with the others, and in particular the recommendations of the Education Committee of the General Medical Council published in *Tomorrow's Doctors* 2002. (Benchmark report for Medicine, para 4)

In such professional courses it is obviously important that graduates allowed to practise the profession are known in the market to have at least certain basic professional competences. This is the case in about half the subject benchmarks so far defined and in most of these, further professional qualifications are needed before the graduate is allowed to practice.[6]

As already noted it is the 'new' subjects where benchmarking has had most impact. It has made a useful contribution to the legitimation of subjects such as media studies and many branches of management, which began to appear to a significant extent in the British higher education curriculum only during the explosive expansion of the early 1990s. The committee chairs from these 'new' areas interviewed by Wisby, 'were preoccupied with the impact that further expansion might have on the status of their subject areas'.

In summary, the subject benchmarking process can be seen as an attempt to assure the higher education community itself and its stakeholders that after the rapid

[6] About 25 per cent of first-degree graduates continue with further academic or professional study immediately after graduating.

transition to mass higher education in the early 1990s there is still a real sense in which first-degree graduates from one university are equivalent to those from any other university in the country. The influential Council for Industry and Higher Education has published a manual based on the Benchmarking reports summarising for employers the main attributes that they can expect graduates of various subjects to possess (Kubler and Forbes 2005). The intention of their report is to 'help make explicit what has often been implicit' (p. 1). However, their study leaves implicit any opinions employers may have about whether the knowledge and skills of graduates from different institutions are equivalent.

Benchmarking is also considered to be important in the burgeoning market for foreign students, which also grew at an extremely rapid rate in the late 20th and early 21st century. Universities are using them as authoritative statements of what knowledge and skills students can expect to acquire when studying for a first degree in particular subjects. Chandler, a member of the benchmarking group for sociology has noted that:

> ...benchmarks have implications for promotional literature. They may also assist graduate students to articulate skills...and promote themselves in the graduate market place. Hence there is an element of competition in the way in which subject groups benchmark their areas as they enable prospective students and potential employers to compare the learning outcomes of different disciplines (Chandler 2001, p. 56).

The success of the subject benchmarking panels in persuading senior representatives from across the whole higher education sector to subscribe to common statements of the aims and broad content of nearly 50 subjects in higher education provides some evidence that there is still a sense of common purpose and standards in UK universities in a wide range of subjects, which students can use when choosing universities and employers can use when recruiting graduates. Whether the employment market for graduates or the student recruitment market will permanently accept this conclusion remains to be seen. Although there is much discussion of diversity of contemporary mass higher education in the UK there is very little formal consideration of differences between degrees from different institutions. This is left implicit.

The benchmarks also, unsurprisingly, make no attempt to resolve the challenge of the equivalence of first degrees in different subjects. Clearly subject content is different and the substantive meaning of the most widely used words, knowledge, skills and understanding vary widely between the subjects. Apart from this the generic skills graduates are expected to have acquired vary considerably between different disciplines and subjects. In the last analysis what higher education and its myriad of individual courses offer continues to depend on the professionalism and integrity of individual teachers and course teams. The market, or the various markets that impinge on UK higher education will be the final arbiter between individual graduates, their courses and their *almae matres*.

In the last analysis subject benchmarking has been a formative and developmental exercise and not, as many academics feared when they were initiated, primarily a regulatory mechanism. Benchmarks are a useful addition to the information available to external stakeholders such as potential students and employers of graduates. They are also useful for course designers, particularly in 'new' subject areas in reminding them of what the general view of their peers is about what such a course should set out to do. Metaphorically a subject benchmark is a combination of menu, basic recipe book and public health manual. These can help to avoid disappointments, particularly important in an increasingly litigious society, but it is what the skills of the chef does with the basic recipe that makes the difference between whether the establishment can charge $10 or $100 for a dish. In a market driven higher education system, the national and international university and course rankings, which are appearing in the British media with growing frequency and detail will probably be more influential in the long run than standardised subject benchmarks.[7]

Postscript

Subject benchmarking has become an integral component of the light touch quality assurance arrangements in Britain. Benchmarks have been embraced by influential parts of the higher education community. In November 2004 the QAA announced the establishment of a 'Recognition Scheme' which aims to involve new discipline areas and to recognise the work of subject communities themselves to define the aims and scope of degree courses in their subjects.

Three more subjects have so far been added to the list on bachelor degree benchmarks, making 50 in total and 20 of the benchmarking statements have been reviewed and up dated. Work is in progress to compile benchmark statements for the proliferation of 2-year foundation degree courses which are growing rapidly.

After a short period of consultation four master's degree benchmarks have been added, business and management, engineering, pharmacy and physics. The main purpose of these is to explore what the holder of a masters degree should be expected have learned over and above the bachelor degree. This is not an easy task as many masters degrees in the UK are in practice conversion courses enabling students to obtain a specialist qualification in an area different from the one in which they obtained their first degree and are often taken some years after the student has first graduated. Of those already prepared only business and management falls clearly into this category.

[7] However, in the Autumn of 2005, under pressure from the government a consultative document was circulated by UUK and SCOP (the bodies representing the heads of higher education institutions) making proposals for the classification of degrees and attempting to ensure that common principles are applied across the whole of higher education for the benefit of students choosing universities and employers recruiting graduates.

The purpose of master's degrees in business and management is not surprisingly clearly deemed to be vocational in nature. It includes advanced study of organisations, their management and the changing external context in which they operate, preparation for a career in business and management by developing skills at a professional level, or as preparation for research or further study, development of the ability to apply knowledge and understanding of business and management to complex issues, and the enhancement of lifelong learning skills and personal development so as to be able to work with self-direction and originality and to contribute to business and society at large.

While there is little doubt that the main function of the subject benchmarking scheme is to assist in the development of new subjects and of new universities that are able to offer degrees for the first time, they have been taken on board by even the longest established universities. Cambridge University for example in its instructions to faculties and departments in drawing up new or revised programmes of study exhorts them to

> demonstrate that your Faculty/Department has taken notice of the Benchmark Statement for your subject in drawing up the Programme Specification. One convenient way of doing this is by cross-referencing the skills mentioned in the Benchmark Statement to the ones that you include on the Programme Specification. You could do this just by putting an asterisk or letter by the benchmark skills that occur in your Programme Specification. You may want to look at the Benchmark Statement first to check on the language used. You may find that your Benchmark Statement uses a slightly different categorisation of skills than the one given above. If this is more appropriate for your course, there is no reason why you should not use it, as long as it covers the same ground. (Cambridge 2007)

At Birkbeck College in the University of London

> The published benchmark statements will form a point of reference for internal and external reviewers for making judgements about the appropriateness of standards in any programme of study. The relevant benchmark statement should also be borne in mind when the learning outcomes of a programme, or part of a programme, are being specified
>
> ... The question of how Schools are reacting to these benchmark standards is incorporated into the internal review process. (Birkbeck 2006)

At Liverpool University course teams are required to

> consider the appropriate subject benchmark statement(s) and the QAA's Framework for Higher Education Qualifications ....
>
> A statement should be given about the key (transferable) skills that a student will be given the opportunity to develop through the programme. The skills should be mapped against the generic skills identified in the relevant Subject Benchmark Statement(s) and against the modules that enable the development of the skills, and there should be an explanation of how the skills are assessed. (Liverpool 2006)

There are similar statements in the course approval manuals in nearly all UK universities.

Appendix: Use of Certain Key Words per Thousand Words in Subject Benchmarking Reports

	Knowledge	*Creative*	*Critical*	*Analysis*	*Independent*	*Evaluation*	*Original*	*Understanding*	*Reflect*	*Employment*	*Skills*	*Synthesis*
Chemistry	6.5	0.0	0.0	2.6	0.0	1.7	0.0	0.9	0.0	0.9	16.1	1.3
History	4.6	0.0	3.4	1.5	1.0	0.6	0.5	3.5	1.3	0.0	5.4	0.4
Law	3.6	0.2	1.5	1.0	1.7	1.4	0.2	2.3	0.9	0.0	3.4	1.0
Total pilot subjects	**4.1**	**0.2**	**1.9**	**1.3**	**1.4**	**1.2**	**0.3**	**2.5**	**0.9**	**0.1**	**5.1**	**0.9**

	Knowledge	*Creative*	*Critical*	*Analysis*	*Independent*	*Evaluation*	*Original*	*Understanding*	*Reflect*	*Employment*	*Skills*	*Synthesis*
Accounting	10.3	0.0	2.0	4.8	0.8	2.0	0.0	7.5	0.0	0.4	7.9	0.0
Classics and ancient history	3.6	0.3	0.4	1.2	0.3	0.4	0.0	1.6	0.6	0.0	5.2	0.0
Archaeology	6.9	0.0	2.4	4.6	0.3	0.6	0.0	7.5	2.4	0.3	4.8	0.2
Architecture etc.	6.0	0.6	1.1	5.5	0.1	5.3	0.0	7.7	1.5	0.4	8.9	1.3
Computing	2.7	0.7	0.5	3.2	0.0	2.1	0.0	2.8	0.4	0.4	3.6	0.3
Earth sciences and environmental sciences and studies	4.6	0.0	2.2	1.7	1.3	1.9	0.0	4.6	0.2	0.4	7.4	0.7
Economics	5.6	0.0	1.5	8.7	0.0	1.5	0.0	5.9	0.0	0.9	9.0	0.0
Educational studies	6.0	0.0	3.0	1.7	0.0	1.0	0.2	9.0	5.5	0.2	3.5	1.5
Engineering	6.9	1.4	1.0	4.5	0.5	0.5	0.1	7.5	0.0	0.3	5.2	0.4
English	7.1	1.0	7.1	2.4	1.3	1.3	0.3	2.1	1.6	0.3	6.3	0.0
General business and management	5.4	1.0	1.7	1.3	0.7	2.0	0.0	3.4	2.4	1.0	9.1	0.3
Geography	5.9	0.2	2.5	3.8	0.7	2.2	0.2	6.8	2.0	0.4	7.2	0.7
Hospitality, leisure, sport and tourism	4.7	0.3	3.0	4.6	0.6	1.4	0.4	5.2	2.1	1.0	4.6	0.6

	Knowledge	*Creative*	*Critical*	*Analysis*	*Independent*	*Evaluation*	*Original*	*Understanding*	*Reflect*	*Employment*	*Skills*	*Synthesis*
Librarianship and information management	6.5	0.0	2.2	0.4	1.7	5.2	0.0	6.5	0.9	0.4	7.8	0.0
Philosophy	2.9	0.5	1.9	2.2	0.7	1.0	0.0	3.1	0.5	0.0	6.5	0.0
Politics and international relations	4.4	0.0	2.3	7.1	1.1	0.5	0.2	4.4	1.1	0.9	5.1	0.7
Sociology	4.7	0.0	0.6	2.6	0.6	2.6	0.0	8.5	0.6	0.3	6.5	0.9
Religious studies	1.0	0.1	3.8	2.5	0.6	1.1	0.3	2.8	0.8	0.0	2.8	0.1
Social policy and administration and social work	6.7	0.3	3.1	1.0	0.9	2.1	0.0	6.6	2.8	0.1	8.7	0.5
Agriculture and forestry	8.1	2.0	1.4	3.3	0.6	9.2	0.0	9.8	0.6	0.4	5.4	1.2
Anthropology	5.2	0.2	1.4	5.2	0.9	0.0	0.0	2.9	0.5	0.0	4.1	0.2
Area studies	4.5	0.3	3.6	1.5	0.3	1.5	0.0	1.5	1.2	1.2	8.5	1.5
Art and design	4.5	2.2	3.1	1.8	1.3	1.1	0.1	2.7	0.5	0.5	6.9	0.5
Biomedical sciences	11.9	0.4	0.4	2.1	0.9	1.7	0.0	4.7	0.0	3.0	5.5	0.0
Biosciences	6.0	0.0	3.8	3.8	0.9	1.9	0.0	6.2	0.0	0.7	0.9	0.2
Building and surveying	7.2	0.0	0.4	2.6	0.4	5.7	0.0	4.5	0.0	1.5	5.3	0.4
Communication, media, film and cultural studies	6.0	5.0	5.3	4.0	1.4	2.1	0.2	11.2	3.4	0.7	6.7	0.0
Dance, drama and performance	4.9	1.9	4.9	3.0	2.2	1.9	0.6	4.3	0.4	0.0	11.4	0.6
Dentistry	4.4	0.2	0.9	0.9	0.4	1.1	0.0	3.3	0.9	0.4	6.4	0.0
Health studies	4.6	0.4	11.3	9.2	2.9	4.2	0.4	4.2	4.2	0.4	11.7	0.8

	Knowledge	*Creative*	*Critical*	*Analysis*	*Independent*	*Evaluation*	*Original*	*Understanding*	*Reflect*	*Employment*	*Skills*	*Synthesis*
History of art, architecture and design	4.0	2.4	3.1	1.8	1.3	1.1	0.1	2.4	0.5	0.5	6.6	0.5
Languages and related studies	5.5	0.6	1.2	1.3	1.0	1.0	0.0	5.5	0.3	0.7	8.1	0.1
Linguistics	2.6	0.0	1.2	10.4	1.4	2.1	0.0	4.6	0.2	0.0	6.1	0.9
Materials	6.8	0.0	0.9	1.8	0.0	0.0	0.9	6.5	0.0	0.3	5.8	0.3
Mathematics, statistics and operational research	5.4	0.0	0.0	2.0	0.3	0.3	0.0	3.7	0.1	0.9	4.9	0.0
Medicine	3.3	0.2	1.4	0.7	0.5	1.9	0.0	3.5	1.4	0.7	6.0	0.2
Music	3.3	4.2	2.4	3.4	0.7	1.0	0.1	4.2	0.8	0.3	12.4	0.8
Optometry	9.5	0.0	2.0	1.0	1.0	3.3	0.0	4.3	0.0	0.3	9.8	0.0
Pharmacy	5.8	0.6	2.2	1.9	1.3	3.0	0.6	1.5	0.4	0.4	3.7	0.2
Physics, astronomy and astrophysics	2.9	0.3	1.3	3.8	0.6	2.5	0.0	4.1	0.0	0.6	11.4	0.0
Psychology	8.3	0.0	3.3	4.3	2.1	2.6	0.5	4.0	0.2	1.2	9.5	0.0
Town and country planning	4.8	1.3	1.9	1.6	0.0	4.4	0.6	4.4	0.6	0.0	5.1	1.0
Veterinary science	6.7	0.0	0.7	0.9	0.2	0.2	0.0	6.3	0.3	1.4	5.5	0.5
Welsh	4.7	1.1	3.7	3.2	0.5	1.3	0.3	4.2	0.0	0.0	3.2	0.0
All subjects	**5.3**	**0.9**	**2.2**	**3.1**	**0.8**	**2.0**	**0.1**	**5.0**	**0.9**	**0.5**	**6.4**	**0.4**

References

Becher, T., Trowler, P. (2001). *Academic tribes and territories: Intellectual enquiry and the cultures of disciplines*. Buckingham, UK: Open University Press.

Bekhradnia, B. (2003). *20 years of higher education policy: Looking back 10 years and forward to the next decade*. Oxford: Higher Education Policy Institute.

Birkbeck (2006). Academic quality assurance. Birkbeck College, University of London. http://www.bbk.ac.uk/qa/external/subjbench Accessed 12 June 2009.

Brown, R. (2004). *Quality assurance in higher education: The UK experience sine 1992*. London: Routledge Falmer.

Cambridge (2007). A Guide to quality assurance University of Cambridge academic division. http://www.admin.cam.ac.uk/offices/education/quality/221.html Accessed 12 June 2009.

Chandler, J. (2001). Benchmarking the Sociology Discipline. In E. Harrison, & R. Mears (eds.), *Assessing sociologists in higher education* (pp. 41–60). Aldershot: Ashgate.

Chemistry benchmarking report – see QAA website below.

Dearing (1997). *Higher education in the learning society. Report of the National Committee of Enquiry into Higher Education*. London: HMSO.

Economics benchmarking report – see QAA website below.

General Medical Council (2002). *Tomorrow's Doctors*. London: GMC.

Henkel, M. (2000). *Academic identities and policy change in higher education*. London: Jessica Kingsley Publishers.

HEQC (1995). Graduateness: What are graduates? Clarifying the attributes of 'graduateness'. Higher Education Quality Council Quality Enhancement Group. London. http://www.city.londonmet.ac.uk/deliberations/graduates/starter.html Accessed 12 June 2009.

History benchmarking report – see QAA website below.

Jackson, N. (1998). Pilot Benchmarking Study of Assessment Practice in Seven Engineering Departments. In N. Jackson (ed.), *Pilot studies in benchmarking assessment practice*. Gloucester: Quality Assurance Agency for Higher Education.

Kubler, B., Forbes, P. (2005). *Student employability profiles: A guide for employers*. London: CIHE.

Law benchmarking report – see QAA website below.

Liverpool (2006). Programme specification: Notes for guidance. University of Liverpool. http://www.liv.ac.uk/tqsd/qualityframework/ps_gn_06.doc Accessed 12 June 2009.

QAA (website) Quality Assurance Agency Website. http://www.qaa.ac.uk / academicinfrastructure/benchmark/honours/ Accessed 12 June 2009.

QAA (1998). Evaluation of the benchmarking process in Law, Chemistry and History. Cheltenham UK: Quality Assurance Agency. http://www.qaa.ac.uk/academicinfrastructure/benchmark/evaluation/overview.asp Accessed 12 June 2009.

Rae, J. (1895). *Life of Adam Smith*. Reprinted in 1965 by New York: Augustus M. Kelley.

Williams, G. (2004). The higher education market in the United Kingdom. In P. Teixeira, B. Jongbloed, D. Dill, A. Amaral (eds.), *Markets in higher education: Rhetoric or reality?* Dordrecht: Kluwer Academic Publishers.

Wisby, E. (2002). Subject Benchmarking as a 'new mode' of regulating teaching in higher education and its implications for academic autonomy – including a case study of the Sociology Benchmarking Group. PhD dissertation, Sheffield University.

Yorke, M. (2001). *Assessment issues arising from Subject Benchmarking statements*. Bristol: Learning and Teaching Support Network.

Young, M. (1961). *The rise of the meritocracy*. London: Penguin Books.

Chapter 10
Subject Assessments for Academic Quality in Denmark

Bjørn Stensaker

Introduction

Higher education in Denmark is mainly public and consists of eight universities, a number of university colleges, and more specialised and professional higher education institutions (in art, agriculture, etc.). The national ministries of education, and science, technology and innovation approve all public higher education institutions (colleges are the responsibility of the Ministry of Education, while the Ministry of Science, Technology and Innovation is responsible for universities). Private institutions may operate without governmental approval, but then run the risk that their students will not be eligible for the state student grant. The higher education system can be divided into a university sector and a college sector (a binary system). Denmark has merged a number of higher education institutions, first in the college sector where a substantial number of (very small) colleges triggered the government in the late 1990s to stimulate voluntary amalgamations (Gornitzka et al. 2001, p. 16). In recent years, institutions in the university sector went through a similar process, leading up to the present number of universities. However, not all universities underwent a merger with the consequence that this sector consists of both some very large universities and also some quite small universities.

The degree system has three levels: bachelor studies (3 years), master degree studies (5 years), and the PhD degree (additional 3 years). However, within the college sector one can find study programmes that deviate from this structure and can be described as "short cycle" (1–3 years), "medium-cycle" (3–4 years), and "long-cycle" programmes (5–6 years) (Thune 2001, p. 3). As such, the degree system is rather complex with limitations regarding the transfer of credit points within the system, especially between the college and the university sector. Denmark has a system of external examiners which partly comprises teachers/professors from other institutions, and partly labour market representatives. The role of the external examiners is to assure that students are treated fairly and there is an equivalent national level

B. Stensaker (✉)
NIFU STEP – The Norwegian Institute for Studies of Innovation, Research and Higher Education, Oslo, Norway

D.D. Dill, M. Beerkens (eds.), *Public Policy for Academic Quality*, Higher Education Dynamics 30, DOI 10.1007/978-90-481-3754-1_10,

of assessment across schools and institutions (Kristoffersen 2003, p. 26, see also Stensaker et al. 2008).

As in other OECD countries, Denmark during the past 10–15 years has experienced a rapid increase in student numbers. In recent years, the gross intake to higher education has been between 50 and 60% of the relevant age group. Most of these students enrol on long-cycle higher education programmes. Higher education institutions are responsible for admissions, but admission requirements are set by the Ministry. In some programmes, for example in medicine, the Ministry still sets the admission number. In general, student numbers in study programmes vary according to student preferences and choice. There are no tuition fees for national students in the public sector, but tuition is claimed for students outside the EU.

The steering and funding of Danish higher education have changed considerably during the 1990s. The trend has been to delegate more responsibility from the Ministry to higher education institutions (see also Wright and Ørberg 2008). One may claim that the changes in the steering of the sector have stimulated the autonomy of the institutions, even though the power and autonomy of Danish universities have been historically quite strong. However, strategic behaviour and strong institutional leadership have not been a central characteristic of Danish universities. Hence, in 2000 the Ministry of Education launched what may be termed as "development contracts" between the Ministry and the individual institution. The purpose is to agree on more long-term objectives and targets (4-year periods) and to enable the institutions to market themselves better. Later reform efforts have been targeted at turning universities into self-owning institutions, with more emphasis on output funding and accounting (Wright and Ørberg 2008, p. 45).

The changes in the steering of higher education have been followed by a change in the funding of higher education with more emphasis on lump-sum allocations and output measures (Gornitzka et al. 2001, p. 19). This means that the higher education institutions can decide on how to allocate resources internally. The most important output measure (the "taximeter-system") is a combination of different indicators related to student numbers, the cost of studies in different disciplines and subject fields, and the number of credit points and exams taken. Research is funded separately. Four streams of money comprise most of the research funding: a lump-sum from the Ministry, allocations from different domestic research councils, applied research programmes, and some funds from the Danish fund for basic research (DGF).

The described changes in the steering and funding of higher education in Denmark in the past 10–15 years have also had implications for how academic quality assurance is conducted. Traditionally the country had a decentralised system of quality assurance, which left the assurance of academic standards up to the individual institution, with the external examiner system as the key component. In 1992 the Ministry of Education established the Danish Centre for Quality Assurance and Evaluation of Higher Education (EVA[1]), and instructed the centre to *conduct*

[1]The abbreviation EVA is used throughout the document even though the organisation "Centre for Quality Assurance and Evaluation" (Evalueringscenteret – EVC) changed its name and formal status in 1999 to the "Danish Evaluation Institute" (Danmarks Evalueringsinstitut – EVA).

systematic evaluation of all study programmes offered in higher education within a 7-year period. Hence this centre can be interpreted as a more centralised and independent actor in the field of academic quality assurance. Until 2007 this system remained fairly unchanged, before a new act introduced accreditation in the country. However, in this chapter the focus is mainly on an analysis of the previous system of subject assessment, why this was established, and the functioning of this system. The chapter ends with a brief discussion of the possible links between the new accreditation system and the former system of subject assessments.

The Policy Problem Resulting in the System of Subject Assessments

In the spring of 1992 a majority of the parties in the Danish Parliament arrived at a number of compromises on higher education, which led in the following year to a reform of the entire educational system. The stated objectives of the reform were to ensure (Thune et al. 1996, p. 21):

- a higher degree of institutional freedom and autonomy combined with a tightening of each institution's management structure,
- a better balance between supply of and demand for study places,
- the quality of the study programmes according to international standards.

The reform implied a new study structure (the bachelor/master/PhD-system), a new Act on universities, which reorganised the political and managerial governance of the institutions (reducing the number of democratically elected governing bodies and introducing external representation in the academic senate and in faculty boards), an introduction of the taximeter principle (an output-based funding system), and the establishment of a national system for the evaluation of higher education (conducted by a newly established agency for conducting such evaluations [EVA]).

When looking at the stated objectives of the reform, the background for the reform also comes to the fore. First, a huge increase in the number of students that applied for higher education. Second, Denmark faced at that time constraints on public spending, which triggered a focus on the efficiency and the effectiveness of higher education. Third, worries that an expansion of higher education could lead to a lowering of the academic quality. Fourth, the international commitments that Denmark had in relation to the European Union and their student exchange system (Erasmus).

The establishment of a national system for evaluation and an independent agency for carrying out such evaluations are in various ways related to the drivers behind the reform. The establishment of the EVA agency could be described along a number of different perspectives:

(a) The creation of the system of study programme evaluations could be interpreted as a governmental response to perceived needs for more *efficiency and output orientation* in Danish higher education. The share of resources spent on higher education, due to the increasing number of students among other reasons, triggered a need to check how resources are spent and to identify "organisational slack" inside higher education institutions. The systematic evaluation of all study programmes offered in Danish higher education can be seen as an indicator of such an orientation.

(b) At the same time, study programme evaluations could also be seen as an attempt to balance the *centralisation–decentralisation dilemma* in Danish higher education. While major parts of the 1992 reform were intended to give institutions more autonomy, the establishment of a national evaluation system could be interpreted as a means of centralising quality assurance. Thus, the evaluations in this perspective represent the need to maintain control even in a more decentralised system.

(c) Since the evaluations were established with a double purpose of accountability and improvement, it is also possible to see the establishment of the evaluation system in a more *developmental* perspective. The decentralisation of authority and responsibility to institutions meant that the institutional leadership had to take on a stronger and more strategic role. However, this is a role that breaks with the traditions of institutional leadership in Danish universities. The traditional power structure in higher education centred round the departments and disciplines (Gulddahl Rasmussen 1997; Foss-Hansen 1997). The national evaluation system could in this perspective be interpreted as being a "support structure" for the institutional leadership (see also Stensaker 1999, p. 257–258).

(d) The notions "knowledge society" and "knowledge economy" and the role of higher education and research in these developments have had a powerful influence in the political debate on higher education in the last two decades. One of the important elements of the knowledge society is that higher education needs to establish better links with the world of work (Rasmussen 1997a). In the new evaluation system, these links are very visible. Not only are members of industry and society part of the review panels, but graduated students are also, after a few years at work, asked about the relevance of their study programme in relation to their current job. In this perspective, the study programme evaluations could be seen as an instrument for increasing the *relevance* of higher education for the society and the world of work.

(e) Finally, one could also interpret the establishment of study programme evaluations as a form of political accountability. Not only higher education, but also those responsible for higher education at the political level need to be accountable to the larger society. Hence, the creation of a national system for evaluation, and an agency responsible for carrying out such tasks could be interpreted as being an important *symbolic* action, by which politicians can show the public that something "is done to assure quality." One indication of this is that how evaluations should be followed up was almost a non-issue in Denmark in the first few years of the 1990s (Askling et al. 1998, p. 9). Not the outcomes, but the fact that evaluations were conducted seemed, in other words, to be the important thing.

To pinpoint the policy problem in accurate terms is, in other words, somewhat problematic. However, the five perspectives mentioned above cover most of the arguments related to the establishment of the national evaluations in 1992 and can be said to have represented a formidable challenge for the leadership of EVA to balance these various needs and expectations.

Content of the Policy Instrument

The mandate for EVA, provided by the Ministry, instructed the centre that future evaluations had to focus on the study programme level, that both control (accountability) and institutional improvement had to be a part of any procedures launched, and that evaluations were not a voluntary activity for the institutions. However, the results of the evaluations were not linked to funding (Evalueringscenteret 1998, pp. 16–17). The evaluation system was not created on a permanent basis, but was set up for an initial period of 7 years, and on the condition that the system and EVA itself should be subject to an evaluation when deciding whether evaluations should become a permanent activity. The political focus on study programmes can probably be related to the huge number of small higher education institutions in Denmark, and the fear that the institutions could not be trusted as assurors of quality (see also Thune 1996). Also, the systematisation meant that all study programmes were treated equally – a particular feature in the Scandinavian culture (Smeby 1996). EVA was created as an independent body. (See Box 10.1 for EVA's legal and organisational framework.) The Ministry of Education was not to instruct the centre, but the National Educational Councils (NEC) (in humanities, science, social sciences, etc.) were given the right to decide the chronological order of the evaluations, and thus could be seen as the bodies responsible for the initiation of a given evaluation.

EVA's formal mandate was (Thune 2001, p. 7):

- to initiate systematic evaluation of all study programmes in higher education in Denmark including the university as well as the non-university sector (within a 6-year period),
- to develop appropriate methods of the recognition of study programmes making them eligible for governments funding and student loans,
- to inspire and guide the institutions of higher education in aspects concerning evaluation and quality,
- to compile national and international experience on evaluation of the educational system and quality development.

Given the many interests surrounding the establishment of the study programme evaluations, it is perhaps not surprising that active dialogue and consultation with institutional and departmental leadership was and is chosen by EVA as a strategy for organising the individual evaluations. This process is "real" in the sense that EVA often adjusts the organisation and design of a given evaluation after this round of consultation.

Box 10.1 EVA's Former Legal and Organisational Framework

1. Legal Framework

Two legal documents regulate EVA's activities. The most important one is the Danish Evaluation Institute Act. The Ministry of Education has established a set of regulations for EVA that specifies the act. The regulations are as legally binding for EVA as the parliamentary act, but it is within the authority of the Minister of Education to amend the regulations within the framework of the parliamentary act. The legal framework regulates the relationship to the Ministry of Education and specifies:

- EVA's right to initiate evaluations;
- the governance of the agency;
- the distribution of responsibilities with regard to evaluation;
- core methodological principles.

2. Main Stakeholders

Within the field of higher education, the Ministry of Education and the new Ministry of Science, Technology and Innovation (established after the Danish election November 2001) represent the main stakeholders, for example they have to approve the annual plan of action and the budget. Besides these formal relations, EVA has regular contact meetings with the Ministry of Education and is in the process of establishing a network at staff level. In addition to the ministries, EVA has maintained contact with stakeholders from the higher education community. EVA meets with the Danish Rectors' Conference, which represents all universities in Denmark, and EVA's Committee of Representatives, which comprises members from different sectors of the education system.

3. Governance

EVA is an independent institution formed under the auspices of the Danish Ministry of Education. EVA is governed by a board. The board is responsible for the overall supervision of the Institute, including the annual action plan, and appoints the management of the Institute. The appointment of the executive director must be formally approved by the Minister of Education. The executive director manages EVA and is responsible to the board. The board formally approves the appointment of other staff. The board consists of 10 members and a chairman. The Danish Minister of Education nominates the chairman. The 10 members are appointed by the minister upon the recommendation of the ministry's advisory boards. Thus, the board does not automatically include representatives of the higher education institutions. The board is appointed for a 3-year period with the possibility of reappointment.

In addition to the board, a Committee of Representatives is established as a mandatory part of EVA's organisational set-up. The Committee of Representatives comments on EVA's annual plan of action, the annual report and the priority of planned activities. The Committee comprises 27 members. They are appointed by organisations from the following sectors: school proprietors, school associations, school boards and employers; rector's conferences and school managers; management and labour organisations; teachers' organisations and students and pupils bodies. In addition, the Committee of Representatives itself appoints two experts with international evaluation experience.

The board draws up the programme for the next year's activities based on the recommendations of the executive director. The Minister of Education approves the annual plan of action. In addition to the evaluations conducted on its own initiative, EVA may conduct evaluations on the request of authorities responsible for education.

Source: EVA 2002.

The system that was established in 1992 had the following procedures, which basically have been kept unchanged ever since (Thune 2001, pp. 7–8).

EVA conducts a *preliminary study* with the purpose of identifying relevant study programmes to be included in a given evaluation (due to difficulties in knowing the content and profile of some study programmes), and to establish possible criteria/objectives to be used as a mandate for the evaluation. The final selection of study programmes in a given assessment is decided after the preliminary study.

Based on the preliminary study and internal guidelines, EVA prepares the terms of reference. The terms of reference are a formal basis for an evaluation and they have to be approved by the Board before the evaluation process can be started. The terms of reference include the background and purpose of the evaluation, time schedule, the list of higher education units involved in the evaluation, items to be included in the evaluation, the division of responsibilities between the evaluation group and EVA, the general framework for the evaluation, and the methods to be applied (EVA 2002).

Traditionally, a fitness-for-purpose approach has been used in Denmark, emphasising the objectives of a given study programme (EVA 2002, p. 14). This approach has been balanced by including national policy objectives in the mandate when relevant. Hence, each evaluation conducted has a specific mandate. In previously conducted programme evaluations the following items were included:

- the objectives of the programme,
- management, organisation and resources,
- structure of the programme,
- content of the programme,
- practical learning,
- methods of teaching and training,
- lecturers/professors, including pedagogical competencies,
- exams and evaluation of students,
- students entry levels and progression,
- internationalisation,
- relations to other institutions and society,
- quality assurance.

The physical environment and learning resources have also been regularly included (EVA 2002, p.14).

When an institution is selected for a review, it is informed of the evaluation process in writing. It is mandatory that EVA informs the institution on the legal basis for the evaluation, including the rights and obligations of the institution, the purpose of the evaluation, the terms of reference, the members of the evaluation group and expectations regarding the institution's own contribution to the process (EVA 2002).

The unit responsible for offering a given study programme (usually a department) then writes a *self-evaluation report* based on a rather detailed protocol provided by EVA. This protocol usually instructs the department to describe the objectives of the programme, the management and organisation surrounding the programme, content, methods of teaching and learning and to include quantitative information concerning the number of academic staff, applications, drop-outs, completion rates, and so on for the past 3 years. The purpose of the self-evaluation is both to provide the external expert committee with background information and to stimulate "development" in the department. The department is free to choose the organisation of the self-evaluation process, but is advised to include academic staff, administration and students in the process.

In parallel to the self-evaluation, a comprehensive *survey* on the quality of the programmes is often conducted among various users, i.e. students, graduates, and employers (and sometimes external examiners within the subject field). These surveys are outsourced by EVA to private consultancies, poll firms, etc. The purpose of the survey is to provide alternative views and perspectives on the subject field and on each particular study programme, and to indicate the relevance of the study programme to important stakeholders. Due to the fact that these surveys are somewhat expensive, only one group of users is surveyed in each evaluation. The selection is made by EVA. Due to the fact that these surveys are conducted in parallel with the self-evaluation, a given department cannot use the results in their self-evaluation. The timing of the process is usually such that the self-evaluation and the surveys are finished simultaneously for the use of the external expert committee.

The next phase in the evaluation process is the *visit* to the department *by the external expert committee*. Before the visit, the committee meets to discuss the information available, to find areas for investigation in addition to, or supplementing, the checklist used for the self-evaluation, and to plan the visit.

The committee is selected and chosen by EVA, and usually consists of three to five persons with extensive knowledge of the subject field and/or expertise in university governance and management. Committee members must be independent of the institutions involved in the evaluation, which is ensured by a formal statement from potential experts. The potential expert must state whether he/she has been employed, been invited to give lectures or in any other way associated with the programme under review, as well as whether his/her child, spouse or near friend has studied or been employed by the programme (EVA 2002). Due to the fact that Denmark is a small country, where it is difficult to find "independent" experts, and that people from Denmark, Sweden and Norway can understand each other's language, experts are often recruited from the latter two countries. Even if these experts could be regarded as "peers," a committee usually also includes a member (sometimes two) from outside of higher education. Typically, this representative is from business, industry or a public organisation. The expert committee must be approved by the EVA's board.

A visit to a department usually takes place over a 2-day period where representatives of the leadership, academic staff, administration and students are interviewed. Usually, but dependent on the number of study programmes evaluated, the same committee undertakes all the visits. After the site visit, the committee meets two or three times in order to discuss and finalise the evaluation report.

For each evaluation, EVA appoints an internal project team of two evaluation officers and one assistant, which provides secretarial help to the external committee and ensures that the evaluation is conducted as specified in both the terms of reference and EVA's formal regulation. The project team is involved throughout the process: from conducting the preliminary study, and preparing the programme and interview guides for the site visit, up to drafting the report to be discussed and approved by the evaluation group (EVA 2002).

After the visit, and after several meetings in the external expert committee, an *external report* is written presenting an overall analysis of the quality of the programme field at the national level as well as individual analyses of all study programmes included in the evaluation. A draft version of the report is then sent to all departments/institutions involved in a given evaluation, and a closed conference is held where only representatives of EVA, the external expert committee and the departments/institutions participate. The purpose of the conference is both to prepare the departments for the coming conclusions, but also to provide an opportunity to adjust the report for any misperceptions or errors. After the conference, the final report is printed and sent to every participating department/institution, the Ministry of Education, and the relevant NEC. The self-evaluation report, the survey results and the external report are open and accessible to the public. The NEC has the responsibility to follow up the report, for example, by checking how departments/institutions took actions based on the recommendations given in the report. It is important to note that the departments/institutions are not mandated to follow the recommendations, but will be asked questions related to what actions have been taken on the basis of the report.

The typical evaluation conducted by EVA includes up to ten study programmes (occasionally even more) within a given subject field, with a self-evaluation report made for each study programme/department, and with an external visit to every study programme. After the release of the final report, Danish newspapers have traditionally taken an interest in the results, sometimes creating a lively public debate.

Implementation

The implementation process in the first years was not without some turbulence. When the study programme evaluations and EVA were proposed by the government, the Rectors conference, the umbrella body of Danish universities, was invited to participate in the development and design of the new centre. The Rectors conference, however, turned down the invitation from the Ministry. The argument used was that the Rectors conference preferred external evaluations, if necessary, to be carried out by the Ministry itself, and not by an independent body (Evalueringscenteret 1998, p. 16). This scepticism from the universities' side can be related to both fear of loosening up established ties between the Ministry and the institutions (the established power structure), but also to protest since the institutions had to cover part of the expenses related to the evaluations themselves (the self-evaluation).

Due to the initial scepticism towards what the evaluations would bring, and the consequences of the evaluations, the procedures surrounding the evaluations were delicately designed (Askling et al. 1998, p. 11). Because of the fear of being seen as just a cover for ministerial and political agendas, or to be perceived as "soft" towards the institutions, the procedures surrounding the evaluations had a focus on methods, systematisation and standardisation (almost unprecedented for quality assurance agencies) as the way to gain legitimacy and respect (Askling et al. 1998, p. 12). Using EVA's own words: "the method is developed with the aim of uncovering the quality of a study programme through a concrete, transparent and trustworthy process" (Evalueringscenteret 1998, p. 25, authors translation). The fact that the mandate for any given evaluation is developed after a pre-study of the subject field, that rather detailed instructions are provided for the self-evaluation process (the protocol), that data is collected from current and graduated students and employers, and that a conference with the involved parties is held before publication to check for any errors or potential overlooked problems, are all indications that the evaluation process is designed to be as robust and solid as possible. Since private consultancy/poll firms were hired to collect and analyse the data from graduated students and employers, not only "methodological" triangulation was obtained (the use of different methods to shed light on a phenomenon), but also "observer" triangulation (the use of different actors to observe the phenomenon). Not surprisingly, all this information sometimes created very lengthy evaluation reports (up to 200 pages, see Thune 2001, p. 8).

Over the years, the evaluation procedure has been incrementally changed and developed. Some of the elements of the evaluation procedure have also evoked debate which sometimes has triggered adjustments. Issues that have been raised are, first, related to the protocol for the self-evaluation. In the early years, this protocol was very detailed with the potential effect that it was perceived as less relevant for initiating more developmental processes at the department level (Askling et al. 1998, p. 16). Thus, over the years the protocol has been revised with the purpose of providing a broader framework for the self-evaluation instead of being a "questionnaire" to be answered. More open, reflective questions to be answered have been included in the protocol (especially related to how departments have established routines and systems for quality assurance).

Second, a debate has developed on the ability of a given evaluation to relate to the needs of the individual study programme when the evaluation covers up to and sometimes even more than 10 programmes at a time. Stensaker (1999, p. 259) has, for example, documented how the departmental/institutional perceived benefit of an evaluation drops when more than ten study programmes are evaluated together. This study shows that the perceived benefit related to the self-evaluation process is constant independent of the number of study programme participating in an evaluation, but the benefit related to the external panel visit and external report is perceived as smaller the more study programmes are included. One possible explanation is the capacity problem of the external expert committee. A large number of participating study programmes results in more general recommendations from the committee, and the individual study programme is not considered in the same way as when a given evaluation only covers a few study programmes.

Third, the user surveys have also been criticised over the years. One line of criticism has been directed at the (lack of) competence of private consultancies for designing and analysing useful surveys (Evalueringscenteret 1998, p. 37). Another argument has been related to the timing of the user surveys in the overall evaluation process. Stensaker (1999, p. 263) has, for example, argued that because self-evaluation and user surveys are conducted simultaneously, the departments cannot use the information from the surveys in their own self-evaluation. As such, it could be argued that these surveys have been more related to external (the external expert committee) than internal (departmental) needs. Since it is EVA in cooperation with the given consultancy/poll firm that decides the content of the survey, departments/institutions also miss a chance to put "their" issues on the agenda.

A fourth issue is related to the fact that a number of departments/institutions have been somewhat dissatisfied with how the closed conference works, and how the results of the evaluations have been commented upon in newspapers after the launching of the final report. The time for debate within the conference is limited and therefore the problems/potentials of each study programme have not always been fully addressed. In addition, many of the participating departments are also potential "competitors" in the Danish higher education market and have a more reserved attitude towards openly discussing problems and solutions

(Evalueringscenteret 1998, p. 44; Askling et al. 1998, p. 21). The fact that Danish newspapers, after the launching of the final report, tend to focus upon findings that give departments bad publicity, has also been raised as an issue, but the departments have not suggested shielding the external reports from public scrutiny.

Despite some debates and criticisms, the systematic evaluations were continued according to the plan. By 1999, EVA had fulfilled its mission and produced 62 evaluation reports, conducted over a hundred user surveys of graduated students, employers, etc., and involved approx. 200 experts in the evaluation processes (Evalueringscenteret 1998, p. 30). The question to be asked is, of course, related to the impact of all this.

Impact

When the initial period was over, the Ministry of Education initiated an external evaluation of EVA. This process was conducted much in the same way as an ordinary study programme evaluation. Hence, not only did EVA have to write a self-evaluation report describing and analysing the previous years, but also an external review of the methods, procedures and roles was conducted by experts from other Scandinavian countries (Askling et al. 1998). A user-survey was also conducted, where a private consultancy firm asked representatives of the students, the higher education institutions (rectors, deans, etc.) and other stakeholders about their views on EVA's activities (PLS-Consult 1998). This section is based on this evaluation process, on a separate article where some of the data collected by PLS-Consult was re-analysed (Stensaker 1999), but also on observations from an independent study conducted by an American researcher in the field (Massy 1999).

Concerning methodology, trustworthiness and relevance of the study programme assessments, Askling et al. (1998, pp. 4–6, 26) stated that the conducted evaluations had gained legitimacy, and that accountability and improvement actually were balanced, even if the improvement dimension could have been better highlighted. The rectors of a number of universities have reached similar conclusions (PLS-Consult 1998, p. 14). The high degree of systematisation, the stringent routines associated with each evaluation, and the various sources of data used to evaluate study programmes were mentioned as important factors leading to this conclusion. The fact that the evaluation system created in Denmark did not integrate any performance indicator system into the evaluation is probably a factor leading to a positive attitude from the higher education sector.

The critique from Askling et al. (1998, p. 25) was, therefore, more directed at what they saw was the weak point – the follow-up of the evaluations. Due to the "arms-length" steering principle, the Ministry only checked whether institutions had launched any actions after completion, and EVA had no responsibility for what happened after the publication of the evaluation report (Smeby and Stensaker 1999, p. 6). Thus, follow-up was a responsibility of the institutions themselves, but the external reinforcement for making sure actions were implemented was not great.

On this background, one could expect that the impact of the evaluations at the higher education institutions was limited. A survey of deans and department heads/other leaders at the department level, and a smaller number of interviews with rectors of various higher education institutions gives a rather different picture. Not least, a general attitude was that the evaluation processes had been a positive experience, and had created much discussion and dialogue inside the institutions (PLS-Consult 1998; Massy 1999; Stensaker 1999).

When asked to pinpoint the most beneficial element in the evaluation process, the self-evaluation process was undoubtedly most often mentioned (Stensaker 1999, pp. 259–260). Many respondents also had a very positive view of the visit from the expert panel (PLS-Consult 1998, p. 15). The perceived benefits of the user surveys were somewhat mixed, where feedback from students in general was seen as most beneficial (Stensaker 1999, p. 261). A negative comment mentioned by all respondents was that the external evaluations were time consuming and rather expensive processes (PLS-Consult 1998, p. 14).

Concerning follow-up, slightly over 60% of department heads and other leaders at the department level claimed that recommendations had been followed up to a great extent and that a large majority of the respondents saw the recommendations as useful advices for improvement of the study programmes (PLS-Consult 1998, p. 16). If one studies the areas where changes were most visible, the curriculum structure, examination, teaching methods and the objectives of the study programmes were the most often mentioned areas. Many respondents also claimed that the evaluations in general had triggered decisions and speeded up existing change processes in the evaluated departments (PLS-Consult 1998, p. 18).

In a broader, institutional, perspective, the assessments of study programmes have had less effect. For example, for a number of years the protocol for departmental self-evaluation did not contain questions related to whether departmental systems for assuring and improving quality had been established. Hence, the early assessments of the study programmes contributed little to the establishment of institutional routines for the systematic maintenance and development of the quality of teaching and learning (see, for example, PLS-Consult 1998, p. 17). Furthermore, since the protocol for the self-evaluations were fully developed by EVA, the departments participating in the evaluation often experienced less ownership, and hence less motivation for going into the evaluation process with an improvement orientation (Evalueringscenteret 1998, p. 34). A last point to be made is that study programme evaluations tend to de-couple the institutional leadership from the evaluation process. Even if the rectors often are drawn into the evaluation process, for example, by being interviewed by the external expert committee, one can detect a feeling that the institutional leadership perceives the study programme evaluations as less relevant for them. For example, a majority of the rectors seemed to prefer an evaluation model where research and education were integrated in an evaluation (PLS-Consult 1998, p. 15). Another example is related to the fact that many rectors perceived that the costs associated with the evaluations outweighed the benefits (PLS-Consult 1998, p. 15). This perception changes, however, when those closer to the study programme were asked about the perceived benefit. Almost 80% of the study programme managers answered that they perceived the evaluation as relevant, and providing them with valuable recommendations. In other words, greater distance to the evaluation creates less perceived benefit.

It is perhaps due to this de-coupling of the institutional leadership from the evaluations that Massy (1999, p. 30), analysing the national evaluation systems in Sweden and Denmark, stated that such systems "need not be an exercise in power and control." (One should, however, also bear in mind that the traditional weak role of the institutional leadership in the Scandinavian countries could impact substantially on the conclusions drawn by Massy.) He maintains that one of the success criteria for these systems is that issues of power and control are not allowed to dominate the agenda. Perhaps as an effect of that, Massy (1999, p. 31) acknowledges that the national evaluation system has not transformed higher education. But they have managed to "start universities and departments on the road to becoming learning organizations (...), to become self-conscious about the processes of teaching and learning and how to improve them."

Costs

Costs have – on the national level – traditionally been a non-issue related to the evaluation system in Denmark. For example, the question of whether the country got "value-for-money" when a new system was designed in 1999 was totally ignored. The necessity to maintain a national system of evaluation and to expand it to all levels dominated the agenda. However, as mentioned earlier, at the institutional/department level, complaints have been launched that participating in the study programme evaluations is time consuming and economically burdensome since departments have to pay for the self-evaluation themselves. Finding resources for freeing staff to be included in the self-evaluation process has been perceived as somewhat difficult.

The costs related to funding and running EVA are decided by the Ministry of Education. EVA's annual budget normally summed up to approximately 45 million DKK (approximately 6 million US dollars). This includes the costs for evaluations in the primary school sector (both the primary and secondary levels of education), but so far only a few pilot-projects have been launched in this area resulting in very limited expenditures.

The majority of the budget is covered by an appropriation from the Ministry of Education (approximately 40 million DKK). The rest of the budget, about 4 million DKK, stems from other external sources. This budget covers the salary of 48 full-time employees at EVA and 10–12 student assistants working part-time. EVA estimates that the costs of a typical study programme evaluation (dependent on the number of study programmes included) vary between 250.000 and 500.000 DKK (approximately 33.000–66.000 US dollars). Most of these resources are tied up in travel expenses and honoraria for the external expert committee. A typical honorarium for an external expert, depending on the number of study programmes included in the evaluation, varies between 12.000 and 25.000 DKK (1.400–2.800 US dollars). In addition, the costs of user surveys vary between 50.000 and 200.000 DKK (approximately 6.500–27.000 US dollars). As mentioned before, private consultancies are often the contractors and conductors of these surveys. In the first years, some evaluations became more expensive than budgeted and some were even substantially delayed according to their time schedule, but this problem has in later years been solved with more rigorous steering of each evaluation project (Evalueringscenteret 1998, p. 24).

Comparisons, Conclusions and Future Action

After the evaluation of EVA and the evaluation system in 1998, the Ministry of Education argued that systematic study programme evaluation on the national level should continue, and that EVA had to be a permanent centre (UFK 1998, pp. 33–36). However, the Ministry suggested that activities should be expanded, both concerning scope and practices. Hence, compared to the past, several new elements were at the turn of the millennium visible in the current evaluation system in Denmark.

Not only study programmes in higher education, but education at all levels in the Danish system became an object for evaluation. The background for this expansion in scope was that Denmark experienced some (relatively) low scores in an international survey testing the knowledge level at certain ages in primary and secondary school (the PISA survey), and that, amongst other things, more systematic evaluations were needed as a means to improve quality (Kristoffersen 2003, p. 26).

Concerning methods, systematic evaluations of study programmes in higher education continued, but new forms of evaluations, including thematic evaluations, evaluations of institutions, system evaluations, and audits (evaluation of quality assurance systems) were also introduced (Kristoffersen 2003, p. 27).

The study programme evaluations carried out during the 1990s were still recognisable in the new version. This has probably a lot to do with the purpose of the evaluations, which still maintained the duality of accountability and improvement, and that the established procedures for conducting study programme evaluations at that point were still being used. In an impact-study of some of the evaluations conducted by the new body, findings suggest that the study programme evaluations continued to have an impact on Danish higher education, even in "round two" (Stensaker 2004, p. 38). The respondents still had a positive view of the evaluations, approximately 60% of the recommendations were followed up, and a typical effect was that dialogue, discussions and reflections at the institutions increased afterwards (EVA 2004). The result is interesting in that counter to the predictions of researchers in the field, the system showed no signs of a diminishing impact of the national evaluation system as a consequence of institutions learning the "tricks of the trade" the second time around (Jeliazkova and Westerheijden 2001, p. 1). A possible explanation could be that evaluations that stimulate dialogue, discussions and reflection may not be so vulnerable for that sort of institutional "strategic behaviour." As neatly said by a US observer: "The strength of the Scandinavian evaluation philosophy lies in its strategy of creating discourse rather that laying down regulations" (Massy 1999, p. 29).

However, the type of discourse that Scandinavian evaluations seem to contribute to, varies somewhat between the countries. In Sweden, where institutional audits (evaluations of institutional work to secure and improve quality) for a number of years were the dominant evaluation type on the national level, the discourse (measured in the number of recommendations made) has traditionally centred around management and strategy issues (Stensaker 2000). Those experiencing the effects of these audits noted that the audits did not address quality issues on the "shop floor." Whether the quality of a given study programme or department is good or bad, in other words, cannot be detected by using this method. Contrary to this, the Danish evaluations address quality issues related to the teaching and learning process more directly. Most recommendations in the study programme evaluations in Denmark have addressed issues related to teaching, curriculum and pedagogy, and organisation, management and strategy issues relating to quality have been less emphasised (Evalueringscenteret 1998, p. 26).

That being said, it is important to remember that even if the focus of the Danish evaluation system has been on study programmes, it should still be seen as a "fitness-for-purpose" process and not directed at securing academic standards per se. The study programmes in Denmark have traditionally not, as has been the case in the UK, been checked against predefined academic criteria developed for every subject area, and neither have the same disciplinary orientation as the programme assessments that have been carried out by the VSNU in the Netherlands (Brennan and Shah 2000, p. 64). A reason is probably the existence of the external examiner system in Denmark (see also Stensaker et al. 2008). In this way, one may argue that Denmark in the period of analysis actually had two parallel systems for academic quality assurance on the national level. That these two systems in the past have been poorly linked has surprised external observers (see Dill 2002).

Dill (2002, p. 28) has also noted that the Danish study programme evaluations have only to a limited extent been focused on assuring the quality of new study programmes established in Danish higher education. As such, the Danish study programme evaluations have been more oriented towards "general" accountability in a system characterised by a relatively high degree of stability than targeted at quality assurance in a fast moving, more competitive and open higher education market. For example, the study programme evaluations have been difficult to utilise as "consumer guidance" for students. It can also be questioned whether they actually address political needs for information about higher education (Rasmussen 1997b, p. 257). The "accountability" generated by the evaluations have, in other words, been directed more to "the environment" rather than to a specific interest group.

On this background, one could ask whether the absence of "revolutionary change" in the Scandinavian higher education systems during the last 10–15 years (Finland excepted) may be one of the factors for the "success" of the Danish (and Scandinavian) evaluation approach. According to Massy, it is the "soft pedalling of the accountability agenda" that has provided the Scandinavian evaluations with high degree of legitimacy (Massy 1999, p. 33). Is it the absence of the need for radical political change of the higher education system that has provided a basic condition for the “soft pedalling”? Along the same line, it should also be noted that the Ministry of Education in Denmark has managed to follow up its own strategy and intention concerning the evaluation system, and that the legitimacy of the evaluation system probably also can be related to the fact that the Ministry have not interfered in the evaluation process. In other words, the arms-length strategy has been allowed to work according to the intentions.

Here, a contrast can be made to the Netherlands where the Inspectorate of Higher Education was to take action if a programme experienced a poor assessment by the VSNU. This inspectorate, which reported directly to the Dutch Ministry of Education, could then instigate its own investigation potentially resulting in a deletion of the programme from the list of recognised (funded) programmes. Even if programmes were not closed as a result of this procedure during the 1990s (Brennan and Shah 2000, pp. 63–64), it is an indication of a policy design more oriented towards control than dialogue.

A comparison can also be made to the UK subject assessment system where the design of the subject assessments in the 1990s was more oriented towards grading different programmes (accountability/control) than initiating improvement processes inside the institutions. According to Brennan and Shah, this relates directly to the “assessment methodology, in particular the summative judgement on a three-point scale” (Brennan and Shah 2000, p. 92). A typical effect of this latter methodology was the assignment and redistribution of programme status and reputation, while a more typical effect of the Danish assessment methodology is that of changing institutional cultures and creating positive attitudes towards teaching and learning. Even if there were national funding schemes available in the UK for starting up quality development projects in teaching and learning at the institutions, the access to these resources was again dependent on a good assessment result (Brennan and Shah 2000, p. 94). Hence, to become better you should be rather good in the first place.

In retrospect, one may argue that the many policy problems highlighted in section two continue to exist in the new system in Denmark. Worries about efficiency, centralisation, control, development, relevance and how to deal with these problems are as present in Danish higher education today as in the later decades. The main difference is that it seems easier to articulate views on these issues today, and that the different actors trust more the motives and engagement of each other. The unfortunate accountability-improvement debate that in many countries, like the UK, dominated the policy agenda and consequently lowered the trust in the evaluation system was, after the first years, resolved through practice in Denmark (Askling et al. 1998) and has not appeared since. Thus, Denmark is a good case for illustrating that external evaluation can balance accountability and improvement (see also Stensaker 2003, p. 157, see also Danø and Stensaker 2007).

However, as in other countries, there has recently been an increased focus on international comparisons and on international developments within the field of quality assurance in Denmark. Accreditation is, for example, one of the new elements introduced in Europe in recent years (Faber and Huisman 2003), a method that recently also was introduced in Denmark. In 2007 the Danish Parliament granted the implementation of an accreditation system in Denmark, and established also a new agency for accreditation of study programmes in the university sector (ACE Denmark). EVA is in the new system responsible for accreditation of study programmes in the college sector. In the new system, a separate accreditation council will decide on whether accreditation should be given based on the evidence provided by EVA and ACE Denmark. However, also other independent accreditation agencies can in the new system operate in Denmark, although all formal decisions on accreditation are taken by the accreditation council.

The introduction of accreditation, and the establishment of the new agency and the new accreditation council, could be interpreted in several ways. First, one could argue that this development is a natural consequence of a more "mature" quality assurance system constantly building and expanding existing activities. In this perspective the new accreditation system could be seen as filling the blank spots of the old system, not least when it comes to recognition of new study programmes (see above). Second, as indicated the new accreditation system is part of broader European developments in which the method of accreditation has spread rapidly. The background for the interest in accreditation is related to a growing interest in internationalising Danish higher education (Wright and Ørberg 2008), driven both by the Bologna process and Danish ambitions of strengthening the country as a knowledge economy in a more global perspective.

What consequences this new scheme will have for higher education in Denmark is at this point unclear. While one could argue that the accreditation system represents a more radical break with past policy-making, and may damage the level of trust that has been built up through the old scheme, this should not be taken for granted. Although accreditation does represent a change with respect to the amount of feedback given to institutions (yes/no), the method of investigation may not necessarily change that much. If the future brings primarily positive and few negative comments concerning the outcomes of the accreditation process, the previous dialogue-based contact between the evaluators and those evaluated may be continued even in the new system. Some years ago EVA's former director Christian Thune specified the challenge of Danish quality assurance in the following way: "how [can] a well-functioning (domestic) and established system of ("fitness-for-purpose") study programme evaluations be translated or reinterpreted in the light of the (international) trend towards accreditation" (Thune 2001, p. 17)? One possible answer is that a new name has been provided to an old procedure.

References

Askling, B., Nordskov Nielsen, L., Stensaker, B. (1998). *Mellom fag og politikk. En granskning av Evalueringscenteret og det danske evalueringssystemet for høyere utdanning*. København: Undervisningsministeriet.

Brennan, J., Shah, T. (2000). *Managing quality in higher education: An international perspective on institutional assessment and change*. Buckingham, UK: OECD, SRHE & Open University Press.

Danø, T., Stensaker, B. (2007). Still balancing improvement and accountability? Developments in external quality assurance in the Nordic countries 1996–2006. *Quality in Higher Education*, 13(1), 81–93.

Dill, D. D. (2002). Through the looking glass: Comments on the Nordic Pilot Project on Mutual Recognition. In Lindeberg, T., Kristoffersen, D. (Eds.), *A method for mutual recognition. Experiences with a method for mutual recognition of quality assurance agencies*. Helsinki: ENQA Occasional Papers 4.

EVA (2002). *Mutual recognition: The Danish report*. Copenhagen: The Danish Evaluation Institute.

EVA (2004). *Effektundersøgelse. Redegørelse*. København: Danmarks evalueringsinstitut.

Evalueringscenteret (1998) *Selvevalueringsrapport*. København.

Faber, M., Huisman, J. (2003). Same voyage, different routes? The course of the Netherlands and Denmark to a "European model" of quality assurance. *Quality in Higher Education*, 9(3), 231–241.

Foss-Hansen, H. (1997). Universitetssektoren på vej mod år 2000. Den splittede omverden og behovet for helhed. Rasmussen, I. P., Jakobsen A. (red.), *Universiteter i dag. Politikk – kultur – ledelse*. Fredriksberg: Samfundslitteratur.

Gornitzka, Å., Maassen, P., Dahl Norgård, J. (2001). Nasjonal prioritering og arbeidsdeling i høyere utdanning – internasjonale erfaringer. Oslo: NIFU skriftserie 9/2001.

Gulddahl Rasmussen, J. (1997). Styring av forskning og undervisning på universiteter. Forandringer i de valgte mellomledernes funktioner og placering. Rasmussen, I. P., Jakobsen, A. (red.), *Universiteter i dag. Politikk – kultur – ledelse*. Fredriksberg: Samfundslitteratur.

Jeliazkova, M., Westerheijden, D. F. (2001). A next generation of quality assurance models. On phases, levels and circles in policy development. Paper given at the CHER-conference in Dijon, 2–4 September.

Kristoffersen, D. (2003). Denmark. *Educational evaluation around the world: An international anthology*. Copenhagen: The Danish Evaluation Institute.

Maassen, p. A. M., Stensaker, B. (2003). Interpretations of self-regulations. The changing state-higher education relationship in Europe. In Begg R. (Ed.), *The dialogue between higher education research and practice*. Dordrecht: Kluwer academic publishers.

Massy, W. F. (1999). Energizing quality work: Higher education quality evaluation in Sweden and Denmark. National Center for Postsecondary Improvement, Stanford University, Stanford.

PLS-Consult (1998). Undersøkelse af effekterne af uddannelsesevalueringer. København.

Rasmussen, p. (1997a). Universitetstradisjonen i krise. Rasmussen, I. P., Jakobsen, A., (red.), *Universiteter i dag. Politikk – kultur – ledelse*. Fredriksberg: Samfundslitteratur.

Rasmussen, p. (1997b). A Danish approach to quality in higher education. In Brennan, J., de Vries, P., Williams, R. (Eds.), *Standards and quality in higher education*. London: Jessica Kingsley Publishers.

Smeby, J. C. (1996). Disciplinary differences in university teaching. *Studies in Higher Education*, 21(1), 69–79.

Smeby, J. C., Stensaker, B. (1999). National quality assessments systems in the Nordic countries: Developing a balance between external and internal needs? *Higher Education Policy*, 12(1), 3–14.

Stensaker, B. (1999). User surveys in external assessments: Problems and prospects. *Quality in Higher Education*, 5(3), 255–264.

Stensaker, B. (2000). Quality as discourse: An analysis of external audit reports in Sweden 1995–1998. *Tertiary Education and Management*, 6(4), 305–317.

Stensaker, B. (2003). Trance, transparency and transformation: The impact of external quality monitoring on higher education. *Quality in Higher Education*, 9(2), 151–159.

Stensaker, B. (2004). Epilog: effekter af evaluering – en relativisering. *In Effektundersøgelse. Redegørelse*. København, Danmarks: Evalueringsinstitut.

Stensaker, B., Brandt, E., Solum, N. H. (2008). Changing systems of external examination. *Quality Assurance in Education*, 16(3), 211–223.

Thune, C. (1996). The alliance of accountability and improvement: The Danish experience. *Quality in Higher Education*, 2(1), 21–32.

Thune, C. (2001). Quality assurance of higher education in Denmark. In Dunkerly, D., Wong W. S. (Eds.), *Global perspectives on quality in higher education*. Aldershot: Ashgate Publishers.

Thune, C., Jacobsen, H., Kristoffersen, D. (1996). Establishing an evaluation system in Denmark. In Smeby J. C. (Ed.), *Evaluation of higher education in the Nordic countries*. Copenhagen: Nordic Council of Ministers.

UFK (1998). Kvalitet og ansvar. Fremtidens kvalitetsoppgaver inden for de videregående uddannelser. København: Undervisningsministeriet.

Wright, S., Ørberg, G. (2008). Autonomy and control: Danish university reform in the context of modern governance. *Learning and Teaching*, 1(1), 27–57.

Chapter 11
Education Quality Audit as Applied in Hong Kong

William F. Massy

Introduction

The Government of Hong Kong made substantial investments in higher education during the decade beginning in the mid-1980s. These investments more than doubled the fraction of school-leavers attending postsecondary institutions, to just under 20%, and the number of institutions grew accordingly. The two traditional universities, the Hong Kong University and the Chinese University of Hong Kong, broadened and deepened their degree offerings. The region's two polytechnics increased their production of bachelor's degrees, reduced subdegree enrollments, and eventually achieved university status as the Hong Kong Polytechnic University and the City University of Hong Kong. The newly founded University of Science and Technology, opened in the early 1990s, soon became a force to be reckoned with across Asia. Hong Kong's liberal arts colleges became full-fledged universities: Hong Kong Baptist University and Lingnan University. With the advent of the Institute of Education, the Hong Kong University Grants Committee (UGC) was responsible for eight postsecondary institutions by the year 2000.

Hong Kong's universities are self-accrediting. As such, they can set their own standards and curricula without outside intervention. Absent self-accrediting status, institutions must get their courses approved by the Hong Kong Council for Academic Accreditation. Achievement of self-accrediting status emancipates an institution from detailed regulation and makes it a substantially autonomous entity. Each UGC institution has its own Council; manages its own finances, procurement, and physical plant; and employs its own academic and non-academic staff outside the civil service system.

Funding from government comes as a block grant whose size is determined by the UGC, with most remaining money coming from tuition. UGC funding, which comprises about 80% of funding, is built up from notional allocations for teaching (68%), research (22%), and performance and role-related factors

---

W.F. Massy (✉)
Stanford University, Stanford, CA, USA

D.D. Dill, M. Beerkens (eds.), *Public Policy for Academic Quality*, Higher Education Dynamics 30, DOI 10.1007/978-90-481-3754-1_11,

(10%). The teaching component depends on a model driven by student numbers differentiated by field of study, level (bachelors, masters, etc.), and mode of attendance (part-time vs. full-time). Tuition rates and student numbers have historically been regulated, but the degree of regulation is declining. The research component is determined mainly by a Research Assessment Exercise (RAE), which will be described presently. The UGC reserves the right to adjust its funding allocations according to judgment and does so regularly – for example, the results of audit are said to "inform funding," though not in a formulaic way.

Prior to the nineteen nineties, the UGC's approach to quality assurance consisted of institutional visitations in which a broad range of university operations was reviewed during a 2- or 3-day period. The visits were not unlike institutional accreditation visits in the United States as they were being conducted at the time. The reviewers, which generally included most or all UGC members, sought to familiarize themselves with the institution's governance, priorities for use of resources, quality of faculty and staff, research and scholarship, and academic standards. But while the agenda was broad, the evidence obtained was not particularly deep. UGC members were able to form impressionistic conclusions, but it was hard to drill down into particular areas – especially the quality of education as actually delivered to students. The institutional visits' shortcomings became increasingly apparent as the number and variety of institutions grew, and mitigating these shortcomings became an important objective for the UGC.

The rise of research in Hong Kong exposed additional shortcomings. Research was viewed as important for the region's economic development and, also, as crucial for the development of top-flight universities. Research growth was spurred by the Research Grants Council, which the UGC created and funded circa 1990. All the UGC institutions sought to appoint and promote research-active academic staff, who in turn demanded investments in research infrastructure, increased numbers of students taught by research, and often reduced teaching loads.

The large research investments made measuring research activity and outcomes a high priority for the UGC. This led to the triennial RAE, which was implemented circa 1993 and continues to this day. The RAE measures the publications and other scholarly work-products of academic staff and assesses the degree to which each staff member is "research active." Because research activity as measured by RAE drives more than 80% of the UGC's notional research allocation, it became an enormously important incentive for both the institutions and staff members.

The Policy Problem

The growth of postsecondary education in Hong Kong coincided with the rise of academic quality assurance around the world. Country after country came to realize that the traditions upon which universities had relied for centuries to assure quality could not cope with dramatic increases in participation rates and huge investments in research. The UGC was quick to recognize this problem. It understood the need for QA in both teaching and research from the outset, but moved first to establish the RAE because it needed to direct its investments and also because the task appeared more tractable.[1]

Quality assurance for teaching and learning emerged as a top priority as the RAE's influence on academic priorities became apparent. The UGC joined the International Network of Quality Assurance Agencies for Higher Education (INQAAHE), and in 1994 this author, the UGC member who had headed the original RAE, was asked to research QA for teaching and learning and make recommendations about the way forward.

Stripped to their essentials, the available approaches fell into three categories. The first, rooted in US-style accreditation, sought to determine whether an institution's governance processes and resources were sufficiently robust for it to be *capable* of educating students at degree level. The UGC believed that its institutions passed this test: after all, as funding agency it was already analyzing the schools' finances and making institutional visitations. The second approach, practiced in Denmark, the Netherlands, and in the Higher Education Funding Council of England's subject level assessments, used external assessors to evaluate the delivered quality of education ("external assessment"). The third approach, developed by the UK's Academic Audit Unit (AAU) and practiced in Australia, New Zealand, and Sweden, viewed quality assurance as an institutional obligation and audited the degree to which institutions were discharging their responsibilities ("academic audit").

The UGC's policy problem was how to discharge its obligation to Government and the public to assure the quality of teaching and learning without disempowering the institutions, infringing their autonomy, or spending too much in relation to the results achieved. However, the Committee wanted to do more than assure traditional academic standards: it wanted to use the QA process to spur improvement in teaching and learning. The policy problem's urgency was underscored by institutional diversity, which meant that "quality" had to be defined differently in different places, and evidence that the RAE was diverting staff attention from teaching and learning at all institutions.

The Committee made its decision based on the principle that quality assurance is intertwined with quality improvement, which is unquestionably an institutional responsibility. Furthermore, institutional autonomy and the Committee's history of collegial interaction with the universities favored the "light touch" represented by audit over the more intrusive interventions needed for external quality assessment. Finally, committee members, including this author, were concerned about the high cost of external assessment and doubted whether good evaluations could in fact be made. US-style accreditation had been ruled out for the reasons given above,

[1] Because the RAE measures number of academic staff whose work meets preset quality standards, it can be viewed as combining QA with measurement of the amount of activity. For teaching, the analogous quantity measure is student numbers. One needs a separate QA exercise for teaching because the relation between student numbers and quality standards is not automatic. For more discussion on the RAE see French et al. (1999, 2001).

which left academic audit as the method of choice. Two rounds of audit have been conducted since this decision and ways of integrating a further round into a general institutional review framework now are being considered (Massy 1997; Massy and French 1999a, Massy and French 1999b).

Rather than adopt the UK's original audit approach, which was judged to be insufficiently improvement oriented, the UGC set out to invent its own methodology. (We knew little about the Swedish and New Zealand approaches, which in any case were in their infancy.) This chapter describes academic audit as developed and used in Hong Kong. The method was named "Teaching and Learning Quality Process Review" (TLQPR) to avoid perceived negative connotations associated with the word "audit," but I have come to prefer the term "Education Quality Audit."

The Hong Kong experience has led to implementations in Missouri and Tennessee and the descriptions that follow will draw upon this experience.[2] It also is worth noting that different lineages of academic audit are developing around the world. They differ from the Hong Kong audits in scope (e.g., education quality or all academic operations) and the definition of "quality work" (described in the next section), but not in fundamental approach.[3] Referring to the Hong Kong, Missouri, and Tennessee implementations as education quality audits calls out their lineage while differentiating them from the other types of academic audit.

Content of the Policy Instrument

Education quality audits can best be understood using the flowchart in Fig. 11.1. The chart consists of three elements: inputs, teaching and learning processes, and learning outcomes. The forward-facing arrows depict how inputs energize teaching and learning processes, which then produce learning outcomes. But what is most relevant to audit are the backward-facing or "feedback" arrows. To produce education quality, teachers must consistently measure the quality of outcomes, contrast it with their objectives, and then adjust the processes as needed to fix problems or effect improvements (arrow A). Process adjustments also can result from self-reflection and comparisons with best practice inside and outside the university (arrow B). Finally, process adjustments may trigger changes in the type, amount, and quality of needed inputs (arrow C). The performance of processes without feedback, which are said to run "open loop," is sure to degrade over time. Decades of experience in quality assurance in a wide variety of fields demonstrate that feedback is essential for maintaining quality.

---

[2] Education quality audit was significantly improved during the second Hong Kong round, and again in the Missouri and Tennessee implementations as described in Massy, et al. (2007). Because this chapter is a policy analysis and not a case study, I will describe the current state of the art rather that the method as originally implemented. Areas where the method has changed materially will be noted, however.

[3] See, for example, Harvey (1999); Dill (2000); Meade and Woodhouse (2000); Massy (2000); Wahlén (1998).

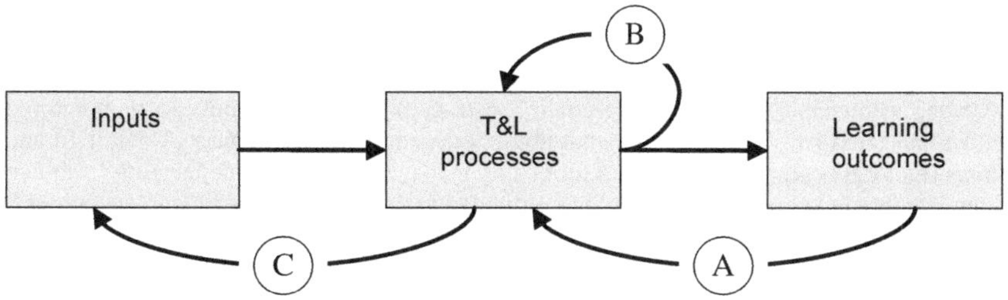

Fig. 11.1 The production of quality education

In a complex environment like education, getting and interpreting the feedback and acting on it requires more than casual effort. Faculty who assess learning carefully and apply what they've learned to improve their teaching do better than those who don't. Likewise, faculty who spend time reflecting on their teaching and thinking about how to improve it tend to produce more learning than their colleagues. The same is true for departments. Those that stress learning assessment and reflection on teaching processes generally produce better teaching. Furthermore, they build a "culture of quality" that triggers a self-perpetuating cycle of improvement. Reflective and evidence-rich feedback processes also help departments optimize their use of inputs and, where necessary, to make the case for additional resources.

The Audited Activity: "Education Quality Work" (EQW)

Feedback is the key to effective quality assurance. For example, one can measure learning outcomes and then take corrective action if quality falls below standard. Or one can measure perceptions about teaching and learning processes, as in student course evaluations, and then take corrective action if the evaluations are unsatisfactory. No feedback means no corrective action and thus no QA. And to get ahead of our story slightly, feedback without a goal or standard to compare against is largely useless.

The examples also illustrate the tight connection between quality assurance and improvement: in each case the corrective action represents an effort to improve. The activities required to set standards, assess outcomes, and take corrective action – in other words, to create and use the feedback loops – have come to be called "Education Quality Work" or EQW for short.[4] We shall see that EQW gets performed at the department level, at the level of a school or faculty, and at the level of a campus or institution.

[4] Sweden's National Agency for Higher Education coined a term for describing the subject matter of academic audit. The term translates to English roughly as "Education Quality Work" (EQW). Massy (2003, 2004) uses the term "Education Quality Processes" (EQP). However, designating EQW as EQP invites confusion with teaching and learning processes. In a recent development, Massy et al. (2007) extended the ideas of EQW to include research. They call the enhanced process "Academic Quality Work" (AQW).

EQW can be defined as:

> Organized activities dedicated to improving and assuring educational quality. They systematize a university's approach to quality instead of leaving it mainly to unmonitored individual initiative. They provide what higher education quality pioneers David Dill and Frans van Vught call "...a framework for quality management in higher education ... drawn from insights in Deming's approach, but grounded in the context of academic operations." (Massy 2003)[5]

Hong Kong's approach to education quality audit examines EQW rather than the inputs, teaching and learning processes, and learning outcomes that most observers view as being the only determinants of education quality. The auditors determine whether systematic feedback processes exist and, if so, what kind. They ask whether the processes make systematic use of evidence, and whether the evidence is robust or circumstantial. They ask whether faculty members and departments compare the evidence with policy objectives and their own clearly stated goals and, if so, whether they act promptly and decisively to correct discrepancies. Education quality audits evaluate the *maturity* of institutions' EQW. Thinking broadly, all lineages of academic audit can be said to evaluate the maturity of "quality work" somewhere in the institution.

Audit's focus on quality work has positive implications for institutional autonomy and academic freedom. For example, auditors do not substitute their judgments about the quality and quantity of inputs or the appropriateness of teaching and learning processes for those of institutional leaders and faculty. What they do is ask whether those judgments are characterized by careful reasoning and informed by good evidence. Nor do they try to measure learning outcomes. They ask whether the local academics are measuring outcomes adequately and whether they use the information systematically to improve teaching. Getting a good audit score depends on having evidence, including evidence from learning assessments, and then using it systematically. However, getting a good score does *not* depend on matching the reviewers' preconceptions about educational content, teaching methods, or the "right way" to assess learning. It is sufficient that the respondent's judgments flow logically from evidence, that they take account of established policy, and people have exercised due diligence in making them.

The proposition that audits of EQW are sufficient for education quality assurance depends upon two fundamental assumptions.

1. Most professors want to teach well. Unless stymied by resource constraints or driven by incentives that discourage investment of time in teaching, they will use feedback to effect improvement – especially if the feedback has been produced by a collegial process.
2. Most professors have only sketchy knowledge of EQW and, therefore, of how to generate and use feedback. They are trained as content experts, and while most have acquired an understanding of conventional teaching and assessment

[5] References within the quotation are van Vught (1994) and Dill (1992).

methods they have little experience with organized quality improvement and assurance activities.

While exceptions can be found at both the individual and institutional level, informed and objective observers generally agree that these assumptions do in fact characterize modern universities.

The above implies that better EQW will pay off for teaching, i.e., that new tools for the improvement of teaching will in fact be put to good use. Furthermore, because EQW includes student learning assessment, better assessments will improve the stock of information about education quality – information that is eagerly sought by external quality assurance agencies and the public. Audit spurs better EQW and vets EQW maturity. It also can vet the efficacy of information about education quality supplied by the institution to the public. The bottom line here is that external quality assessment isn't the only way to get good information about education quality to Government and the public. Audit also can do that job better and, as argued later, there is reason to believe more effectively.

All academic audits involve two basic steps: (1) the entity being audited prepares a self-evaluation of its quality work and (2) the audit panel reads the self-evaluation, visits the entity, and prepares a report. These two steps do not differ from most other types of evaluation. What is different is the *content* of the self-evaluation and of the conversations that take place during the audit visit. The differences reflect audit's improvement orientation as well as its focus on quality work.

Consistent with its improvement orientation, education quality audit elicits *structured conversations* among auditors and auditees about how the EQW quality principles described in the Appendix are being applied across the five focal areas.[6] "Conversations" are important because the complex issues of teaching and learning quality are best addressed through dialog. "Structure" is important because the auditors must cover all the relevant topics, gauge quality process maturity, and produce a meaningful report.

But however structured audit's basic design, the conversations themselves are free flowing and collegial. Respondents are encouraged to "come as you are" and standup presentations are held to a minimum. This approach has several advantages. First, the auditors and auditees learn about EQW from each other, which spurs improvement. Second, the auditors learn whether the auditees' descriptions of their EQW activities are "for real" – the gloss one can sometimes hide behind in a PowerPoint or document breaks down in deep conversation. Third, conversation blurs the distinction between accountability and improvement. Auditees learn that

[6] A primer on education quality work as audited in Hong Kong, Missouri, and Tennessee can be found in the Appendix. It provides brief descriptions of the "focal areas" of EQW (the subjects that audit should cover), some principles by which the efficacy of a respondent's EQW can be judged, and a maturity scale for summarizing the audit results. The material is important because understanding the education quality audit as a policy instrument requires an understanding of EQW itself. Agencies that adopt education quality audit may wish to substitute their own materials, but having some kind of standards to audit against is essential.

what they are accountable for is a sincere effort to improve, not adherence to rigid standards. They want to hold up their end of the conversation, and this introduces a degree of self-accountability.

Approaching audit through structured conversation mitigates the problem attributed to the original UK academic audits: that their focus on formal policies and documentation resulted in a bureaucratic "paper exercise." Policy statements and other written materials should be present in the audit rooms for reference before, during, or after the audit conversations. However, the dialog should be more concerned with respondents' attitudes, behavior, and command of quality processes and principles – i.e., their EQW maturity – than written policies and paper trails. The Hong Kong auditors stressed that "it's what you're doing that matters," not the precision of your documentation.

Implementation

Hong Kong's education quality audits involved six distinct steps: (i) the initial design process, (ii) onsite briefings or workshops for prospective auditees, (iii) the auditees' self-studies, (iv) the audit visits, (v) preparation of the audit reports, and (vi) a meeting to debrief the exercise and share exemplary practice after the audit round was completed. Steps iii–v encompass the two "basic steps" introduced above. However, all six steps are crucial for a successful implementation.

Initial Design

The exercise began with a detailed design for how EQW concepts would be introduced, what would be included in the self-study, how the audit visit should be conducted, and how the report would be written and promulgated. The UGC felt the auditee institutions should participate in the design process, so each campus appointed members to a Consultative Committee that worked with UGC members and staff through both the first and second audit rounds. The Committee included people responsible for quality assurance and improvement on their respective campuses. They contributed valuable insights about emergent design ideas and provided a reality check on the result. Most became enthusiastic supporters of quality improvement and of education quality audit, and they helped transmit this enthusiasm to their colleagues within the institutions.

Onsite Briefings

Each institution's introduction to EQW, before the first audit round, began with a 2–3 h briefing by the chair of the UGC's audit team and a few of his colleagues. The briefing occurred about 9 months before the audit visit. It described quality process concepts and principles, the institutional self-study, and the audit visit and report. The session was open to all faculty and staff, and participation often numbered in the hundreds. In addition to launching the self-study, the briefing sought to focus attention on EQW and initiate self-reflection and improvement.

The UGC secretariat followed up on the briefing with written guidance notes describing the self-study and arrangements for the audit visit. The team chair and a member of the Secretariat visited each campus a second time about 4 months before the audit to finalize the arrangements. The briefing and follow-up visits were omitted in the second audit round because people were familiar with the exercise and the requisite activities already were being conducted on the campuses.

Self-Study

Doing the self-studies stimulated institutions to reflect on their EQW and begin working on improvements prior to the audit team's arrival. As in other quality assurance regimens, the self-study reports helped orient the audit team before its arrival on campus. Members could request additional information and/or supporting documentation before the audit visit.

Initially the institutions were free to structure the self-study reports as they wished and include appendices of any length – which soon overwhelmed the audit team. The UGC responded by putting a 20-page limit on the self-study reports and discouraging voluminous appendices. Lists of relevant documents were included, however, so team members could conveniently request the ones they wanted.

Audit Visit

The visits were conducted by intact teams of 18 members in Round 1 and 10 members in Round 2.[7] The large size in Round 1 was due to inclusion of one member of the Consultative Committee from each institution. Eight UGC overseas academics also served, along with two overseas academic quality experts who were not UGC members. The second-round team was similar except that it included the consultative committee chair but not other members. Ten is still a fairly large number of auditors, but the size was dictated by the need for division into subgroups as described below.

The audit visits lasted between 1½ and 2 days depending on the size of the institution. More time might have been desirable, but the limit was dictated by the availability of the overseas UGC members and experts. In the event, the amount of time available did prove sufficient.

[7] In Round 2, additional two-person sub-panels addressed research postgraduate programs and continuing education.

A typical visit schedule follows.
Day 1

- Executive session (60 minutes). Team members compared notes on the self-study, look at documents, and plan their queries.
- Opening plenary with the institution's president, chief academic officer, and other senior officers (45 minutes). The president gave an opening presentation not to exceed 15 minutes. Questioning by team members generally addressed institutional priorities and policy issues raised by the self-study.
- Plenary with the institution's Quality Assurance or equivalent committee (45 minutes). There was no opening presentation. Questions generally involved institution-level QA policies and procedures.
- Plenary with students (30 minutes). The group often consisted of representatives from student government and/or institutional student-faculty committees. Questions addressed perceptions about education quality, whether students were involved in quality assurance, whether the problems they identified were addressed promptly, and whether prompt feedback on resolution was forthcoming.
- First set of small-group sessions as described below (90 minutes)
- Second set of small-group sessions (90 minutes)
- Executive session to recap the day (30 minutes)

Day 2

- Third set of small-group sessions (90 minutes)
- Plenary with the deans of schools (60 minutes). Questions generally addressed the deans' familiarity with education quality processes and principles, and their role in the institution's self-regulation of quality.
- Executive session to recap the morning and plan the audit report (90 minutes)
- Exit conference with the opening plenary group (30 minutes)
- Executive session to recap the exit conference (15 minutes)

The time allocations varied depending on institutional size and complexity.

Panel size allowed for six replications in each of the three small-group sets: for a total of eighteen separate meetings. (The sub-panels had three people in Round 1 and two in Round 2 – two members proved sufficient.) About two-thirds of the sessions were with departments; the rest were with schools and special-purpose entities like educational technology and teacher development units. Most respondents were faculty, but students always were included. Numbers ranged from half a dozen to as many as twenty people. The 90-minute sessions were divided three ways: about 70 minutes with the whole group, 10 minutes with the students separately, and 10 minutes in an executive session. The students tended to be fairly quiet in the general session but opened up when asked separately, "You heard the faculty – is this how things really look to you?"

The small-group sessions were the most important part of the audit visit. Conversations at the grass-roots level allowed panelists to get past the formalities of policies and procedures and find out what was really happening on the ground. The multiple replications also provided data about interdepartmental and inter-school variance – which often contradicted the positive face put on by institutional leaders and quality assurance committees. Moreover, the grass-roots conversations proved almost impossible to fake. Faculty in departments that had embraced quality processes would back up their remarks with a rich mosaic of examples, whereas those whose experience was limited to lip service would soon sputter into generalities. The subgroups noted good and bad examples of quality work and assigned capability-maturity scores for subsequent discussion with the full audit panel.

The auditors also tested quality processes further up in the institution's academic hierarchy. For example, they quizzed deans and their associates about EQW in their schools and, in particular, what they were doing to improve weak-performing departments. The teams observed considerable variation in the deans' knowledge and attitudes. Some deans were aware that certain of their departments needed improvement and were working to achieve that, whereas others didn't know and still others knew but didn't believe they were responsible for effecting change. Such observations were usefully provocative in our subsequent plenary sessions with the deans and institution-level leadership. One of the points pressed by audit is that everyone in the hierarchy, from president to individual professors, should take education quality seriously. Deans, provosts, and presidents should join with quality assurance committees in reinforcing the quality message at every opportunity. They should take all needed steps to assure and improve departmental EQW.

The desire to detect variance conditioned the selection of which departments and schools to visit. The institution made nominations, but the panel chair and UGC Secretariat always added their own selections. Sometimes these were based on hunch or insider knowledge, sometimes simply by a desire to span a range of disciplines while visiting multiple departments within a given school. The selections were announced about a month before the audit visit. This meant all departments and schools had to participate in the institution's preparation for audit and that the ones selected could not over-prepare. Selected units were asked to table a one- or two-page "talking paper" to guide discussion of their quality processes but otherwise no special preparation was required.

Audit Report

The reports described each institution's education quality processes and, importantly, what it was doing to improve them. They did not grade or rank the institutions, but careful reading does reveal a rough ranking. (Links to the reports for both rounds can be found at www.ugc.edu.hk.) The team chair wrote all the reports in Round 1 but workload dictated that a professional secretary (a retired professor at one of the institutions) do the job in Round 2. The Secretariat sent the report drafts to the institutions for correction of significant factual errors, but no attempt was made to vet the draft with the individual units visited. Hence examples of good and bad practices at the grass-roots level were not identified as to unit. The reports were written in non-technical language in order to make them as accessible as possible.

The UGC viewed the institutions as owning the reports but required publication in both English and Chinese along with whatever comments the university wished to make. The press took a keen interest, and some reports turned out to be lightening rods for discussion. This was positive on the whole, since it highlighted the importance of education quality and quality work for the general public as well as for the institutions.

Debriefing Session

The Consultative Committee convened a Region-wide meeting to air comments on the Round 1 exercise and share exemplary practices identified during the audits. The meeting was attended by several hundred faculty and staff from the eight institutions. The testimony was mainly positive, and many exemplary practices were described (Massy and French 1999a). Presentation of such practices served a dual purpose: to propagate the specific practices and to illustrate quality process concepts through the vehicle of examples.

Impact

Four kinds of evidence about impact can be identified: (i) testimony from auditors and, especially, auditees; (ii) changed institutional behavior; (iii) external evaluation of the audit process; and (iv) evidence from a subsequent audit.

Testimony from People Involved in the Audits

The testimony from participants tended to be positive. The debriefing session after Hong Kong's Round 1 included favorable testimonials from a cross section of university respondents. The closer a person was to the audit processes, whether as an auditor or an auditee, the more likely it was that his or her opinion would be favorable. There were concerns about the number of different reviews being conducted by the UGC, but few people having first-hand experience with audit complained that the time spent was not worthwhile.

The UGC received numerous reports to the effect that "we should have been asking ourselves these questions all along" and "at last someone cares about education quality." One should recognize that these comments came from converts, but the participants' conversion was by no means preordained. Most heartening was confirmation that the structured conversations were in fact meaningful to auditors and auditees alike.

Changed Institutional Behavior

While the Hong Kong audits did not rank the institutions, the UGC was sensitive to whether the institutions were taking their quality work seriously. It had been made clear from the beginning that the audit results might "inform funding" but not in a formulaic way. Fortunately the linkage hardly ever had to be demonstrated, but when it was the reason was poor performance *and* a lack of demonstrated willingness to improve. Removing a small increment of funding solved the problem in short order. Having made its point and obtained reports that the situation was improving, the UGC restored funding to its previous level. The problem did not reoccur.

The example teaches an important lesson about audit's capacity to further an agency's accountability and improvement goals simultaneously. The Hong Kong audits are oriented primarily toward improvement, but their role as an accountability tool also was recognized from the beginning. Because the concept of quality work was largely undefined prior to the first exercise, the UGC did not penalize institutions for EQW immaturity. However, it acted decisively in when good-faith improvement efforts were not forthcoming. Exemplary EQW might well be rewarded, and conversely, in future audit rounds when the nature and importance of EQW has become clear to all.

External Evaluation

As a quasi-governmental agency accountable to the public, the UGC commissioned a formal review of the first audit round. The review was performed by the Center for Higher Education Policy Studies (CHEPS) at the University of Twente in the Netherlands, which fielded a team of international experts.

The CHEPS team interviewed audit participants in all the UGC institutions to ascertain their views on the exercise and its impact. The team's major conclusions follow. (Recall that Hong Kong's name for audit is "TLQPR.")

- The overall major conclusion is that 'TLQPR was the right instrument at the right time', because the review was a positive stimulus to institutional attention to teaching.
- There were clear achievements with respect to the first goal of TLQPR – *to focus on teaching and learning as the primary mission of higher education institutions*. The signal given by TLQPR that teaching and learning were as important as research was seen by many as the prime impact. (The team also noted that the

substantial funds associated with the RAE continued to drive incentives in that area.)

- With respect to the second goal of TLQPR – *to assist higher education institutions in their efforts to improve teaching and learning quality assurance processes* – there appeared to be institutionalization of quality management procedures in some institutions, especially those that did not have such procedures in place before. A number of examples of innovation in existing quality procedures were also found. Sustainment of present efforts and new initiatives would be helped by the signal that there would be a second round of external reviews of teaching and learning processes (a signal that was given).
- The TLQPR has certainly contributed to achieving the third goal – *accountability of the UGC and the higher education institutions to society* – firstly, through the review process itself and secondly through the publication of the reports and the institutional progress reports. However, in the eyes of the higher education community, the press coverage of the reports was not seen as a balanced reflection of the process. (Westerheijden et al. 1999)

Evidence from a Subsequent Audit

The most telling evidence about the impact of education quality audit came from Hong Kong's second audit round. Once again an audit team visited each of the Region's institutions for an in-depth review of education quality work. Team members who participated in both rounds concluded unanimously that the institutions had made great progress. All demonstrated systematic EQW with substantial traction at every level from departments to the central administration. Some institutions had gone so far as to institute "internal education quality audits" to maintain momentum between the UGC's visits. Several with poor results in Round 1 scored well in Round 2 and no institution regressed. The degree of progress across the Region surpassed the UGC's most optimistic expectations. While improved quality processes don't guarantee improvements in delivered educational quality, it is hard to believe that such improvements are failing to materialize. The one disappointment from the second round concerned student learning assessment. While progress had been made, most departments still had a long way to go. The evaluation and improvement of student learning assessment remains a high priority for subsequent exercises.

Cost

The numbers of audit visits and the types and numbers of visitors are the main cost drivers for any onsite quality evaluation program. Audit offers significant advantages on both dimensions. A quality assurance agency need mount only one audit per institution as opposed to the separate evaluations of each department needed for subject-level evaluations: for *n* institutions and an average of *m* departments per institution, this means only n as opposed to $n \times m$ visits. Second, auditors need not be expert in any particular discipline, which greatly simplifies team selection.

The external cost of audit depends on local circumstances and design details, but a sample calculation of the staffing requirements for Hong Kong-style audits is possible.

Per-visit cost drivers

- Audit teams usually run in the range of 6–10 members, with larger numbers for more complex venues that require more small-group sessions.
- An audit requires from 1 to 2 days on site and about an equal number of days for visit preparation and reviewing the draft report.
- Team chairs require 2 or 3 additional days for report writing and first-time auditors require a day of training.

Sample calculation
(Assume five institutions with 2-day visits and an intact team of eight auditors.)

Auditor training:		8	person days
Visits:	$5 \times 2 \times 8 =$	80	
Preparation and review:	$5 \times 2 \times 8 =$	80	
Report writing:	$5 \times 3 =$	15	
Total		183	person days
Travel and sustenance for site visits and the training day:		86	person days

Institutional costs are harder to calculate because it is never clear which activities should be attributed specifically to audit and which to things that should be done anyway. For example, faculty should attend quality process workshops and strive for improvement whether they will be audited or not. The fact that the audit stimulates such behavior should be viewed as a benefit, not a cost. The same is true for the reflection that informs a self-study. The actual writing up of the self-study, direct preparation for the audit visit, and the visit itself might reasonably be counted as a cost of audit, yet even these activities confer benefits. The key point is that audit addresses practical issues associated with important day-to-day activities rather than requiring large amounts of bureaucratic make-work.

Comparisons

While detailed comparisons of the Hong Kong education quality audits with other quality assurance methods are beyond the scope of this chapter, a brief summary may prove useful.

Variants of Audit

Academic audit started with the UK's AAU, which was organized by the Committee of Vice Chancellors and Principles circa 1990 – in part as a counterweight to the

Higher Education Funding Council of England's external assessment initiatives. The emphasis was on quality assurance policies and processes and not so much on improvement. Two criticisms were leveled at the AAU's approach, both of which were discussed earlier in this chapter. The first was that the audits were overly concerned with formalities and documentation. The second was that audit doesn't provide external assessments of education quality. Citing duplication of QA effort in the UK, the government merged audit with external assessment in a new quality assurance agency (QAA in the mid-1990s). The QAA adopted external assessment as its primary methodology but changed to audit circa 2001. Both audit and assessment have played a role, and the situation continues to evolve.

Another academic audit lineage runs from the UK through New Zealand's Academic Audit Unit to the Australian Universities Quality Agency (AUQA). Australia's audits are comprehensive. They go beyond education quality to include governance, research, management, access, and support services among other things. The preparation and review processes are very well organized and the Agency staff provides strong support to the audit teams. The reports are taken seriously by the institutions and by the public. The Australian process provides a good model for agencies that wish to pursue a comprehensive audit approach. AUQA also is developing a good practices database.

Denmark's new audit program was triggered by legislation in 2003 that requires universities "systematically to develop and improve the quality of their processes for teaching and learning." The approach draws on UK, Australia, and Hong Kong as well as the Danish Evaluation Agency's own rich history of subject-level external assessment (see Chapter 10 and Massy 2000). Agency staff again play a strong role: for example, they gloss the self-evaluation documents, provide auditors with suggested questions, and draft the panel's report. The University of Copenhagen and The Technical University of Denmark received audit visits during 2004 and reports have been issued.

The first education quality audit application in the United States was begun in 2001 by the University of Missouri System. The approach was to apply audit at the department level. One department on each of the university's four campuses was audited in 2003 with good results, and a successful second round has been completed. The second U.S. application was by the Tennessee Board of Regents, which audited department-level EQW on 13 of its 19 campuses during 2005. This project also appears to have been successful. The methodology used in Missouri and Tennessee is a direct lineal descendent of the one developed Hong Kong. These applications are described extensively in Massy et al. (2007).

Other variants of audit also are emerging in the United States. The Senior College Commission of the Western Association of Schools and Colleges ("WASC," a regional accreditor) piloted institution-level audit at the California State University at Fullerton in 1999 but has yet to roll out the method in a major way (WASC 1999). Other regional accreditors have adopted elements of education quality audit, as have subject-level accreditors like the Teacher Evaluation Council (TEAC) and the Association of Collegiate Schools of Business (AACSB). While not related to higher education, it is interesting to note that the Education Commission of the States (ECS) adopted education quality audit as the core concept for its second-generation accountability proposal for primary and secondary education (ECS 2003).

External Assessment

The strongest and most recurring criticism of audit is that it doesn't provide an "objective third-party assessment" of education quality. We heard this criticism from people in the Hong Kong Government and the Hong Kong Council for Academic Accreditation, for example, and one runs into it frequently in discussing academic quality assurance around the world. The RAEs conducted in the UK and Hong Kong represent strong forms of external quality assessment, so why not do the same for education?

The answer flows from the difficulty of assessing education quality. Research and scholarship can be judged by artifacts, especially peer-reviewed publications, which external assessors can review at whatever length they desire. Even this is not easy, given that there is not just one but four kinds of scholarship (Boyer 1991) and the observed differences in what's considered "quality" across disciplines (French et al. 1999, French et al. 2001), but it is vastly easier than assessing the delivered quality of education.

Traditional external assessments focus on the quality of faculty, staff, and infrastructure, and the degree to which the breadth and depth of curricula meet generally accepted standards for the degree being granted. But while these are necessary conditions for quality, they are not sufficient. A curriculum that passes the "generally accepted" test may not fit the needs of the student segments served by a particular institution, or it may be poorly taught. Hence quality assurance agencies are focusing more and more on outcomes measures.

But measures of what? Decided by whom? How implemented? How used to effect improvement? Programs in which third parties assess education quality are costly and inspire institutional resentment. Furthermore, the history of the so-called assessment movement in the United States, which demands that institutions assessment learning outcomes but ignore the other aspects of EQW, does not inspire confidence. I have written extensively about these matters (see Massy 2003, 2004) and will offer only a brief summary here.

External assessment compares the "educational production function" (the three boxes in Fig. 11.1) against a predetermined standard for what's good in the circumstances. Education quality audit compares EQW (the feedback arrows in Fig. 11.1) against a standard. Both methods consider learning outcomes. External assessment makes its own outcomes quality determination whereas audit reviews the institution's measurements. This might seem like a small difference but it has profound consequences.

First, external assessment is inherently confrontational. Hence it is difficult to approach quality in a collegial way, let alone combine quality assurance and improvement goals in the same exercise (Trow 1994). Education quality audits as defined in this chapter are inherently collegial. They are designed from the ground up to combine improvement with accountability.

Second, external assessors bear the burden of proof for getting valid quality measures. Institutions can hardly be expected to search out and lay bare their shortcomings. That's the assessor's job. The complexity and subtlety of higher learning makes this a difficult burden to bear, all the more so when quality is defined in terms of the institution's own goals – goals they can interpret retrospectively when their quality is challenged.

Auditors, on the other hand, need only ascertain whether institutions and faculty are doing good job of EQW. Because the standards for effective EQW don't vary across institutions and disciplines, the auditors' task is far simpler. The burden of proof shifts to the respondent – for example, to convince the auditors that effective feedback loops exist and are being used? Countless examples will be brought to the auditors' attention if the answer is "yes." However, a negative answer means no examples will be forthcoming and respondents will stumble repeatedly in their dialog with the auditors.

It is my strong conviction that university-level education is too complex and subtle – and the opportunities for institutions to go into a "compliance mode" and withhold, distort, or simply not produce evidence are too great – to make external assessment the method of first choice for QA agencies. (Assessments may be useful in special circumstances. The UK, for instance, relies mainly on audit but reverts to assessment when an institution fails the audit.) The danger is that, when propagated on a large scale, external assessment will oversimplify the definition of quality and drive higher education toward its lowest common denominator. It also will be very costly. In my view, those who demand external assessment regardless of circumstances are seeking a magic bullet that simply doesn't exist.

My concern about external quality assessment applies with much less force to assessments conducted within the institutions, particularly for departmental majors and organized general education programs. What is difficult or impossible when propagated at the level of a higher education system becomes manageable at the program level. The development of robust and meaningful outcomes-based performance indicators, managed at the local level as part of EQW, is not too much to hope for.

In closing, I must admit that critics who argue that universities cannot be trusted to perform their own quality assessments may have an element of history on their side. However, it is precisely this history that education quality audit is designed to change. EQW as described herein embeds learning outcomes assessment in a complete program of quality assurance and improvement. Education quality audits spur the development of EQW and, after allowing a reasonable period for maturity to develop, they can be used to hold institutions accountable for good EQW. In the long run, when locally based assessment methods have matured and become generally accepted, institutions also can be held accountable for results as measured by locally generated, audited, performance indicators. The road to accountability through local action is longer than through preemption of quality assessment by external agencies, but the results will be more meaningful.

To test the efficacy of this assertion, I suggest that a robust external assessment system be given its own education quality audit. Let the assessors have structured conversations with experienced auditors about the assessment's conclusions and the evidence that supports them, the methods used to obtain the evidence, and the difficulties, if any, encountered during the assessment process. I predict that the external assessors will get less complete evidence and draw less valid conclusions about education quality than faculty within institutions characterized by mature EQW – faculty whose work is, of course, audited.

Appendix. Primer on Education Quality Work (EQW)[8]

Quality Focal Areas

Education quality addresses five key "focal areas" of education quality.[9] In the first focal area the questions center on *learning objectives*. What should students whom the department has taught know and be able to do? (Goals and standards are essential for the use of feedback.) How do the students' educational experiences contribute to their employment success, their capacities as citizens, and their quality of life? Are the specified learning objectives based on the needs of enrolled students rather than the ideal student most faculty want teach?

The next focal area deals with the *curriculum and co-curriculum*.[10] How does the curriculum relate to the program's learning objectives? What is being taught, in what order, and from what perspective? Does the curriculum build cumulatively on the students' prior knowledge and capacity? To what extent does the co-curriculum, those organized experiences outside the classroom, support the curriculum? Does the curriculum meet accepted standards for the degree, to the extent such standards have been articulated by accreditors or other external bodies?

The third focal area centers on *teaching and learning processes*. For example, what methods are employed for introducing students to new materials, for interpreting those materials and answering student questions, for stimulating student involvement, and for providing feedback on each student's work? Is learning active? Is technology being used, and if so, is it being exploited effectively?

---

[8] This Appendix describes what the author considers to be current best practice. A comprehensive statement (including sample audit questions), developed for the Tennessee Board of Regents, can be found at www.tbr.state.tn.us/academic_affairs/acadaudit/audit.htm.

[9] The focal area definitions used by the UGC, called "domains" rather than focal areas, differed slightly from the ones presented here. Those domain definitions were curriculum, teaching and learning processes, student learning assessment, quality assurance, and resources devoted to quality processes. The current best-practice definitions evolved subsequent to (and as a result of) the first UGC audit round. However, given everyone's familiarity with the original definitions, the UGC chose not to adopt the change for the second round.

[10] While the Hong Kong audits did not explicitly address the co-curriculum, co-curricular issues did arise frequently.

The fourth area focuses on *the assessment of student learning*. What measures are used to assess student learning? Are they aligned with the learning objectives? Do they compare beginning and ending performance to ascertain value added? Who is actually responsible for student learning assessment – each individual faculty member? A department committee? Members of the administrative staff, either within or outside the department?

The fifth and final area focuses on the institution's or department's processes for *assuring educational quality*. Can faculty and administrators assure first themselves and subsequently the audit team that the designs for curricula, teaching and learning activities, and student assessments are being implemented as intended? Can they be certain, in short, that curricula and teaching are subjects of robust evaluations?

Quality Principles

To achieve rigor, academic auditors need principles against which to judge activities in each of the focal areas. The principles are still evolving, but the (U.S.) National Center for Postsecondary Improvement (NCPI) developed a starting set that was used successfully in Hong Kong. The principles have their roots in business, health care, and government quality work, but they have been adapted for use in academe. They are analogous to the "Generally Accepted Accounting Principles" (GAAP) used in financial audits. As in finance, academic auditors can ascertain whether respondents know about and practice the principles, and then follow up on shortfalls.

The first principle is obvious but often overlooked: define educational quality in terms of outcomes. The quality of student learning, not teaching per se, is what ultimately matters. The outcomes should pertain to what is or will become important for the students enrolled in the program. Outcomes mandated by institutional or oversight-agency policy should of course be observed, but the entity should have its own outcome goals in any case.

The second principle calls for a focus on process – on how things get done. It becomes important to know, in some detail, how teachers teach, how students learn, and how each approaches the task of assessment.

Third, quality should be everyone's business. Faculty need to demonstrate collegiality in teaching, just as they do in research. The department, as the organizing unit, needs to encourage faculty members to work together, to hold one another accountable, and to bring a broad array of talent to bear on difficult problems. The goal of such teamwork is to make the institution or department a learning organization with respect to education and EQW as well as disciplinary content.

The fourth principle calls for decisions to be based on evidence. Faculty should collect data on student preparation, learning styles, and, where relevant, probable requirements for employment. The data – testimonies by current and former students, and perhaps by employers or the faculty members who taught them in graduate or professional school, along with the numeric data culled from the institution's student record system – need to be analyzed carefully in light of disciplinary standards along with the faculty's own professional experiences. The results should have a direct and demonstrable impact on curricula, learning processes, and assessment methods.[11]

Fifth, coherence should be viewed as a virtue. The goal is to have departments see learning through the lens of the student's entire educational experience. In an ideal curriculum courses build upon one another to provide the desired depth and breadth, and students' educational portfolios should reflect the same coherence.

Sixth, there is a paramount need to identify and learn from best practice. Institutions and departments should seek out examples of good practice and adapt the best to their own circumstances. They should compare well-versus average or poorly performing methods and students, assess the causes of the differences, and seek ways to minimize the variation.

Seventh and finally, continuous improvement should be viewed as not just an important but an attainable priority. Quality should be everybody's business all of the time. While faculty will continue to place strong emphasis on research, they should spend enough discretionary time on educational quality to keep the improvement process moving. The department's as well as the institution's personnel committees need to make the results of such work, along with teaching and research performance, a criterion for promotion and tenure.

The quality principles are couched in terms of departmental performance – that is, what a department should be doing to improve and assure educational quality. Translation to multi-disciplinary programs like general education is straightforward: the committee or other entity responsible for program quality should apply the quality principles. What may seem less straightforward is how the principles apply to a school (a faculty in most places outside the United States) or to the institution as a whole. But the answer is clear. Deans, and the quality committees in their areas, need to make sure that the departments under their purview apply the principles effectively. Chief academic officers should make sure the deans take education quality seriously and hold their departments accountable. Systemwide administrators and external quality agencies need to make sure campus leaders do the same. The audit methodology, described under "Implementation," is designed to help the responsible parties at each level perform these tasks.

EQW Maturity Ratings

Having established the focal areas to be addressed by audit and principles of good practice in each area, the last requirement is a language by which auditors can describe the performance of an institution or department. Fortunately, by the time of the second Hong Kong audit round, Carnegie-Mellon University had developed its capability maturity model for tracking the prowess of software development teams.

[11] The Tennessee audits make use of principles for the use of evidence in teaching and learning prepared by the Senior College Division of The Western Association of Schools and Colleges (WASC).

Somewhat unexpectedly, the categories and definitions, along with the model's conceptual approach, helped the Hong Kong auditors gauge the relative maturity and hence systematization of the quality processes in its eight universities. The Hong Kong auditors used the capability maturity scale in their internal deliberations but did not make their ratings public. Since then, however, the UGC has begun experimenting with maturity ratings for institutions and focal areas based on the published audit reports. Departmental maturity levels were discussed publicly in Missouri and they are integral to both the self-study and auditing processes in Tennessee.

The capability maturity scale's zero point is *no effort* at all. The department or institution being evaluated does not have organized educational quality processes. Quality and quality assurance remain in the hands of individual professors. Next comes *firefighting*. The entity responds to problems, but mostly with ad hoc methods. The five focal areas described above are not covered systematically, and the quality principles receive little attention.

Mid-point on the scale is occupied by *informal effort*. The entity can report individual initiatives and experimentation with the principles in one or more focal areas. Coverage remains spotty, however, and the entity had yet to become a learning organization with respect to its educational quality processes.

The fourth point on the scale is reached when the entity's quality process show evidence of *organized effort*. The entity plans and tracks quality process initiatives in all five focal areas. Emergent norms encourage investment in the quality principles. Methods for gauging performance are under development.

The true winners are departments or institutions that have reached the scale's terminus: *mature effort*. The quality principles have become embedded in the entity's culture, and the idea of regular improvement in all five focal areas has become an accepted way of life. The entity recognizes the planning, tracking, and performance evaluation of quality processes as important elements of peer accountability and collegiality, and it has developed appropriate and feasible performance indicators.

The Hong Kong experience suggests that auditors, and also most chairs, deans, and provosts for that matter, can use the scale to evaluate the maturity of an entity's EQW. Departments, schools, and institutions with immature education quality work can be encouraged or spurred to do better. Fully mature entities can be celebrated – one can say that education quality has become "job one" when an entity reaches the mature end of the scale.

Acknowledgement I am indebted to David Dill, Ralph Wolff of the Western Association of Schools and Colleges my colleagues on the Hong Kong UGC and the Region's two audit panels, Steve Graham of the University of Missouri System, and Paula Short of the Tennessee Board of Regents for their help and encouragement in the development and application of the education quality audit method.

References

Boyer, E. L. (1991). *Scholarship reconsidered: Priorities of the professoriate*. Princeton, NJ: Carnegie Foundation for the Advancement of Teaching.

Dill, D. D. (1992). Quality by design: Toward a framework for academic quality management. In *Higher education: Handbook of theory and research* (pp. 37–83). New York: Agathon Press.

Dill, D. D. (2000). Designing academic audit: Lessons learned in Europe and Asia. *Quality in Higher Education*, 6(3), 187–207.
ECS (2003). *Accountability – Next generation models: Quality improvement model*. Denver, CO, USA: Education Commission of the States.
French, N. J., Massy, W. F., Young, K. (2001). Research assessment in Hong Kong. *Higher Education*, 42(1), 35–46.
French, N. J., Ping, K. K., Massy, W. F., Sui, H. F. H., Young, K. (1999). Research assessment in Hong Kong. *Journal of International Education*, 10(1), 46–53.
Harvey, L. (1999). Quality in higher education. Paper at the Swedish Quality Conference, Goteborg, November.
Massy, W. F. (1997). Teaching Learning Quality Process Review: The Hong Kong programme. *Quality in Higher Education*, 3(3), 249–262.
Massy, W. F. (2000). Energizing quality work: Higher education quality evaluation in Sweden and Denmark. Technical report. National Center for Postsecondary Improvement, Stanford University.
Massy, W. F. (2003). *Honoring the trust: Quality and cost containment in higher education*. Bolton, MA: Anker Publishing Company.
Massy, W. F. (2004). Academic audit for accountability and improvement. In J. C. Burke (Ed.), *Achieving accountability in higher education: Balancing public, academic, and market demands* (pp. 173–197). San Francisco, CA: Jossey-Bass.
Massy, W. F., French, N. J. (1999a). Teaching and Learning Quality Process Review: What the program has achieved in Hong Kong. *Quality in Higher Education*, 7(1), 33–45.
Massy, W. F., French, N. J. (1999b). Teaching and Learning Quality Process Review: What has the programme achieved in Hong Kong? Presented at the 5th Conference of the International Network for Quality Assurance Agencies in Higher Education, Santiago de Chile, 2–5 May.
Massy, W. F., Graham, S. W., Myrick Short, P. (2007). *Academic quality work: A handbook for improvement*. Bolton, MA: Anker Publishing Company.
Meade, P., Woodhouse, D. (2000). Evaluating the effectiveness of the New Zeeland Academic Audit Unit: Review and outcomes. *Quality in Higher Education*, 6(1), 19–30.
Trow, M. (1994). *Academic reviews and the culture of excellence*. Stockholm: Kanslersãmbetet.
van Vught, F. (1994). The new context for academic quality. Paper for symposium "University and Society", Vienna, 9–10 June.
Wahlén, S. (1998). Is there a Scandinavian model of evaluation of higher education? *Higher Education Management*, 10(3), 27–41.
WASC (1999). Newsletter 4. www.fullerton.edu/wasc/Newsletter4.htm. Accessed 12 June 2009.
Westerheijden, D., Brennan, J., Dill, D., Shah, T., Verkleij, A. (1999). *A campaign for quality: Hong Kong Teaching and Learning Quality Process Review*. Hong Kong: University Grants Committee.

Chapter 12
The German System of Accreditation

Barbara M. Kehm

Introduction: State Approval and Accreditation

Accreditation was introduced in Germany in 1998 as a procedure to ensure minimum standards in terms of the quality of curricular content and to assess the labour market relevance of newly established bachelor's and master's programmes. Shortly after the Sorbonne Declaration in June 1998, which preceded the Bologna Declaration by 1 year and was then only signed by the Ministers of Education of France, Germany, Italy, and the United Kingdom, the German higher education framework law was revised enabling higher education institutions to introduce bachelor's and master's programmes in a trial phase.

Traditionally, the establishment of new degree programmes at German higher education institutions, including their study and examination regulations, needed the approval of the responsible Ministry of the respective State. This was typically a long and tedious process lasting two and more years in which the Ministry examined

- the compatibility of the proposed new degree programme with respective State planning;
- the availability of resources to establish the new degree programme;
- the compatibility with the examination regulations of the State; and
- the adherence of the proposed new programme to framework regulations for examinations (in particular the envisaged standard period of study, the number of classes in weekly hours per term, and the number of examinations on the subject matter).

In its decision about the introduction of accreditation procedures for newly established degree programmes according to the tiered structure of bachelor's and master's degrees, the Standing Conference of the German Ministers for Education

---

B.M. Kehm (✉)
International Centre for Higher Education Research (INCHER),
University of Kassel, Kassel, Germany

D.D. Dill, M. Beerkens (eds.), *Public Policy for Academic Quality*, Higher Education Dynamics 30, DOI 10.1007/978-90-481-3754-1_12,

and Culture emphasised the functional separation of state approval and accreditation. The responsible State Ministry continued to retain its right to approve of every new degree programme with respect to three dimensions: (a) a guarantee that the programme to be established had sufficient resources; (b) the compatibility of the new programme with the higher education planning of the respective State; and (c) adherence to the structural rules and regulations of the State. While the State continues to approve of resources and legal issues, accreditation was established to assess quality and labour market relevance. All newly accredited study programmes have to undergo a process of re-accreditation every 5 years. The regulations of the Standing Conference also envisage that eventually not only the newly established bachelor's and master's programmes should be accredited but also the already existing traditional *Magister* and *Diplom* programmes which will continue to exist in quite a few cases for the time being because many universities are offering traditional degree programmes parallel to the new degree programmes.[1]

Reasons for the Introduction of Accreditation in Germany

The decision to introduce accreditation as an instrument of quality assurance into the German system of higher education was regarded as an important element of modernisation of the system vis-à-vis growing European and international cooperation as well as competition. The revision of the higher education framework law in 1998 did not only introduce a trial phase for a tiered structure of programmes and degrees but was accompanied by other political decisions as well. The State intended to give up detailed regulation of a number of areas of higher education policy and decision making while at the same time triggering a stronger differentiation of higher education provisions, encouraging profile building and competition among higher education institutions, and allowing more higher education offers by private and also foreign providers.

There were basically four reasons to change the existing system of quality assurance through framework regulations decreed by the Standing Conference of Ministers for Culture and Education of the German States (henceforth shortened to Standing Conference).

First, a new quality assurance system was deemed necessary because the far-reaching changes to a tiered structure of study programmes and degrees according to the bachelor's and master's model could not be based on any previous experiences.

Second, there was considerable criticism with regard to the traditional system of state approval and the opportunity was taken to establish a new quality assurance system which was more in line with international developments.

[1] The reason for this is a constitutional law that all students have the right to finish their degree programme under the same conditions which were in place when they started it so no student can be forced to switch from a traditional to a new programme in the middle of his or her course of study.

Third, if the German States would have continued to approve of newly established bachelor's and master's programmes according to the traditional procedures, the change to the new structure could not be completed until 2010 (Standing Conference 2002).

A fourth reason became more obvious once the reform dynamics had picked up speed and an increasing number of new bachelor's and master's programmes were in the making or being introduced. While minimum standards of curricular content were traditionally tied to state regulation of study and examination procedures, thus determining a canon or core curriculum for every degree, accreditation is based on the consensus of peers about the relevant subject matter. In particular at the master's level, German higher education as well as higher education in most of the other Bologna signatory states experiences a wealth of newly designed interdisciplinary programmes that cannot be closely linked to a single discipline or a clear-cut subject matter and can have no established canon. These programmes intend to prepare their students for jobs and professions in the newly emerging knowledge societies that either did not exist before or are going through a process of professionalisation, i.e. requiring a higher academic form of education and training.

The Structure of the German Accreditation System

Accreditation Council and Agencies

The structure of the German accreditation system consists of two levels. The overarching level is the German Accreditation Council under the guidance of which are the actual accreditation agencies as a second level. This actually reflects the German federal system with a framework responsibility of the federal government to guarantee equality of opportunities and actual responsibility of the States for all matters in the field of education. The Accreditation Council has 17 members: four representatives from higher education institutions (professors), four representatives from the ministries for education and research of the German States, four representatives from various fields of professional practice (trade unions, large companies, public authorities), two student representatives, two international experts and (in a consultative and advisory function) one representative from the accreditation agencies. Its responsibilities are

- accreditation of accreditation agencies;
- monitoring the work of the accreditation agencies and their periodical re-accreditation;
- definition of standards, procedures and criteria for accreditation procedures.

In its first few years of existence the Accreditation Council also engaged in accreditation of bachelor's and master's degree programmes, thus providing a valuable source for pilot schemes and field knowledge. Since 2003 however, the

Accreditation Council is no longer involved in the actual accreditation of degree programmes.

Its legal status was originally somewhat unclear but associated to the Secretariat of the Standing Conference which also provided its budget. Since 2004, the Accreditation Council has been given the legal status as a foundation. However, it still remains unclear how legally binding the decisions of the Accreditation Council are. Finally the Accreditation Council is a member of the international networks for quality assurance, in particular in INQAAHE (International Network for Quality Assurance in Higher Education) and in ENQA (European Network for Quality Assurance in Higher Education). In this context the Accreditation Council negotiates cooperation agreements with foreign accreditation agencies for mutual recognition of accreditation decisions and degrees (Schade 2005, p. 129).

The second level of the German accreditation system consists of the accreditation agencies (Kehm 2005). Only agencies accredited by the Accreditation Council are allowed to accredit degree programmes. However, higher education institutions and departments can choose to get an additional accreditation for a particular degree programme from an international accreditation agency (e.g. a highly reputable American agency) in order to market this as an additional quality feature. There are altogether six accreditation agencies in Germany which are either regionally active and then for all subjects and disciplines or nationally active and then only for specific subjects or subject groups:

- Agency for Quality Assurance Through Accreditation of Study Programmes (AQAS): general accreditation with focus on North-Rhine Westphalia and Rhineland Palatinate;
- Accreditation Agency for Study Programmes in Computer Sciences, Natural Sciences, and Mathematics (ASIIN): subject-specific accreditation;
- Accreditation Agency for Study Programmes in the Field of Health Care and Social Work (AHPGS): subject specific accreditation;
- Institute for Accreditation, Certification, and Quality Assurance (ACQUIN): general accreditation with focus on Bavaria, Thuringia, and Saxony;
- Foundation for International Business Administration Accreditation (FIBAA): subject-specific accreditation in Germany, Austria, Switzerland, the Netherlands, and Sweden;
- Central Evaluation and Accreditation Agency (ZEvA): general accreditation with focus on Lower Saxony and Hesse.

The legal status of these agencies varies, the majority are non-profit organisations. The composition of the decision-making bodies reflects that of the Accreditation Council minus representatives of the State, i.e. higher education institutions, students, and professional fields/employers. In some agencies, representatives of the trade unions play a role; in others representatives of the respective ministries have an advisory function or observer status.

Accreditation in Germany follows some basic principles as well as formal and subject related criteria that are the same in all agencies and procedures. There are four premises on which accreditation is based:

- quality assurance;
- proof that the curriculum of a given degree programme can be studied in the envisaged standard period of study;
- enabling diversity;
- creation of transparency.

Accreditation Procedure

The accreditation procedure is carried out in three steps. In the first step the higher education institution sends an application for accreditation to an agency. In some German States the application is examined by the responsible ministry in terms of its compliance to the respective state planning. The agency examines the application in terms of completeness of forms and information and in terms of the question whether the study programme to be accredited is conceptualised in such a way that it fulfils the basic requirements at a formal level. The agency then determines the costs for the accreditation and agrees with the higher education institution about an appropriate schedule.

The second step starts as soon as the higher education institution has assured the agency that it is willing to pay the costs for accreditation. Then the application is examined in more detail, an audit team is proposed to the responsible accreditation commission within the agency and peer reviewers are nominated. The institution has a right to propose peer reviewers as well. The peer review consists of an on-site visit. Usually the group of peer reviewers also includes representatives of the respective professional field or of employers of graduates in that subject and a representative from the ministry. A representative of the agency deciding about the accreditation acts as a rapporteur of the visit. During the visit the review group talks to the dean, to the academic staff responsible for the programme, the teachers and to student representatives. The report is based on the accreditation application, the results of the visit and ends with a recommendation concerning the accreditation. The institution then receives the report as well as the recommendation and can comment on the report (feedback).

The third step consists of the finalisation of the report and its recommendation, which is then submitted to the responsible accreditation commission of the agency. The commission takes the final decision. It can be a clear yes or no or a yes under conditions the fulfilment of which are examined during re-accreditation. Typically, a study programme which has to fulfil further conditions will be accredited for a shorter period than the regular 5 years, after which a re-accreditation must be carried out.

It might be interesting at this point to take a closer look at the actual accreditation procedure, at the issues the accreditors look at and how they determine academic quality.

The application for a first accreditation of a study programme consists of two main parts. The first part provides context information about the faculty or department as whole, e.g. how many students, how many degree programmes, how many teachers, how many graduates per year, number of rooms and other information about infrastructure. The second part describes the study programme itself, which is supposed to be accredited. The guidelines for this part vary somewhat among the accreditation agencies, but as a rule they include a detailed description of the study programme, its targeted students, the teachers, the modules, the potential jobs available for graduates of that programme, and the key competences or skills included in the curriculum. The accreditation agency checks the application for completeness and consistency on a formal level while the peers check the content and the quality.

The peer review also consists of two parts. First there is a preliminary analysis of the application among the reviewers during which open questions are noted and topics for the interviews during the on-site visit will be determined. The second part is the actual on-site visit, which again consists of several steps (Reuke 2005, p. 148):

(a) Starting discussion with the dean or department head, sometimes representatives from the central institutional management are participating as well. Focus of the discussion:

  - development planning of the institution,
  - importance of the subject in the overall institutional context,
  - profile and perspectives of development of the subject in the view of the central level management,
  - situation of study and teaching in the department or faculty;
  - staff planning,
  - cooperations,
  - perspectives of development,
  - infrastructure,
  - communication and coordination in the department or faculty,
  - role of the study programme to be accredited in the department or faculty,
  - measures and instruments of quality assurance.

(b) Discussion with the person responsible for the programme. Focus:

  - educational goals,
  - curriculum,
  - course of study,
  - teaching content and teaching methods,
  - advice and counselling offered to students,
  - organisation of examinations,
  - study success,

- labour market relevance of programme (employability of graduates),
- marketing of the programme

(c) Discussion with the teaching staff of the programme. Focus:

- curriculum,
- course of study,
- content and methods of teaching,
- advice and counselling offered to students,
- provision of staff development courses.

(d) Discussion with students in various stages of the study programme and with representatives of the students' union. Focus:

- educational goals and study programme,
- organisation and course of studies,
- examinations,
- advice and counselling offered to students,
- study conditions (e.g. access to library, books, computers, state of classrooms, laboratories etc.),
- opportunities for temporary study abroad,
- opportunities for work placements and internships,
- excursions.

(e) Reviewers are taken through a guided tour of the department or the institution. Focus:

- This should offer opportunities to discuss open questions of the reviewers individually with members of the department or the central level and ask additional questions or request clarifications.

(f) Closing talk and possibly a first oral feedback with academic staff responsible for the programme and the dean or head of department.

Re-accreditation is a more simple procedure and also accompanied by an application. It can be carried out by an external evaluation agency or organisation which has been recognised by the agency. The price for re-accreditation might be somewhat but not considerably lower than the price for accreditation, which means that there is a continuing extra financial burden on the institutional budget. An application for re-accreditation must include the following information (Reuke 2005, p. 150):

- description of the current curriculum and explanation of possible changes in comparison to the first accreditation;
- a list of all academic staff involved in teaching the programme with short CVs;
- proof that possible conditions connected to the first accreditation have been fulfilled;

- summary of results of an evaluation of the study success of students, including their success on the labour market;
- results of examinations, final theses (statistics) plus examples of very good and just barely passing theses;
- possibly answering additional questions of the reviewers;
- existing examination regulations;
- a table providing an overview of the modules.

The procedure for a first accreditation described above is an ideal type and would require the reviewers to be on-site for at least a whole day if not longer in order to carry out all the talks with all the target groups. In reality the on-site visit is often much shorter, e.g. half a day, so that the various group discussions take place not individually but with other groups present. Typically the reviewers will only ask to speak to the dean and the students individually. A "cluster" accreditation, i.e. an accreditation of several programmes in a given department or faculty in one go, will basically mean that the reviewers have 1 or 2 h per programme.

German Accreditation Specifics

The framework regulations for the introduction of accreditation in Germany issued by the Standing Conference include a few specifics which are worth mentioning because they are a reflection on the traditional German system of higher education and unique in the sense that the systems of accreditation which have been introduced in recent years in other European countries as well don't have these features.

Although the German *Diplom* degree differentiation between universities and universities of applied sciences (*Fachhochschulen*) was given up with the introduction of bachelor's and master's programmes; i.e. both types of higher education institutions are allowed to offer bachelor's as well as master's programmes, another form of differentiation has been introduced. All accreditation applications for master's programmes have to include information on whether the programme is "research oriented" or whether it is "application oriented." This is examined in detail during the accreditation procedure.

One would assume now that universities will tend to offer research-oriented master's programmes while universities of applied sciences will tend to offer application-oriented master's programmes. But they often don't. For universities of applied sciences it was the long-hoped-for loophole for academic drift. They finally wanted to become more similar to universities. That has led to considerable protest from universities. In fact, the nine biggest technical universities[2] in Germany reacted by forming an association, the "TU 9 Group," and declaring publicly that

[2] Technical universities have the highest number of study programmes awarding the degree of *Diplom* which is the usual one subject study programme in most of the technical and engineering sciences, but also in economics and social sciences. Basically the *Diplom* was/is the degree in all professional subjects. Universities of applied sciences also award or awarded a *Diplom* but with the

there was no such thing as a bachelor's degree in their subjects and that they would only award master's degrees. If students wanted to leave the university after 3 years of study they would be given a bachelor's degree but not be recognised as properly trained engineers and properly trained engineers were the only graduates that they were producing.

When these debates flared up experts expected that they would die out sooner or later and indeed, this is what happened. Bachelor's and master's degrees have also been introduced by now in technical universities. However, there are still strong feelings about the value of a bachelor's degree, especially in engineering and some of the pure sciences (e.g. in physics). On the surface we find compliance to the new rules and regulations but a more in-depth analysis shows[3] that many academic staff in universities still reject the idea of a bachelor's degree in these subjects. But German accreditation introduces a second form of differentiation at the master's level. All master's programmes have to be classified according to the question of whether they are "consecutive," "stand alone" or "continuing academic education" programmes. Consecutive master's programmes are programmes in any given subject that follow more or less on top of a bachelor's programme in the same subject so that students have a choice whether to finish their studies after a bachelor's degree or go right into the master's programme.[4] However, most master's programmes, regardless of the type, now have defined criteria for admission because there is a widespread consensus that a smaller proportion of students than traditionally will and should be accepted into the master's level. Still, many of these master's programmes have been derived with only some curricular change from the previous, i.e. traditional, long cycle university study programmes which generally finished with a degree equivalent to a master's degree. In order to implement the bachelor's and master's structure, a traditional programme was often just cut in half and, with a few additional contents and qualifications added, the first half became defined as a bachelor's programme while the second half became defined as a master's programme. The second category of master's programmes which I have defined as "stand alone" – following Stefanie Schwarz and Don Westerheijden (2004) in this – are newly developed programmes which frequently do not have a matching bachelor's programme as a basis. They are often particular specialisations which

---

additional letters FH in brackets behind the degree (for *Fachhochschule*) to denote the difference between the two types of institutions.

[3] For example the analysis of the implementation of bachelor's and master's degrees in physics carried out in INCHER-Kassel on behalf of the European Physical Society since 2008.

[4] For non-German readers it should be mentioned here that the idea of defining a master programme as "postgraduate" is not very widespread in Germany. The traditional university degree programmes all finished at the level of a master degree from which the issue of getting a doctoral degree was clearly separated. This also holds true for the majority of graduate schools or graduate centres emerging currently in Germany. Most of them are clearly targeted towards doctoral students and apart from a few exceptions which have explicitly adopted the American model, German graduate schools do not accept students aiming for a master's degree. Thus, selection for a doctorate takes place on the basis of master's degrees not bachelor's degrees and doctoral candidates are expected to be familiar with research methodology and theory of their subject or field of expertise.

have newly emerged in a given subject or field or interdisciplinary programmes. Frequently students from a broader range of bachelor's programmes than only one particular subject are accepted into such "stand alone" master's programmes. The third category of master's programmes are continuing education programmes; i.e. they are targeting students with two or more years of professional practice who want to upgrade or update their qualifications. Typically those programmes require tuition fees and have done so for quite some time while the general introduction of tuition fees (for bachelor's as well as master's programmes) has been politically decided only recently and started mostly from 2007 onwards. All East German states and two West German states have decided not to introduce tuition fees.

The final specific is that a market has been created for accreditation in Germany. The accreditation agencies are actually non-profit organisations, and the price they demand for the accreditation of a given degree programme basically covers only the actual costs that are incurred, but the agencies compete against each other for customers. That also includes the more regionally oriented agencies which will recruit their customers predominantly from two or three of the German States, but they are not restricted to their dominant geographical area of activity.

Accreditation Statistics

According to the accreditation statistics from January 2009 there are altogether 13,791 degree programmes on offer at German higher education institutions. Of these programmes altogether 9,712 (70.4 percent) have been converted into bachelor's (5,322) and master's (4,390) programmes. In January 2009 the agencies had accredited altogether 4,115 study programmes, among them 2,195 bachelor's programmes, 1,889 master's programmes and 31 study programmes with traditional German degrees (Diplom/Magister). More than half of all accredited degree programmes were only provisionally accredited and have to fulfil further conditions.

This status points to at least at two problems which will be analysed in the next section:

- Since 2004 the reform dynamics have gained momentum so that the agencies can't keep up the pace and are clearly lagging behind. Accreditation is currently a serious bottleneck for the implementation of reforms with regard to the changes in the degree and study structure.
- More than half (57.6 percent) of the newly established Bachelor and Master programmes have started without proper accreditation. This might give rise to student complaints and legal problems.

Problems of Implementation

Concerning the implementation of a system of accreditation in Germany five problems are quite obvious (Teichler 2006; Schade 2004).

The *first* problem is the fact that accreditation considerably lags behind in the face of the ongoing reform dynamics. Many new study programmes have been developed and are starting or have started already without being accredited. Not only is the process of application for accreditation immensely work intensive for those responsible for the programmes, e.g. for the application alone a dossier has to be put together comprising statistics that are often not readily available and descriptions and arguments have to be submitted that expand the paper work to 40 pages and more. The visit of the reviewers, the discussion of the report and recommendations, and the possible fulfilment of conditions for final accreditation eat up valuable time of academic teachers and researchers. Furthermore, the accreditation agencies themselves are more and more often in need of reviewers. Applicants can propose subject specific reviewers for the accreditation of their programme but whether the proposed reviewers are willing to find the time to do the job is another question altogether. In the beginning many well-reputed professors might have accepted a request from the accreditation agency to act as reviewer out of curiosity. But the process is time consuming and more often than not there is no honorarium attached to it. As a consequence the agencies – at least in some subjects – are desperately seeking subject specific experts willing to do the peer review job. Thus, it can happen that applicants are confronted with reviewers whom they do not consider as their proper peers. For example, a newly established or designed study programme at a well reputed university might be confronted with reviewers from the *Fachhochschule* sector who then start prescribing how to design the programme properly. That will not only cause a reaction of outrage but there are already the first cases of universities which have decided to refuse to go through such an accreditation, in particular as the costs of accreditation have to be borne by the institutions themselves.

The *second* problem which can be observed is the fact that the accreditation agencies are independent in their judgement while the Accreditation Council is not. The Accreditation Council is neither allowed to decide about structural guidelines and regulations nor about the assessment of programmes without agreement of state representatives. As Serrano-Velarde (2006, p. 9) put it, "the history of accreditation (in Germany) is to be read as a constant fight for organisational independence against regulative attempts on behalf of the federal states." The Accreditation Council is in several ways dominated by the policy of the Standing Conference of the Ministers for Culture and Education of the German States (Standing Conference). The Standing Conference is responsible for educational policy making and planning in Germany and is the main funder of the Accreditation Council. In addition, representatives of the Standing Conference constitute more than half of the members of the steering committee of the Accreditation Council. The Council is supposed to take up the resolutions of the Standing Conference and forge them into legally binding framework regulations for the work of the accreditation agencies. The agencies themselves are constantly fighting for their independence and thus the Accreditation Council finds itself in an uncomfortable position between the rather powerful political body of the Standing Conference and the relatively autonomous accreditation agencies.

The *third* problem is a consequence of the second one: Accreditation in Germany is not a complete substitute for state approval of curricula. Although the states and the Federal Ministry by setting up a system of accreditation have reduced close state control in this field there continues to be a double structure of accreditation and state approval of degree programmes.

As mentioned in the previous section, a *fourth* problem is constituted by the fact that accreditation lags behind the establishment of new degree programmes and many programmes have started without accreditation. A few cases have occurred in which programmes were started and the accreditation process was interrupted. In order not to have wasted students' time (and possibly money), the accreditation was stopped mid-way, new discussions were started, and before it could come to a definitive rejection by the peer reviewers and the responsible commission in the accreditation agency, the reviewers formulated a number of conditions to be fulfilled by the university and the person responsible for the programme. This suggests initiating programmes that subsequently are not accredited not only runs the danger of being taken to court by the students, but also that the legal implications of a possible rejection of accreditation is not yet properly clarified. So far no accreditation agency has rejected a new degree programme outright, instead the accreditation process was usually stopped and requirements were formulated which had to be fulfilled before a new accreditation procedure was started.

The *fifth* problem in the implementation process has to do with costs (see below). Accreditation has sometimes been characterised as being a successful money-generating machine. Accreditation as part of quality assurance services should therefore be regarded as a market in the sense of a "contextualised zone of interaction" (Serrano-Velarde 2006). The higher education institutions have to come up with the money for accreditation themselves. No additional government funding is being provided. That has led to cost cutting measures by the higher education institutions. The most favoured form in this respect is the so-called cluster accreditation. Usually any given department or faculty in the sense of the basic organisational unit of an institution of higher education offers several degree programmes. In addition, a faculty may consist of several departments or a department of several subject groups each of which offers more than one degree programme as well. In those cases in which a department or faculty decides to change all its existing degree programmes into the new bachelor's and master's structure at once, there will be several new programmes to be accredited. An application is then prepared and submitted to the chosen accreditation agency requesting the procedure be organised in such a way that all programmes can be accredited at the same time, preferably in the framework of one peer review as well. The group of peer reviewers must therefore be somewhat larger and include specialists for the subject matter of the programmes to be accredited. Naturally, the institution will ask for a reduction of the price per programme accreditation and the request is normally granted. The problem here is that the peer review and in particular the on-site visit is cut short for the individual programme because the accreditation of several programmes within the framework of a "cluster accreditation" will not take much longer or be more intensive than the accreditation of a single programme. Accrediting several programmes at once therefore tends to be a more superficial procedure.

The Impact of Accreditation: Strengths and Weaknesses

Although the system of accreditation is still relatively new in Germany – apart from the pilot period systematic accreditations have been carried out since 2004 only – a few strengths and weaknesses can be identified at this point in time.

It should certainly be noted as a strength that Germany for the first time has established a comprehensive and external quality assessment system geared towards the organisation of teaching and learning. Accreditation in Germany looks at the existence of minimum standards comparable to European criteria and provides the new programmes with one of the preconditions to achieve state approval, i.e. "the right to exist" as it is frequently formulated in the relevant literature.

However, apart from the problems enumerated in the previous section, which could also be interpreted as weaknesses, one of the main problems identified so far is that there is a multitude of quality assessment procedures emerging in Germany (accreditation, state approval of curricula, external evaluation, internal evaluation) that tend to overlap and make the whole quality assurance system higher education overly complex, in some cases even sending out contradictory signals (Teichler 2003, 2006; Schade 2004, 2005). Schade (2004), in particular, emphasises that there is no national institution to coordinate evaluation activities in teaching and learning – just a wealth of local and regional initiatives – and that there is no link between these decentralised evaluation activities and accreditation (ibid., p. 191). The super-complexity (Teichler 2003) of the emerging quality assurance system in German higher education does not only make it onerous for the departments and faculties involved – all this gathering of data and writing of reports which all need to be detailed and put together in a different way for the different occasions – it is also not very economical. The time and cost factors will soon become too high for the institutions. Schade (2004, p. 191) points out that "if the strict division of evaluation and accreditation were to be maintained, there would additionally be a danger that the quality assurance system could disintegrate into two parts: one for comparability and the other one for quality improvement." In the long run, therefore, Germany needs to develop an integrated system of quality assurance in higher education (ibid., p. 193). There are currently discussions going on whether a change to institutional accreditation, or "process or system accreditation" as it is called in Germany, might be a solution to this problem.

The decision taken at the national level to allow universities as well as universities of applied sciences (*Fachhochschulen*) to establish master's programmes has led to a blurring of boundaries between these two distinct types of higher education institutions. However, the signals produced by the guidelines of the Standing Conference of the Ministers for Culture and Education for accreditation are somewhat contradictory. On the one hand, the distinction between research-oriented and professionally oriented master's programmes could be interpreted as reproducing the institutional distinction at another level. On the other hand, all new degrees (bachelor's as well as master's degrees) have to provide proof of the competences and skills they offer to enable graduates a smooth transition into the labour market. Due to the phenomenon of academic drift, the universities of applied sciences are including in their bachelor's and master's programme curricula more academic and theoretically oriented elements in order to provide opportunities for their graduates to continue studying at a university or get access to a doctoral programme, while the universities are forced to include more practical and labour-market-oriented skills and qualifications into their curricula in order to provide proof that their graduates have the required competences for the transition into the world of work.

At the same time current higher education reforms in Germany at a more general as well as national level aim at a higher degree of institutional differentiation. This differentiation no longer follows the traditional divide between universities and universities of applied sciences but along the lines of competition for excellence, i.e. a more strictly vertical differentiation according to research excellence in the university sector. The German "initiative for excellence" (Kehm 2006) selected a number of German universities on the basis of a competitive bidding procedure for considerable extra funding provided by the Federal Ministry and the German States in three categories (graduate schools, clusters of excellence and institutional development concepts). The aim is to support nine German universities that successfully bid for an institutional development concept and turn them into elite universities. The process has been highly contested, but its impact on the vertical differentiation of university reputation can be already felt. It thus remains an open question whether standardised procedures of accreditation will contribute to this development or be an impediment. The impact of this trend on accreditation has been little discussed up to now.

A final weakness is the unfamiliarity of the labour market with the qualifications of the graduates with the new degrees. There are a few initiatives to welcome bachelor's graduates on the labour market and the signals coming from potential employers of higher education graduates are generally positive. But the fact remains that bachelor's graduates from universities and master's graduates from universities of applied sciences are unknown species and given the inherent conservatism of personnel managers and recruitment departments the transition might not be all that smooth for all graduates. Until now the number of graduates entering the labour market with the new degrees is still too small for large-scale surveys and analyses of the transition period but there are plans to study this issue closely. Concerning this problem it could also be stated that the absorption of higher education graduates into the labour market has basically always been more supply led than demand led in Germany and the link between higher education and the world of work has always been more a loose coupling than a tight one in order to provide necessary flexibility. Therefore, it can be assumed that once the potential employers become more familiar with the qualifications of the new degree graduates the problem will likely disappear.

Who Bears the Costs of Accreditation?

As mentioned in previous sections, all new degree programmes which are established according to the bachelor's and master's structure must be accredited and the higher education institutions have to come up with the costs themselves. In a country in which there yet are no tuition fees and higher education institutions are almost exclusively state funded this constitutes a heavy burden on the institutional budget. In particular, the institutions have experienced reductions in their state funding for several years now and accordingly believe themselves to be in a permanent financial crisis. There is limited evidence on how much the institutions are actually paying for the accreditation of their programmes and it is difficult to get accurate information on prices.

Depending on the subject the accreditation of single degree programme will incur costs for the institution at a level between €8,000 and 15,000. Prices per programme accreditation will be somewhat less in the framework of a "cluster accreditation." However, these latter prices are negotiated between the higher education institution and the accreditation agency. Since institutions are free to choose an agency this leaves room for competition and bargaining. It should also be kept in mind that a medium sized German university (medium-sized is between 15,000 and 30,000 students) might offer between 60 and 100 different degree programmes. Since the traditional degree programmes ended at the master's level, the majority of these programmes will now be turned into one bachelor's and one master's programme plus additional "stand alone" master's programmes. So the change of all existing degree programmes into the bachelor's and master's structure at a medium-sized university can easily incur costs of 1 million US dollars and more. A rough estimate by the president of one of the medium-sized German universities is that accreditation costs amount to about 10 percent of the overall institutional budget.

In the majority of those German states which have introduced performance contracts between the responsible ministry and the individual institutions of higher education,[5] the introduction of bachelor's and master's degrees and programmes are part of the performance indicators. However, as a rule the institutions do not receive any additional financial means, e.g. an incentive, but rather agree to the changes in the form of a self-commitment. Such a contract will stipulate among other things that the respective higher education institution will change all or a negotiated number of its existing degree programmes into the bachelor's and master's structure in a given period of time. Generally the introduction of the new degree structure takes place gradually so that the costs will not become too high in a given year.

In the year 2000, the German Science Council, an important buffer body making policy and planning recommendations in the field of higher education, was given the task by the German states and the Federal Ministry to include private higher education institutions into the accreditation system. There are currently 53 private and 44 church affiliated higher education institutions in Germany compared to 333 public or state approved higher education institutions. The proportion of students studying at private and church affiliated institutions of higher education as compared to all students is only 3.3 percent. However, in contrast to the accreditation of programmes in the public sector it was decided to have institutional accreditation in the private sector. In addition, institutional accreditation by the Science

[5] Performance contracts are now more commonly used to determine the annual budget provided by the responsible state for a given higher education institution located in that state.

Council follows a different procedure than programme accreditation by the agencies. Institutional accreditation is guided by the principle that not individual degree programmes are assessed but the core object of assessment is the question whether the institution as a whole has an adequate quality assurance system in place. If that is answered positively in the process of accreditation then the institution has earned the authority to set up any degree programme it wants. This approach is actually supported by quite a number of individual actors in the field of programme accreditation and evaluation and some of the accreditation agencies as well. Institutional accreditation, in Germany often called "process" accreditation in contrast to "programme" accreditation, would certainly reduce the current complexity of the system and its procedures but it might also require changing the system of additional state approval of programmes and its strong links to educational planning at state level. The advantage of institutional accreditation would be that the costs would be considerably lower than they are for programme accreditation. The Science Council has proposed to price an institutional accreditation between 22,000 and 35,000 US dollars.

European Developments and Models for the German System of Accreditation

Accreditation is one of a number of quality assurance instruments. Certainly the concern about quality in higher education is not new. Quality has become a focus of Western European higher education policy since the mid-1980s (Schwarz and Westerheijden 2004, p. 6) and now has become a global phenomenon. The quality assurance instruments, however, have varied over time and preferences have also been influenced by the distinctive cultures of institutions and systems. In Europe accreditation of degree programmes has been introduced only recently, that is after the Bologna Declaration of 1999. In order to make the European higher education systems more attractive to students from non-European countries, to make the systems more competitive and to increase intra-European student and staff mobility, the Bologna Declaration aims to create a European Higher Education Area by the year 2010. One of the most important elements of this common space is the creation of a comparable and transparent degree structure at the higher education institutions characterised by two main cycles of studies: undergraduate and graduate (leading to a bachelor's and a master's degree, respectively).

Westerheijden (2005, p. 98) observes that after the Bologna Declaration quite a number of national governments in the European countries decided with notable speed to have all their study programmes accredited. He goes on to state that in this process four possible alternatives were excluded (ibid., p. 98):

(a) Quality assessments of study programmes as they had been implemented during the 1990s in most Western European countries because the reports did not provide sufficiently transparent information.

(b) Quality assessments at institutional level because they did not provide information about the actual quality of study programmes.
(c) An accreditation system like in the USA because this was based on a general institutional accreditation in the sense of minimum standards for institutional recognition and complemented by an accreditation of programmes in highly professionalised fields (like medicine, nursing, law, engineering and teacher training).
(d) The establishment of quality assurance bodies at the European level because there was no need to introduce an additional layer of bureaucracy in the European higher education area and thus lessen national sovereignty.

At the first Bologna Follow-Up Conference in Prague in 2001,[6] the European Network of (national) Quality Assurance Agencies (ENQA[7]) was given the responsibility for the further development of quality assurance in the Bologna Process. ENQA had originally emerged from a pilot project funded by the European Commission. Today, ENQA has 42 members and is a professional network of quality experts rather than a political body (see Schwarz and Westerheijden 2004, p. 6). The membership fee is €2,000 per year. Although ENQA quickly took over a central position for quality assurance and accreditation, in particular with regard to the definition of standards and procedures, the representatives of national governments insisted that mechanisms of quality assurance and assessment remained at the national level. However, at the second Bologna Follow-Up Conference in Berlin in 2003, the Ministers of the signatory states decided that all Bologna signatory states should establish a national accreditation system or an equivalent by 2005 (Westerheijden 2005, p. 99).

Thus, accreditation was introduced in Europe later but spread much faster than evaluation. Only Greece has not yet introduced a national accreditation system, although it is in the process of establishing one. There are already committees working to determine the respective procedures.

Despite the attempts, through ENQA, to develop comparable structures of quality assurance mechanisms at the European level, current accreditation systems in Europe continue to be relatively diverse. In some countries the unit of analysis is the study or degree programme (e.g. in the Czech Republic, in Germany, Hungary, Italy, the Netherlands, Norway, Poland, etc.) in other countries it is the institution of higher education (e.g. in Austria, in the Czech Republic, in Norway, and Sweden). In some countries the state retains the right to approve of new degree programmes in

---

[6] The Ministers of Science and Education of those countries having signed the Bologna Declaration meet every 2 years in the framework of so-called follow-up conferences in order to discuss progress made and possibly add to the reform agenda or correct it if developments go in an unwanted direction.

[7] The original name of ENQA was: European Network of Quality Assessment Agencies. It is now: European Association for Quality Assurance in Higher Education.

addition to accreditation, in others it does not. In some countries several accreditation agencies are operating, in others there is only one national accreditation agency. In some countries evaluation and accreditation are similar processes carried out by one and the same body or agency, in other countries evaluation is carried out either by the institutions of higher education themselves or by national/regional bodies not necessarily identical to those responsible for accreditation.

By and large, however, according to Westerheijden (2005) two basic approaches to accreditation can be distinguished in Europe: centralised accreditation systems and open accreditation systems.

Centralised accreditation systems can be found predominantly in the Central and Eastern European countries but also in Norway and in Spain, though the accreditation system is different in the latter two. In the majority of Central and Eastern European countries accreditation is directed at study or degree programmes. As a rule a national accreditation agency – often established by the government – establishes a set of standards for the input (infrastructure, teaching staff, curriculum, planning, etc.) of all programmes in a discipline or field of knowledge. The standards are typically defined by consulting the academic oligarchy. Each programme is controlled whether it conforms to the standards or not by the national agency. Output, in particular employability of graduates which is high on the European agenda is not part of accreditation. This approach has led to conformity rather than diversity of study programmes. In Spain and Norway there is only one accreditation agency as well which is responsible for the procedure and the process of accreditation. However, in contrast to the accreditation in the majority of Central and Eastern European countries it is more output oriented. Both agencies have also been involved in the formulation of a European qualification framework and the so-called Dublin Descriptors,[8] from which national qualification frameworks will be derived (Westerheijden 2005, p. 102).

Open accreditation systems can be found in particular in the Netherlands and in Germany. In both countries there is one national agency or body responsible for the regulation and coordination of accreditation procedures. But both bodies are not monopolistic. Quality or accreditation agencies having been recognised by the national body can carry out accreditations and also include foreign accreditors in the procedures. Institutions or degree programmes are free to choose by which agency they want to be accredited. This freedom includes the right to be accredited by an accreditation agency from another country (multiple accreditation). In fact, some German programmes, especially MBA programmes and a few engineering programmes have gone through accreditation procedures by European (e.g. by the European Foundation for Management Development) and foreign agencies (e.g. American agencies among others) in order to increase their reputation and market value. However, there is also market regulation insofar as the national body, the

---

[8] The Dublin Descriptors (named after the city where the Joint Quality Initiative – an informal group of experts for quality assurance from a variety of European countries – met to formulate them) are a short list of competences which can be expected from bachelor and master graduates and from Ph.D. candidates independent of their particular field of study.

Accreditation Council in Germany, decides which agencies are to be recognised in the market (ibid., p. 104).

To summarise it can be said that the development of a German system of accreditation has been strongly influenced by policies and developments at the European level. Westerheijden (2005, p. 99) notes that ENQA was established at the first Bologna Follow-Up Conference in Prague in 2001 and continues to act as the "spider in the net" for the further development of quality assurance in the framework of the Bologna Process. While ENQA became a central contact point for quality assessment and accreditation agencies at the national level, the ministers also made sure that the mechanisms of quality assurance remained a national affair. In other words we find here a process of "policy transfer" (Pratt 2004) which is not simply the copy of an existing model but rather the use of one or more possible models put forward by ENQA as stimulus for innovation (ibid., p. 112). The fact that quite a number of the Bologna signatory states decided to introduce a system of accreditation was strongly related to a higher degree of competition among higher education institutions within the European higher education area and the need to provide more transparency to increase recognition of study achievements abroad. While ENQA assumed the role of an elite network of experts and brokers with international experience – which national agencies either joined or could consult with in establishing the national accreditation systems – the national quality assurance systems established had to incorporate into their procedures and criteria the standards developed at the European level. In addition, national governments and in Germany the responsible state ministries referenced the European level to overcome possible resistance to the implementation of the new policy of accreditation.

Conclusions

The German accreditation policy is a significant attempt by the government to ensure the quality of education in German universities; however, the policy has imposed quite high transaction costs on the higher education system. This is not so much due to the fact that the policy transfer has been "coercive" and accreditation has been implemented against the wishes of the higher education institutions; on the contrary, the German Rectors' Conference was an influential participant in the establishment of the Accreditation Council. It has more to do with the fact that the whole system of quality assurance in higher education, of which accreditation is only one, albeit an important element, is decentralised and its various elements are not properly linked. The German States can influence the accreditation business through their membership in the agencies. The accreditation sector itself is rather incrementalist and not very coherent. In addition, accreditation is not properly integrated with other quality assessment and assurance activities. This leads to less transparency, possibly even to less recognition within Germany, and to a fragmentation of the quality assurance system, which then requires even more control, audit and accountability.

But transaction costs are high on the European level as well. First of all, as Serrano-Velarde (2006) rightly points out, accreditation and evaluation is an emerging market in which not only quite a bit of money can be made but a market which is prone to international (or European) competition of agencies as well. Second, Van Damme (1999) in his report on behalf of the European Association of University Presidents noted as early as 1999 that this market might be open to fraud and the operation of agencies of dubious quality. ENQA then proposed to establish a European register of recognised accreditation agencies which should be allowed to operate in all of the Bologna signatory states (ENQA 2005). This proposal was turned down by the Ministers at the Bologna Follow-Up Conference in Bergen in 2005 because they wanted to keep authority and control in the field of higher education quality assessment and assurance in their own countries and not transfer this power to a supra-national agency (Serrano-Velarde 2006). Thus and thirdly, at the European level as well we have differing quality assurance and accreditation systems in each of the Bologna signatory states, a situation which is not conducive to increased transparency, mobility and recognition.

But the situation has a further policy implication since no overarching conventions can be found at the European level and that is the fact that national accreditation markets basically tend to become "closed shops" and national governments will determine the rules of the game. Even where international accreditation is possible – and there are a few cross-border agreements in place – accreditation by a foreign agency will either have to be done in addition to national accreditation for which there is little incentive due to the extra costs involved, or foreign accreditation agencies will have to seek recognition of national governments (in the case of Germany of the national Accreditation Council) in those countries in which they wish to operate and play the game according to the national rules. For the time being, however, the creation of a European market for quality assurance and accreditation which would include an element of European consumer protection seems to be out of reach (Serrano-Velarde 2006). Whether this will prevent the achievement of a European Higher Education Area altogether must currently remain an open question.

Finally it should be mentioned that the impact of the German "excellence initiative" and its emphasis on vertical stratification of the university sector will lead to changes that the system of accreditation will eventually need to take into account. These changes will include more selective admissions for programmes in those universities that profit from the initiative. These universities will only want the very best and most promising talent. Other universities might react with local access restrictions too in order to uphold their reputation. This might also affect recognition of degrees. Will students coming from a regional university be able to change into the "elite sector" in order to continue their studies if they move from one location to another? Will students with a master's degree from a university of applied sciences be accepted into doctoral programmes of universities? It is unclear yet whether these changes will lead to a differentiation of accreditation and evaluation procedures as well.

Acknowledgements I thank Kathia Serrano-Velarde for providing me with her (as yet unpublished) paper and thus contributing not only additional information but also some of the ideas and perspectives which have been included in this analysis. However, all possible mistakes and misinterpretations are those of the author.

References

Accreditation Council (2009). Accredited study programmes – Statistics. http://www.hochschulkompass.de/kompass/xml/akkr/akkr_stat_a.htm. Accessed 23 January 2009.

Bologna Declaration (1999). The Bologna declaration of 19 June 1999. Joint Declaration of the European Ministers of Education. http:/www.bologna-berlin2003.de/pdf/Bologna_; declaration.pdf. Accessed 4 June 2006.

Bretschneider, F., Wildt, J. (Eds.) (2005). *Handbuch Akkreditierung von Studiengängen. Eine Einführung für Hochschule, Politik und Berufspraxis*. Bielefeld: Bertelsmann.

ENQA (2005). Standards and guidelines for quality assurance in the European higher education area. Helsinki: ENQA. http://www.bologna-bergen2005.no/Docs/00-Main_doc/050221_ENQA _report.pdf. Accessed 4 June 2006.

HRK (Ed.) (2006). Statistische Daten zur Einführung von Bachelor- und Masterstudiengängen. Sommersemester 2006. Statistiken zur Hochschulpolitik 1/2006. Bonn: HRK. http://www.hrk.de/download/dateien/HRK_Doku1_2006_Sose2006.pdf. Accessed 8 June 2006.

Kehm, B. M. (2005). Akkreditierungsagenturen in Deutschland. In: F. Bretschneider, & J. Wildt (Eds.), *Handbuch Akkreditierung von Studiengängen. Eine Einführung für Hochschule, Politik und Berufspraxis* (pp. 131–144). Bielefeld: Bertelsmann.

Kehm, B. M. (2006). The German 'Initiative for Excellence' and rankings. *International Higher Education*, 44(summer), 20–22.

Pratt, J. (2004). The accreditation model and policy transfer. In J. Pratt (Ed.), *The 'Accreditation Model': Policy transfer in higher education in Austria and Britain* (pp. 107–127). Oxford: Symposium Books.

Reuke, H. (2005). Das Akkreditierungsverfahren. In F. Bretschneider, & J. Wildt (Eds.), *Handbuch Akkreditierung von Studiengängen. Eine Einführung für Hochschule, Politik und Berufspraxis* (pp. 145–150). Bielefeld: Bertelsmann.

Schade, A. (2004). Shift of paradigm in quality assurance in Germany: More autonomy but multiple quality assessment. In S. Schwarz, & D. Westerheijden (Eds.), *Accreditation and evaluation in the European Higher Education Area* (pp. 175–196). Dordrecht: Kluwer.

Schade, A. (2005). Der Akkreditierungsrat. In F. Bretschneider, & J. Wildt (Eds.), *Handbuch Akkreditierung von Studiengängen. Eine Einführung für Hochschule, Politik und Berufspraxis* (pp. 125–130). Bielefeld: Bertelsmann.

Schwarz, S., Westerheijden, D. (Eds.) (2004). *Accreditation and evaluation in the European Higher Education Area*. Dordrecht: Kluwer.

Serrano-Velarde, K. (2006). The politics of quality assurance. Paper presented at the 3rd EUREDOCS Conference, 16–18 June 2006, Kassel.

Sorbonne Declaration (1998). Sorbonne Joint Declaration on harmonisation of the architecture of the European higher education system. http://www.bologna-berlin2003.de/pdf/Sorbonne_; declaration.pdf. Accessed 8 June 2006.

Standing Conference of the Ministers of Culture of the German States (2002). Decision about the future development of quality assurance in German higher education. Bonn, 1 March. http://www.akkreditierungsrat.de/KMK_Qualitätssicherung.pdf. Accessed 3 June 2006.

Teichler, U. (2003). Die Entstehung eines superkomplexen Systems der Qualitätsbewertung in Deutschland. In E. Mayer, H.-D. Daniel, U. Teichler (Eds.), *Die neue Verantwortung der Hochschulen* (pp. 213–215). Bonn: Lemmens.

Teichler, U. (2006). Akkreditierung in Europa. In G. Simonis, & T. Walter (Eds.), *LernOrt Universität. Umbruch durch Internationalisierung und Multimedia* (pp. 131–151). Wiesbaden: VS Verlag für Sozialwissenschaften.

Van Damme, D. (1999). Internationalisation in quality assurance: Towards worldwide accreditation? Report commissioned for the IAUP XIIth Triennial Conference in Brussels, 11–14 July 1999. http://www.ia-up.org/. Accessed 3 June 2006.

Westerheijden, D. (2005). Akkreditierung im europäischen Kontext. Neuere Entwicklungen der Qualitätssicherung im europäischen Hochschulraum. In F. Bretschneider, & J. Wildt (Eds.) *Handbuch Akkreditierung von Studiengängen. Eine Einführung für Hochschule, Politik und Berufspraxis* (pp. 97–111). Bielefeld: Bertelsmann.

Chapter 13
The Accreditation and Quality Processes of the General Medical Council in the UK

Lee Harvey

The British Accreditation Context

In the UK, most of the professions are controlled to a greater or lesser extent by a professional or regulatory body. In most cases, these bodies, and there are more than 100 of them, have some input into professional education. However, the degree to which control is exercised over the profession and over the training of professionals varies enormously from one body to another.

Accreditation of programmes in the UK provides approval and recognition of academic and vocational awards. Accreditation is vital if a higher education institution wishes to run courses that offer awards controlled by professional or regulatory bodies. The approach and extent of accreditation varies from the recognition of courses as representing industry standards of training (such as those accredited by the National Council for the Training of Broadcast Journalists) to the complex and tightly constrained legally binding procedures of the General Medical Council (GMC).

There is a considerable difference between voluntary regulation by a professional body and by a regulatory body. The former is regulation by a body representing, in the last resort, the interests of the professionals. The latter is regulation by a non-membership body, established by law to protect the public. The GMC is such a body.

Regulatory Body

A regulatory body is created by government to regulate qualifications or training for a particular occupation. Unlike professional bodies, regulatory bodies do not offer membership to practitioners and do not see themselves as serving practitioners in the first instance. The Teacher Training Agency, for example, claimed, when it was established that it served:

L. Harvey (✉)
Copenhagen Business School, Copenhagen, Denmark

D.D. Dill, M. Beerkens (eds.), *Public Policy for Academic Quality*, Higher Education Dynamics 30, DOI 10.1007/978-90-481-3754-1_13,

> the educational community generally. Specifically we serve the Secretary of State as agents of the Education Act within the remit we were given in that Act... and we serve the colleges and schools that are involved with teacher training. (TTA 1995)

Regulatory bodies exercise control over the profession in various ways. They are external watchdogs at one step removed from the profession. Regulatory bodies usually:

- control entry to the profession by specifying the required knowledge and competence;
- maintain a register of practitioners, inclusion on which is required for continued practice;
- enforce a code of practice determined to be in the public interest.

Most, but not all, regulatory bodies are established by statute and have their powers defined by statute. However, not all UK regulatory bodies maintain a register of practitioners. However, the regulatory bodies in the areas of medicine and health do have statutory powers and maintain a register of practitioners. These include the:

- General Medical Council
- General Dental Council
- Health Professions Council
- Nursing and Midwifery Council
- The General Osteopathic Council
- General Chiropractic Council
- General Optical Council

Health and medicine are, thus, regulated by separate bodies in the UK.

Role of the General Medical Council

The GMC is a regulatory body established under the *Medical Act* of 1858. Its by-line until recently was 'Protecting patients, guiding doctors'. Following events described below, the byline is now 'Regulating doctors, ensuring good medical practice'. The purpose of the GMC is 'to protect, promote and maintain the health and safety of the public by ensuring proper standards in the practice of medicine'. The GMC is formally established as a registered charity with strong and effective legal powers under the Act, designed to maintain the standards the public have a right to expect of doctors.

The governing body, the Council, has had its membership cut from 35 to 24 (12 lay and 12 medical members, all appointed by the Appointments Commission) replacing the previous majority held by the profession. The GMC regulates the medical profession in its entirety: it accredits educational provision and registers practitioners. Without accreditation, the education is effectively valueless as it will not lead to registration. Without registration a qualified medic cannot practice.

Formation

The GMC was formed as a result of the 1858 Medical Act.[1] The Act was passed after 18 years of parliamentary debate on the reform of the medical profession. The debate mainly centred on the abolition, or at least the restriction of, unqualified practice. Seventeen previous Medical Bills had been introduced to Parliament between 1840 and 1858. Each of these had encountered problems because of conflicting interests and views.

The idea behind the 1858 Act was that anyone needing medical treatment should be able to distinguish between qualified and unqualified practitioners, which, at that time, was not always the case. There were 19 separate licensing bodies that conferred professional titles. The tests that doctors were expected to pass differed widely in nature. The various bodies used cheaper licences and easier examinations to attract candidates. To help to counteract these problems, The Medical Act of 1858 authorised the establishment of the GMC and the publication of the Medical Register.

The GMC began to take responsibility for medical education 28 years later as a result of provision in the Medical Act of 1886. Following the 1886 Act, applicants for registration had to pass a qualifying examination. These examinations were administered by any licensing body or university authorised to grant medical qualifications. It became the GMC's responsibility to oversee the standard of the institutions and the examinations they offered. The GMC's role was to make sure that the standards were 'sufficient' to guarantee the knowledge and skills needed for efficient practice. If it appeared that any examination was 'insufficient', the General Council was bound to inform the Privy Council. They could, if necessary, order that the examination should no longer be deemed a qualifying examination.

GMC's Responsibilities for Medical Education

The GMC's current responsibilities are an extension of those established in 1886. Its current responsibilities for medical education are set out in the Medical Act 1983, which it fulfils via a statutory Education Committee. The statutory duties of the GMC include the following:

---

[1] Many thanks are due to Joanne Lowe, Registration and Education Directorate, General Medical Council, for providing me with the detail in this section through e-mail correspondence.

1. To determine the extent of the knowledge and skill required for the granting of primary medical degrees in the UK.
2. To ensure that the universities provide medical undergraduates with the teaching and learning opportunities necessary to acquire that knowledge and skill.
3. To determine the standard of proficiency required of the graduating medical student.
4. To ensure that the examining bodies maintain this standard at qualifying examinations/assessments.
5. To determine the patterns of experience that must be undertaken by trainees during the Pre-Registration House Officers (PRHO) year (internship year).
6. To specify the form of the certificate to be completed by universities confirming that the required experience has been gained by trainees during the PRHO year.

According to the Act, 'The Education Committee shall have the general function of promoting high standards of medical education and co-ordinating all stages of medical education.' The Committee has specific responsibilities for undergraduate medical education delivered in the medical schools and for the first year of practice after graduation (the 'PRHO year'). The GMC sets the outcomes that students and PRHOs must achieve and it quality assures the medical schools and the providers of PRHO training to ensure the outcomes are achieved. As such, the Education Committee of the GMC has 'the power to visit universities to make sure that undergraduate teaching is appropriate and to inspect examinations to make sure that the standards expected at qualifying examinations are maintained and improved' (GMC 2005a). The GMC, on the basis of the work of the Education Committee, is statutorily obliged to makes recommendations to the Privy Council about whether a university should be added to or removed from the list of institutions that can award a registerable UK medical degree (Sections 8 and 9 of the Act).

The GMC is responsible for registering medical graduates who are able to work as doctors in the UK. Without registration, a person cannot practice medicine. The education of doctors is a continuous process. It starts formally at medical school where courses normally last 5 years (four for graduate entrants). It continues through initial training as a new doctor with limited responsibilities. (This is the year most doctors spend as PRHO). Next comes initial training as a fully fledged doctor followed by training in the particular specialty chosen by the doctor. Throughout their careers, in addition to, and following on from their formal training, doctors keep themselves up-to-date through continuous professional development.

Control of Medical Education

In essence, the GMC not only accredits but controls medical education in the UK. If a school is not accredited, the qualification is effectively useless in the UK. The GMC (2008) web site states:

The following schools claim that the course of study undertaken substantially or wholly in the UK leads to a Primary Medical Qualification awarded by an overseas University. The GMC does not register graduates who have been awarded primary medical qualifications in such circumstances nor does it give any entitlement to book or sit the PLAB test.

The list below contains institutions that we are currently aware of – it may not be exhaustive. Therefore the GMC accepts no liability for the reliance placed on these institutions or for any action or decision taken.

European College of Medicine, London (ECM)

Grace University School of Medicine, London

London College of Medicine

London School of Medicine

London Medical School

School of Health and Neural Sciences, Nottingham

American International School of Medicine, UK satellite campus

St. Christopher's College of Medicine, Luton

Kigezi International School of Medicine, Cambridge

Medical College London, Montserrat

New Schools

The only way new undergraduate medical programmes can be established is by the establishment of new medical schools. This requires the approval of the Privy Council, in the last resort, which will only be granted on receipt of a petition from the GMC, something it has, until recently, not done for decades.

The Education Committee (which is ultimately responsible to the Privy Council) has power to petition the Privy Council to add to the list of universities entitled to award registerable primary degrees in medicine and surgery in the event that new medical schools are established. The Committee would only proceed in this way if it were satisfied that the arrangements for teaching/learning and assessment met the requirements laid down in its *Recommendations on Undergraduate Medical Education*. It would satisfy itself by means of appointing a team to carry out a formal 'visit' (to assess the teaching) and 'inspection' (of the examinations/ assessments) over a 5-year period and would only reach a final decision when the first cohort of students had completed the 5-year course and the qualifying examinations (GMC 1995).

The GMC announced 'an exciting project to establish four new medical schools in the United Kingdom'. The new schools are Brighton Sussex Medical School, Hull York Medical School, Peninsula Medical School (Exeter and Plymouth), and University of East Anglia Medical School. Taking this forward the GMC are required to make sure that the graduates from these medical schools can demonstrate the requirements set out in *Tomorrow's Doctors*.

To do this we are working with the four medical schools in a process that is similar to, but more intensive than, that proposed for the quality assurance of existing medical schools. Our aim is to recommend to the Privy Council that the graduates from these medical schools are given medical qualifications by 2008 that allow them to be registered (GMC 2007).

GMC's Responsibility for Registration and Licensure

The other way in which the GMC exercises ongoing control over the profession is via registration of doctors. Without registration doctors cannot practice medicine. The system of registration through which the GMC regulates is evolving and attempts to link re-registration more firmly to continuing professional development and performance appraisal and periodic review.

From 1 April 2005 every doctor wishing to practise medicine in the UK needs not only to be registered, but also to hold a licence to practise. When this change was instituted, the GMC noted that 'The privileges currently conferred by law on doctors registered with the GMC will from 1 April 2005 be restricted to those holding a licence to practise .... Doctors will be required to satisfy the GMC, on a regular basis, that they are up to date and fit to practise. They will do this using evidence derived from their medical practice. This process, known as revalidation, will be a condition of a doctor's continued licensure with the GMC.' (GMC 2004b)

This is a central element of the reform process, which the GMC described as 'the most ambitious since the GMC was set up in 1858' (GMC 2004d). The process began with publication of *Good Medical Practice* (GMC 1995), which 'signalled a different approach to medical regulation' (GMC 2004d).

Instead of describing what doctors should not do, it sets out the principles that they should operate by and the standards of medical practice that every patient has a right to expect. This concept of regulation by the medical profession in partnership with the public is at the heart of all the changes now taking place – from the reform of our governance, which led to the establishment of the new Council in 2003, to the reforms of our Fitness to Practice procedures and new arrangements to revalidate the license to practise from 2005. Reform of governance marks a fundamental change in how we approach medical regulation and has implications for everyone using medical services in the UK (GMC 2004d).

The intention is that all licensed doctors will face revalidation after 5 years (the initial revalidation from the start of the new process in 2005 will be staggered; thereafter it will be a quinquennial review). The purpose of revalidation is to ensure that patients can have confidence that their doctors are competent and abide by high ethical standards. The proposals clearly reveal the patient focus of the GMC and of government policy and the degree to which the GMC controls the medical profession. This, as will be shown, lies at the heart of the educational accreditation process.

The revalidation process requires doctors to gather and present evidence drawn from medical practice, throughout the 5-year period to show that they have been practising in accordance with the standards of competence, care and conduct set out in *Good Medical Practice*. The exact nature of this process was put on hold following the Shipman Inquiry's[2] fifth report. The Department of Health issued a

---

[2] Dr Harold Shipman was a mass murderer who killed unknown numbers of patients. Despite a higher-than-average death rate amongst his patients, it took many years before he was apprehended. He subsequently apparently committed suicide while in custody before being brought to trial.

statement, in the wake of the fifth report, which included the decision to review the GMC's proposed new system of revalidation. This has led to a last minute postponement of the intended launch of licensing and revalidation from April 2005. Given little alternative, the GMC have 'warmly welcomed the review'. They noted (GMC 2007):

> The whole purpose of revalidation is to create public confidence that all licensed doctors are up to date and fit to practise, and if there are ways of improving the revalidation model we have proposed, we would of course want to include them in our plans.

Doctors, regardless of specialty or type of practice, must, in effect, maintain a portfolio of their medical practice over the revalidation period (which will be independently validated), reflect regularly on his or her standards of medical practice and satisfy the GMC that there are no significant unresolved local concerns about his or her fitness to practise (GMC 2007, 2004b).

The GMC view emphasises the need to: connect the different parts of the regulatory environment, ensure patients and public are involved, make information about registration accessible and meaningful to patients, doctors and employers, adopt a risk-based approach to regulation and distinguish resolving complaints from decisions about fitness to practice.

Funding

The majority of the GMC income comes from fees paid by registered doctors amounting, in 2007, to £60.4 million of the £63.5 million total income (GMC 2007). Most of the £73 million expended in 2007 arose from assessing fitness for practice (£49.1 million) and registration (£11.3 million). The remainder was spent on communications (£4.4 million), standards (£1.2 million), governance (£3.7 million) and education (£3.9 million, which is 5 percent of total expenditure). There is no state subsidy despite being a body established by law.

Political Pressures and Policy Initiatives

The GMC, as a regulatory body, wields a double-edged sword: as accreditor of education and as licence granters to individual practitioners – the latter are to be subject to quinquennial review and strong sanctions. The development of the quality assurance process in both strands has been conditioned by the political agenda, which has shaped aspects of public policy.

The 'better regulation' and the 'choice' agendas, strongly advocated by the government, are supposedly designed to give patients more say. As a result, the GMC has had to sharpen up a system of control and accreditation that has been, hitherto, rather taken for granted.

On the one hand, there is a need for more medics, recognised by expansion in the system and the need to delegate aspects of that expansion, in part, to the providers. There is an implicit trust agenda that the GMC appears to want to develop. On the other hand, there is the consumer choice perspective, which applauds demystification of the profession, getting the 'customer' more involved in decision making and providing alternatives, while, in the area of medicine, in particular, ensuring client (patient) safety.

This is a situation that has been thrown into stark relief by exceptional circumstances (the Shipman affair, and more recently the case of Sir Roy Meadow[3]) and the scandal of 'dirty' hospitals and 'superbugs' that blight patient care. The background to the GMC reforms has been performance indicators, league tables, 'naming and shaming' and a continuous political battle over who can best provide for the nation's health. During 2003–2004, the GMC was embroiled in two other public inquiries,[4] apart from that of Shipman. The second phase of the Shipman Inquiry, chaired by High Court judge Dame Janet Smith, included an examination of monitoring and disciplinary and regulatory procedures and the handling of complaints. It turned the spotlight on the GMC's past and current procedures and the reforms then being planned. In giving evidence, the GMC explained the thinking behind the reform programme, which has been approved by Government and Parliament, and described its proposals for a gateway that would make it less confusing for members of the public to register concerns or complaints about any aspect of the profession. The intention of the reform programme was to 'inspire public confidence in the profession' (GMC 2004d, p. 6).

The pressure on the GMC to increase its own accountability and that of every practitioner in the profession has made it hard to disentangle the pressures on changes in educational evaluation from those impacting on re-registration and ongoing monitoring of practitioner competence.

During the first decade of the century the new quality assurance procedures have detached from the broader registration issues by being located on a micro-site within the GMC's website. The proposed and initially implemented approach is outlined below. Throughout, the aftermath the Shipman affair has created uncertainty over the proposed changes. Although there are adjustments to the quality process the key aspects of the approach initiated in 2005 are set out below. As will also be outlined in the conclusion, the GMC's approach makes limited use of the QAA subject review (QAA 2000) and, in effect, the latter was a temporary diversion from the main issue. However, it does reflect QAA concerns in emphasising the need for robust internal as well as external quality assurance mechanisms, as reflected in the *Principles of Good Medical Education and Training* (GMC 2004b).

The 2005 quality enhancement process is supposed to maintain stringent control of medicine while giving more responsibility to education providers. The proposed changes, that encourage guided self-reflection, more delegated responsibility but ultimately more external control are intended to walk this unenviable tightrope. The

---

[3] Professor Sir Roy Meadow (June 2005) is accused of professional misconduct because of allegedly giving misleading evidence in the trial of women who were initially convicted of killing their babies and then subsequently cleared. This is a high-profile case in the British media. (See BBC 2005.)

[4] The cases of Clifford Ayling and Richard Neale.

approach, as will be shown below, for medical accreditation is a type of controlling quality audit.

Tomorrow's Doctors

The GMC sets standards to describe the knowledge, skills and attitudes that new doctors should have. The GMC is also required to ensure that these standards are met before registering people as doctors. The latest standards are set out in *Tomorrow's Doctors*, first published by the GMC Education Committee in December 1993, revised in 2002 (with a publication date of 2003). The guidance advocated the development of a curriculum comprising a core component and special study modules.

The main areas covered by *Tomorrow's Doctors* are:

- curricular outcomes: the principles of professional practice, outcomes;
- curricular content, structure: the scientific basis of practice, treatment, clinical and practical skills, communication skills, working environment, medico-legal and ethical issues, disability and rehabilitation, the health of the public, the individual in society;
- delivering the curriculum: supervisory structures, teaching and learning, learning resources and facilities, student selection, student support, guidance and feedback;
- assessing student performance and competence, principles of assessment, assessment procedures, appraisal, student progress;
- student health and conduct: confidentiality for medical students, the responsibility of medical students to protect patients, the responsibility of other doctors to protect patients, the responsibility of universities to protect patients;
- putting the recommendations into practice: what the law says about undergraduate education, UK law, European Union law.

However, despite prescriptive content, the GMC does not prescribe educational approaches to pedagogy. Institutions can adopt, for example, a problem-based approach or a more traditional didactic approach to delivery, or anything in between, subject to quality controls. The GMC considers the diversity of approaches to delivering undergraduate medical education in the UK to be one of the reasons why it is held in such high regard abroad. Having said that, there appears to be a growing pressure towards more student-centred pedagogy.

Quality Assurance Process

Traditionally, the GMC adopted an inspectorial approach to checking standards. However, the GMC is moving away from inspectorial approaches to accrediting courses to quality assurance approaches with more emphasis on a process of continual engagement and continual improvement. The quality assurance process was reformed following two rounds of informal visits (the first 1995–1998 and the second, 1998–2001) to all established medical schools. The new process was trialled in three volunteer medical schools (Aberdeen, Birmingham and Liverpool) in 2003 and 2004.

In the past, statutory duties have been met by carrying out a range of different activities, including:

1. Inspection of qualifying examinations held by UK universities with medical schools. Last inspected between 1982 and 1994.
2. Informal visits to UK universities with medical schools. The last round, designed to consider the implementation of the GMC's recommendations on undergraduate medical education (*Tomorrow's Doctors*) and on the PRHO year (*The New Doctor*) took place between October 1998 and April 2001.
3. Written monitoring. This took two forms: first, summaries of how the universities have addressed the recommendations in GMC reports of the informal visits. Second, information requested annually from universities about the primary medical qualifications they award.

The new quality assurance process will ask each medical school how they are meeting the standards set out in *Tomorrow's Doctors*. Furthermore, medical education is also subject to evaluation by the Quality Assurance Agency for Higher Education (QAA), discussed below.

Previous Approach

The GMC undertook a series of visits between March 1995 and March 1998 to then existing 25 medical schools in the UK to monitor their response to the recommendations in *Tomorrow's Doctors*, which, inter alia, had identified an initial implementation period of 5 years from the date of publication. However, due to the length of the undergraduate course, there was a flexible approach to those schools that had not begun the process of curricular reform prior to the publication of the new guidance. Nonetheless, every medical school had to show the GMC evidence of real progress towards meeting the goals and objectives laid down by the Committee. The GMC published summary reports and an overview of the visits to show progress towards the recommendations in *Tomorrow's Doctors*, to identify obstacles to change and examples of good practice (GMC 1999).

The visits, in the late 1990s, were conducted by teams with an appropriate range of medical expertise and knowledge. All teams comprised a leader and two or three visitors, who were usually either members of the Education Committee or members of Council. To facilitate consistency between the visiting teams, only two leaders were appointed and they undertook, between them, to lead all 25 visits, collaborating on two of them to ensure similarity of approach.

Two-to-three months before a visit, schools were asked to complete a questionnaire and return it with supporting documentation, which formed the basis of the 2-day, on-site visit. The first day was devoted to the undergraduate curriculum and the second to the final year of basic medical education (the pre-registration house officer year).

At each school, the visitors had meetings with key staff involved in teaching and in implementing the undergraduate curriculum, and met students drawn from each stage of the course. Following each visit, a report was prepared for the GMC Education Committee, setting out the findings of the visiting team, including areas of good practice and suggestions for change. The reports were treated as confidential to the institution visited, although the GMC requested permission to share with other institutions information about the good practice identified by the visitors. A version of these reports is now available on the GMC website.

Approximately 1 year after a visit, the GMC wrote to the schools to ask how they had addressed the recommendations made in the visiting team's report. Their responses were considered in detail by the Sub-Committee on Assessment and Monitoring (SCAM) and then reported to the Education Committee. The information so obtained was taken into account when considering the sequence of the second-round of visits to medical schools, which began in autumn 1998.

The GMC requested written submissions about progress from schools that would not be visited until later in the cycle. This was done to ensure that the GMC were properly informed about the work being undertaken in these institutions. These submissions were reviewed in detail by SCAM and provided useful background information for the more extensive informal visits themselves. Where the GMC felt uncertain about aspects of the information provided, they organised short, on-site visits to obtain clarification.

In October 1998, the GMC began a series of visits to universities with medical schools, and associated postgraduate deaneries, to monitor the implementation of *The New Doctor* (published 1997), which set out expectations for the training of PRHO as well as identifying the components of a high-quality PRHO post. These visits also monitored further progress on undergraduate medical education, *Tomorrow's Doctors*. This involved 23 visits, which were completed in April 2001.[5]

At around the same time, the Quality Assurance Agency for Higher Education (QAA) was undertaking its subject reviews of medicine. The GMC and the QAA have different agendas. The GMC agenda is about control, accountability and accreditation. The QAA agenda is about reviewing the quality of provision at a subject level, supposedly on the basis of fitness for purpose. It explored six aspects of provision: curriculum design, content and organisation; teaching, learning and assessment; student progression and achievement; student support and guidance;

[5] Aberdeen, 2000; Belfast, 1998; Birmingham, 1999; Bristol, 2000; Cambridge, 2001; Cardiff, 1999; Dundee, 1999; Edinburgh, 2000; Glasgow, 1999; Leeds, 1997, 1998; Leicester, 1999; Liverpool, 1999; London: Barts & Royal London, 2000; Guys, Kings & St Thomas, 2000; Imperial, 2000; St George's, 2000; Royal Free & University College, 2001; Manchester, 2000; Newcastle upon Tyne, 1998; Nottingham, 2000; Oxford, 2001; Sheffield, 1998; Southampton, 1999.

learning resources; quality management and enhancement. In essence these dimensions are tested by exploring the extent to which the student learning experience and student achievement, within each aspect of provision, contribute to meeting the objectives set by the subject provider (QAA 2000).

Reviews of the quality of the educational provision in medicine were carried out by teams of subject specialists.[6] QAA conducted eleven reviews in conjunction with a visit by the GMC. Visits to students working in placements in hospitals, community health-care providers and general practice were also carried out. Student groups ranged in size from four students on a part-time postgraduate programme to over 1,000 full-time students at the undergraduate level. Some part-time postgraduate programmes also recruit large numbers of students. The reviewers approved all of the undergraduate provision that they evaluated. The GMC's sphere of interest is the medical schools. This does not entirely overlap with the QAA who have undertaken reviews of these schools plus the University of Derby's pharmacy programmes. While co-operating in these QAA reviews, it seems that the GMC made little use of them in their own processes.

Although these evaluations were often undertaken in partnership (or at least simultaneously) the reporting was entirely separate with minimum of cross-referencing. For example, in the QAA Report on the University of Leeds (1998), the GMC is mentioned just three times: first to say that 'the review visit was undertaken at the same time as the GMC curriculum monitoring visit' (QAA 1998, para 1), although in no way implying they were working together. Second, that 'The School has responded to the recommendations of the GMC set out in *Tomorrow's Doctors* with a measured, evolutionary approach that is designed to ensure that the undergraduate curriculum will take full account of these by 1999' (QAA 1998, para 9) and, third 'The School states that its aim is to deliver the current undergraduate course in the spirit of the GMC's recommendations, but the reviewers consider that this aim is not yet met.' (QAA 1998, para 10), which is not surprising in view of the timetable set out in 1998.

Similarly, the GMC report on the visit had little to say about the QAA:

---

[6] The QAA visits were as follows (as reported on the QAA website (document number)): University of Newcastle upon Tyne, October 1998 (Q6/99); The Queen's University of Belfast, October 1998 (Q69/99); University of Leeds, November 1998 (Q56/99); University of Sheffield, November 1998 (Q60/99); University of Derby, November 1998 (Q85/99); University of Birmingham, January 1999 (Q98/99); Royal Free and University College Medical School, February 1999 (Q192/99); University College London, February 1999 (Q187/99); University of Liverpool, March 1999 (Q149/99); University of Leicester, May 1999 (Q188/99); University College London, Institute of Child Health, October 1999 (Q43/2000); University of Bristol, November 1999 (Q79/2000); King's College – Institute of Psychiatry, November 1999 (Q124/2000); University of Southampton, December 1999 (Q104/2000); St George's Hospital Medical School, January 2000 (Q170/2000); London School of Hygiene and Tropical Medicine, January 2000 (Q198/2000); Imperial College of Science, Technology and Medicine, February 2000 (Q199/2000); Queen Mary, University of London, February 2000 (Q211/2000); University of Manchester, March 2000 (Q319/2000); King's College London, March 2000 (Q290/2000); University of Cambridge, May 2000 (Q310/2000); University of Oxford, May 2000 (Q328/2000); University of Nottingham, March 2002 (Q588/2001).

> The visit lasted 2 days. The first day was concerned with the undergraduate curriculum, and involved us in collaborative working with a Quality Assurance Agency (QAA) team that was simultaneously conducting a review of medicine at the University. The second day focused on the provision made for general clinical training. (GMC 1998, para 3)

A second mention was linked to the sharing of documentation required of the institution:

> Prior to the visit the School provided us with background material including the Self Assessment Document prepared for QAA visitors. Members of the QAA team were supplied with copies of the completed GMC questionnaire that the School had produced for our visit. (GMC 1998, para 6)

The other two mentions indicated that, on the first day, members of the QAA team joined the GMC visitors (not the other way round!) (GMC 1998, paras 9 and 141).

Neither report referred to the other and it seems that despite degrees of overlap these processes seemed to have little synergy. The GMC review explored the PRHO year as well as undergraduate training, the former being beyond the scope of the QAA visit. On the undergraduate front, QAA approved the quality of education in medicine at Leeds although two areas were graded only 2 (out of 4) and a total rating of 18 out of 24. The GMC were also less than impressed, having visited a year earlier and expressed concerns about the speed of evolution of the curriculum. It noted

> 136. There are a number of areas of good practice in the current curriculum. The use of clinical skills centres and bed-side teachers to assist students to develop their clinical skills is praiseworthy. We were also impressed by the student body, a view clearly shared by the NHS managers we met on the second day of our visit.
>
> 137. However, little has changed since our last visit. The current curriculum exhibits a paucity of vertical and horizontal integration, and the pre-clinical and clinical phases are still clearly discernible. While plans for revising the curriculum are being developed little has been achieved, and a major effort is required to ensure that the changes proposed will be implemented in line with the intended timetable.
>
> 138. The School needs to consider whether its supervisory structures are appropriate for implementing change. Staff are clearly making great efforts to develop plans for the new curriculum, and it is vital that effective mechanisms are in place to secure the desired outcome. We look forward to hearing how implementation of the new curriculum is progressing in a year's time.

The GMC and the QAA reports noted similar areas for improvement, although the former was more detailed and somewhat more directive. Given that the QAA subject reviews have ended (and there was never any certainty about a second round or any follow-up that involved sanctions) and that the GMC is a powerful regulatory body that is not going away, it seems certain that the University of Leeds will have focused on the specifics of the GMC visit outcomes rather than that of the QAA.

However, there is a general concern on the part of the GMC to ensure that appropriate quality assurance processes are in place and that medical schools are equipped to deal with external monitoring of any kind. For example, the GMC report on the

University of Birmingham (GMC 2004e) noted, albeit tucked away under ‘other issues’ in para 69, that:

> There were a number of issues emerging from the 1999 GMC and Quality Assurance Agency visits, including assessment, integration and identification of core, which had not been fully addressed yet by the School. Visitors were not sure whether the mechanisms were in place to respond sufficiently rapidly to legitimate outside criticism.

Indeed, the *Principles of Good Medical Education and Training* (GMC 2004b, paras 26–33) note that

> There must be rigorous and evidence based quality assurance (QA), both internal (IQA) and external (EQA), to ensure that standards are being maintained, curricula are being continually reviewed and good practice is being shared.

It adds that QA processes should be able to show that they add value. Repeating earlier advice (GMC 2002), it states that QA processes should be efficient, valid, reliable, convenient, fair and focused, with a ‘clear statement of QA responsibility for the different aspects of each programme’. QA processes should ensure that the students and doctors provide information and opinion on their education, training, supervision and clinical experience. The processes should be transparent, flexible, reflective and evolve ‘in response to diversity and innovation or constructive criticism’. Furthermore,

> The EQA should confirm the evaluation of processes and outcomes of the IQA, and build upon them. QA should support the creation of common data sets and a resource for innovative practice and the sharing of information.

A New Approach

The GMC has been developing its new approach for half a decade, taking into account the visits that began in 1998. However, the approach to evaluation and accreditation of undergraduate and PRHO education is tied up closely with changes to overall regulation of the profession and the attempt to introduce a more ostensive lifelong learning, continuous professional development element.

In making its reforms to the quality assurance of education, the GMC intended a shift to a more continuous engagement that emphasises dialogue rather than checking and that empowered institutions encouraging more reflection and engagement with quality as opposed to accreditation issues. Furthermore, the new quality assurance processes are also expected to engage with the government’s widening participation agenda. However, the process has been embroiled in the wider politics that have impacted on the GMC regulatory process as a result of high-profile cases that have, with the aid of a good deal of negative media reporting, led to a public ‘crisis of confidence’ in the medical profession. This at a time when the GMC is going through a lengthy process of structural reform GMC 2000a, b, 2002.

The new approach was proposed in 2002, piloted over the next 18 months and encoded on the new GMC *Quality Assurance of Basic Medical Education Extranet* (GMC undated). The aims of the process are to

1. Make sure that the outcomes in *Tomorrow's Doctors* are met.
2. Identify examples of innovation and good practice.
3. Identify, discuss and resolve issues of concern.
4. Identify changes that need to be made and a timetable for their introduction.
5. Promote equality and diversity in medical education. (GMC 2004a)

The objectives of the process are to

1. Monitor changes to curricula, assessments and staffing through information received annually from each school.
2. Make sure that medical schools tell the GMC about any new courses they are developing and seek formal approval for these.
3. Allow issues of common concern in undergraduate medical education to be identified, discussed and resolved, thereby contributing to the ongoing review of *Tomorrow's Doctors*.
4. Produce evidence-based visit reports on whether schools meet the requirements in *Tomorrow's Doctors*.
5. Identify examples of good practice for widening participation in medical education.
6. Provide evidence that will allow the Education Committee to make a recommendation to the Privy Council whether a university or institution should be added to or removed from Section 4 of the Medical Act 1983 that allows them to award a primary UK medical qualifications. (GMC 2004a)

The pilots, during 2004, to the volunteer sites of Aberdeen, Birmingham and Liverpool were somewhat curious in that they surprised the volunteer schools by their focus, without clearly identifying, at least in a public document, what they revealed about the proposed new process.

Professor William Doe, Dean at Birmingham Medical School in the published letter of response to the Final Report of QABME Visits to Birmingham Medical School for 2003–2004, stated:

> The School has found the QABME visits to be a worthwhile and beneficial process and has given us the opportunity to critically assess our existing provision. We were, however, a little surprised at the extent to which our own educational processes were reviewed, given that, when we volunteered to be a pilot site, we believed that it's purpose was, primarily, to assist you in optimising your new processes. (GMC 2004e)

Reading between the lines one might infer a similar reaction from Professor Mike Greaves, Head of School of Medicine, at Aberdeen Medical School:

> The School was, of course, aware of the likely recommendations and areas that the Education Committee would ask to be considered further from your earlier drafts and also

> from discussion with the visiting panel in June. We have, therefore, already been considering the issues raised.... We have enjoyed taking part in your pilot process and have found the process both challenging and refreshing. The visits and the production of supporting documentation has encouraged us to re-evaluate our curriculum and already accelerated progress in particular areas. Above all, the input from the visiting team was always both stimulating and enjoyable. (GMC 2004f)

Professor Anne Garden, head of school at Liverpool Medical School, provided somewhat more feedback on the process itself in her response of December 2004. She stated:

> I thought that the QABME process worked well – certainly the visitors were very thorough in their duties – although we did feel somewhat over-visited! However, at all times they were very professional and the process was carried out in a robust yet supportive way as I think the report bears out. I certainly have no regrets that we volunteered to be a pilot site. (GMC g2004)

She did have two major concerns, though:

> 1. There should not be mention made in the report of things that have not been discussed with the School. The obvious example is the external examiners reports – what we do with them and what changes have been made as a result of them. We could have easily provided evidence about that – but were never asked.
>
> 2. Evidence should be available for comments made. Again the examples would be the comments about the variability of the clinical sites and the feedback to students. I have tried hard since the visits to find out the basis of these comments to no avail. This is unhelpful – it may be these things are true and no-one is brave enough to tell me to my face – in which case I will not be able to put it right and it will catch us again next time – or it may be that this was a single unhappy student or at worst a small group who are unhappy but not representative of the whole – in which case it probably should not have been in the report. When I did QAA (not that I am saying everything we did in those visits was exemplary) we were not permitted to put anything in the report that could not be 'triangulated' – we should apply the same rigour to our comments. (GMC 2004g)

The reference to the QAA subject reviews, in which Professor Garden was an assessor, is a rare case in the medical evaluation literature suggesting that QAA provides an example of good practice.

It is not at all clear how the pilots informed the QABME process nor indeed what the relation is between the QABME process and the *Principles of Good Medical Education and Training: Draft for Consultation,* issued in August 2004, which does not directly mention QABME, although it does refer to quality assurance, both internal and external. Furthermore, the report, *Quality Assuring Undergraduate Medical Education – An Overview,* issued in February 2005, neither mentions the pilots nor responds to, for example, the concerns about over-visiting.

In the new approach, the GMC will arrange a series of visits, over the period of a review year, to the medical school to confirm how they are meeting the standards set out in *Tomorrow's Doctors.* The key objectives are to ensure that the curricular outcomes (attitudes, behaviour, knowledge and skills) are achieved and demonstrated by new graduates and that for the PRHO year, to ensure that systems are in place that allow the GMC to be confident that only those doctors who are fit to receive full registration do so.

The GMC publish their views of the school's areas of innovation and good practice, where the school may wish to consider further developing its work as well as any required action.

To some extent, the new process represents a shift from inspection to audit of internal processes, albeit a tightly controlled audit that can invoke sanctions in the last resort and which, unlike most quality audits, also passes judgement on the adequacy of the standard of medical education and training.

Following the on-site visit, the GMC will ask the school to update information each year and the school will be visited again at least twice in any 10-year period. It is intended that the quality assurance process will be a continuous exercise. The process will involve:

1. Annual requests for written information from universities describing any significant changes from their last return.
2. Regularly (at least twice in every 10 years) the GMC's Education Committee will visit every Medical School. It was proposed that on the designated visit year, there would be a series of site visits to universities focussing on issues identified in the annual returns, this would culminate in a 1-day synoptic visit to universities involving all team members.
3. Regular reports to the Undergraduate Board on the information collated. (GMC 2002; 2007)

Outwith this process, if a school makes significant changes to its curriculum, then special arrangements come into play. Furthermore, the new medical schools face a similar but more intense régime of quality assurance. The visiting cycle is completed every year for the first cohort of students. This will result in annual reports that will allow the Education Committee to gauge the progress of each school, and compare progress across the new schools. The annual report for the final year of the first cohort of students will be the final report that is presented to the Education Committee and sent to the Privy Council with the Committee's recommendation about the awarding status of the medical school concerned.

Information

All universities are asked to provide baseline information before the first cycle of visits starts, using a standardised template. This information, which is updated each year, will be assessed by the office working in conjunction with the Undergraduate Board, a sub-committee of the GMC's Education Committee. The proposed information for *undergraduate education* is the following (GMC 2002):

- A description of how their curriculum meets the requirements in *Tomorrow's Doctors*.
- A description of their assessment system and their internal QA processes.

- External examiners' reports covering the last 3 years prior to a visit, including a list of the issues identified by external examiners during that period and the action taken by the university in response to these issues.
- A report from the student body at the university giving its views on the curriculum and the assessment and QA systems.
- A report from NHS partners commenting on the quality of the university's recent graduates in terms of their attitudes, behaviour, knowledge and skills.
- A copy of their QAA institutional review report.
  This was modified in the 2004 guidance document. Each year all medical schools will be asked to:
- Provide information about how their curricula and assessments meet the requirements in *Tomorrow's Doctors*.
- Identify any significant changes to their curricula, assessments or staffing levels.
- Highlight issues of concern, corrective action taken and proposed solutions.
- Identify examples of innovation and good practice.
- Respond to issues of current interest and debate in medical education including the promoting equality and valuing diversity (GMC 2007).

A standardised QAMBE questionnaire will be used to collect this information.

The changes, from the proposed information requirements (GMC 2002) to the guidance document requirements (GMC 2004c), are significant. The focus in the latter is on conformance to curriculum and assessment, changes in practices as well as highlighting good practice and responsiveness to debates and policy. This contrasts with an initial proposal that not only wanted curriculum conformance but emphasised system procedures and external commentary.

The initial proposal for information relating to the *PRHO year* was that it would include:

- A description of the school's strategy for delivering high quality general clinical training that also explains how the educational and training objectives in *The New Doctor* are being met.
- Information about procedures for quality assuring PRHO posts.
- Information about general difficulties arising in relation to the provision of PRHO posts across the region, and the action taken in response to these issues in the year of the visit.
- Information about specific posts that have caused problems and the action taken to resolve these difficulties.
- Details of how the school liaises with their NHS partners in agreeing educational objectives and service targets.
- A written report from the university's current PRHOs giving their views on the quality of education and training provided.
- Feedback from the university's recent past PRHOs (possibly in the form of a summary report of the exit questionnaires PRHOs complete) giving their views on the quality of education and training they received when PRHOs.

- Copies of any evaluation surveys about their PRHO training that have recently been undertaken. (GMC 2002)

Although this is extensive, if there is no change, on an annual basis, to any of the information requested for basic and PRHO education, then the university would simply provide notification to that effect.

Site Visits

It was proposed to have a series of site visits to universities, focussing on issues identified in the annual information returns. Depending on the range of issues to be covered, there might be up to three or four 2-day site visits undertaken over the course of the year selected for the visit. It is proposed that institutions are visited every 5 years 'unless innovative developments or concerns about provision required an earlier visit' (GMC 2002).

According to one part of the GMC website, these visits will be carried out over an academic year by pairs of visitors looking at particular areas. Another part of the sites states 'Visits will be rigorous and reliable and will be carried out by a small group of trained visitors who will be recruited against competencies'. The visits would involve:

- meetings with university and deanery staff;
- observation of teaching and assessments;
- sampling of student assessment exercises (including, written scripts, portfolios and logbooks);
- meetings with students and PRHOs.
- observation of university and deanery procedures for approving PRHO posts.
- evaluation of university and deanery systems for ensuring that only those doctors who are fit to receive full registration do so. (GMC 2002)

In addition to specific-focus visits there will be a synoptic visit. This will be a 1-day visit to universities involving all team members. This will be undertaken at the end of the academic year in which the university is being visited and will provide the opportunity to draw together and review all the issues considered during that year and for clarification of any outstanding issues prior to publication of the final report. The timetable for visits will be constructed for a 10-year period. Changes to the norm of two visit years per decade will be dictated by pre-set criteria.[7]

[7] These criteria are (1) failure by the university to respond to GMC recommendations within an agreed timeframe, (2) issues arising from assessment of information received via the annual returns, (3) issues arising from other reports that the GMC receive about the university, (4) The development of new and innovative educational and training systems, and (5) The length of time since the last visit.

The 2004 guidance has a different slant on this. The shift was in response to the choice and delegated-responsibility agendas, although as will be shown, the re-emergence of accountability has stalled progress. In the 2004 guidance, site visits are less inspectorial and are part of the information gathering process which has three stages (GMC 2005a, pp. 5–6). Stage 1, collecting information (June to December); stage 2, confirming information (January to July); stage 3, integrating information and making judgements (June to August).

As of 2004, selected schools were contacted in June. GMC administrative staff undertook a preliminary visit to explain about the QABME process. In September, the schools received the QABME Questionnaire requesting information in a standard format, to be completed and returned by 1 October. This information is shared with the GMC Visitors teams who formulate action plans, which will include a series of visits to take place between January to July the following year.

This process will allow visiting teams to collect information, explore issues, and observe parts of the teaching and learning process in a systematic and explicit way. Teams will be provided with practical guidance to help them to collect, confirm and evaluate information so that the process is based on the requirements set out in *Tomorrow's Doctors* and managed consistently across all schools. (GMC 2005a, p. 6)

Visiting teams undertake all three stages and produce a final report on each school that will be submitted to the Education Committee.

Continual Monitoring of Education Provision and Ongoing Practice

The GMC's mantra in all this development is continual improvement, both for education providers, practicing doctors and their own quality processes. They intend to build into the arrangements, systems that will allow them continually to improve the quality assurance process. Clear communication with medical schools is important in making sure that the process works and is amended. In particular, the GMC needs to ensure that medical schools understand the process, provide the information required and give feedback about the process and about individual members of the visiting teams. Schools are, therefore, requested to keep a log of issues about the process, so that concerns and difficulties can be identified, captured and addressed in a managed and consistent way. To this end, the GMC introduced a QAMBE *Monthly Update* available online.

Each medical school, regardless of whether it is being visited in the current cycle or not, will have a named GMC officer as a contact point with the GMC. This individual will work with a named contact at the school to ensure the smooth operation of the process and to facilitate the free flow of information (GMC 2005a, p. 7).

It will be instructive to see how the proposed continuous evaluation works out in practice. In other spheres, notably QAA evaluation of higher education provision in England, the idea of a continuous dialogic audit process has resulted in rather sharp reaction from the potential auditees, on the grounds of unwarranted intrusion and bureaucratic burden. There are already hints of this, in the response to the pilot from Liverpool, mentioned above, about being over-visited.

The key difference in the medical setting may be that the proposed new approach is seen as a lessening of control and that, in any event, the medical schools have little choice but to comply given the stringent powers, backed by legislation, of the GMC. Whether, though, the medical schools adapt to a continuous dialogue on quality and standards, or simply comply with a new inspectorial *régime*, remains to be seen.

Although the GMC comes in for criticism, this tends to be directed at the way it controls doctor re-registration rather than any criticism from the academic community about the accreditation process. It is hard to locate any critiques of either the old system of accreditation of medical education or the proposed new continuous engagement approach. The latter may be because people have been biding their time, waiting to see the outcomes of the first round of the new QABME process.

However, the GMC's own survey of the QABME process suggests a considerable degree of support. Throughout the introduction of the new quality process, the GMC have 'sought feedback from GMC staff, Visitors, Medical Schools, visit observers and other independent organisations such as the Better Regulation Review Group (BRRG)' (GMC 2004a). The survey of Medical Schools was conducted over the telephone and, for Visitors, by e-mail. The hope was that 'this method of approach will encourage communication about the effectiveness of the pilot programme'. The GMC (2004h, p. 2) report notes that survey replies are 'predominantly positive' and 'constructive suggestions or areas of concern have been taken forward as part of the continuous improvement programme'. There were 'no surprise issues arising from the survey'.

Most of the questions were about the specifics of the process, such as the helpfulness and timeliness of information received, the relevance, conduct and feedback of the visits. All of this was reported as satisfactory. One question asked 'Do you think the GMC's quality assurance process is effective at meeting the GMC's statutory responsibilities as set out in *Tomorrow's Doctors*?': 30 percent were neutral while the remaining 70 percent thought it effective.

Overall, the survey results were positive about the process and its ability to meet its basic aim, the monitoring of the curriculum. There was slightly more concern, though, about the reporting, highlighting good practice and new initiatives. To some extent this reflected concerns raised in the responses to the pilots. In all, 30 percent of respondents did not think the report on the process was as fair as it could have been and 60 percent thought that not all their areas of good practice were highlighted through the process. Only 60 percent though that the process was effective 'as a way of encouraging schools to address contemporary concerns including equality and diversity, patient-centeredness, learner-centeredness, interprofessional learning and preparing students for the permanence of change in medicine and social expectations'.

The relative failure to identify good practice and the limited impact on contemporary concerns must raise questions about the nature of the exercise. This is exacerbated by the clear concerns by respondents about the perceived cost of QABME.

The high-profile concerns and the budget expenditure reflect the huge emphasis that is placed on the regulation of the profession, of which the accreditation of initial education is a relatively small part. However, the ongoing monitoring of practitioners itself, it is intended, will henceforward involve a more directive continual learning element, which the GMC will oversee. The intention, subject to further developments and confirmation, is essentially to encourage a self-reflective process on the part of practicing doctors, as far as possible linked into an authorised appraisal process. This would, if carried out as intended, mean that doctors, as noted above, need to maintain a 'folder' outlining their activities and engagement with medical practice, including, in theory, direct feedback from their colleagues and patients via questionnaires.

Conclusion

The GMC is a very powerful and highly scrutinised regulatory body. The recent high-profile case of Professor Sir Roy Meadow is indicative of the pressures on the organisation. Not only does the GMC regulate at all levels of practice, but are also in the process of reforming the ways in which they scrutinise ongoing practice and link that to continual professional development alongside changes in the accreditation and quality assurance of initial medical education. The three elements are interlinked and it is difficult to understand the complexity of the arrangements for medical education independently of the GMC's overall regulatory role. The intention to give medical schools more responsibility for their own quality is to some extent confounded by the need to ensure accountability.

Indeed, when Professor Graeme Catto was elected as the GMC's President in 2001, the primary task was to communicate more widely the changes at the GMC. Professor Catto is reported to have said, 'We have to get the public to understand what it is we're doing,' which would improve the public's perception of the GMC. Further, he said the medical profession itself needed better educating and training, and that improvements in that area were preferable to the GMC policing the medical profession. 'Policing is the least best option. Of course it's essential, but it's an admission of failure to go down that route' (BBC 2001).

Within this context, it is hard to draw comparisons with other forms of evaluation and quality assurance. The GMC are doing more than accrediting in their attempt to quality assure, accredit and control medical education as part of their remit to control the profession while themselves being a powerful regulator but subject to intense public and political scrutiny. The GMC is highly susceptible to government policy and political pressures, is continuously scrutinised and has to be accountable for itself and for the medical profession.

The GMC not only accredits medical education but also regulates the profession. This means that the accreditation of medical education operates within a legislative regulatory framework beyond that of most other quality assurance and accreditation processes. In the United States, for example, the Accreditation Council for Graduate Medical Education (ACGME 2005) is responsible for the Accreditation of post-MD medical training programmes within the United States. However, it does not regulate the profession. Similarly, the Liaison Committee on Medical Education (LCME) 'is the nationally recognized accrediting authority for medical education programs leading to the M.D. degree in US and Canadian medical schools' (LCME 2005). Although sponsored by the Association of American Medical Colleges and the American Medical Association, the LCME has no regulatory function for the profession.

The new QAMBE approaches are not overly concerned with defining quality or even quality assurance. The key role of the process is not to judge academic excellence per se but to ensure appropriate levels of competence of graduating doctors. To that end it requires compliance on several fronts; notably curriculum content, supervision, practical experience and vigilance in ensuring only appropriate students reach the stage where they may achieve registration. This is not so much about fitness for purpose, as the institutions and the programmes of study are not in a position to determine their own mission-related purpose. It is more about quality as transformation – transformation to competent reflective practitioners – and about excellence, ensuring that medical education in the UK maintains its excellent status around the world.

Despite all the problems and scrutiny faced by the GMC through its period of reform, the only expressed concerns about the new quality processes comes from policy quarters, not from the institutions themselves. In the main, the institutions, used to the idea of being subject to the inspection of a statutory regulatory body, seem to be complying, and indeed embracing, the new methodology. Expressed concerns, as noted above, have so far been minimal and relate to too may visits during the audit year and the reservations about the methodology adequately highlighting and sharing good practice and of enabling and encouraging diversity.

The attempt to shift the balance of the new approach from constraint to empowerment – from an inspectoral and controlling methodology to one that encourages more dialogue and delegated responsibility for improvement, external circumstances continue to conspire to emphasise accountability, of both education providers and the regulator itself. This creates uncertainty and has a potentially negative effect on innovation.

In Chapter 10 Bjorn Stensaker examines the evaluations in Danish higher education (undertaken by EVA) and he explores the policy drivers. These are predominantly concerns of politicians about the accountability and standing of Danish education given the increasing institutional autonomy, the balance between supply of and demand for study places, and international comparability of programme standards. These are not unusual concerns of national agencies within Europe evaluating programmes across the sector. Indeed, as noted above, the QAA subject reviews in the UK had similar concerns (institutional autonomy apart) about quality, focusing on six dimensions that purportedly evaluated fitness for purpose.

Impact studies of quality in the general higher education setting are notoriously difficult to do, other than at the level of identifying documentary compliance and measuring the proportion of recommendations implemented. Stensaker's study of the Danish evaluations showed, for example, that 60 percent of recommendations had been implemented. In the UK medical education setting, impact is a rather more deferred concept. Impact as compliance and implementing recommendations is, on the face of it, non-problematic as the institutions are observed directly until such times as progress towards implementation occurs. Sometimes recommendations, particularly relating to the implementation of new curricula can take some time given the length of undergraduate courses (5 years); however, there is no question that such implementation will not take place. The real impact factor that the GMC has to address is further down the line: the competence and (absence of) malpractice of registered doctors.

The other factor that affects evaluations is 'game playing' and familiarity with procedures that potentially render later cycles of evaluations less effective. It seems, in the case of UK medical education, apart from the change in processes which has led to more continuous engagement, there is little incentive for game playing or attempting to short-circuit processes as it is not in the medical school's best interests to be producing incompetent graduates. In short, the schools work closely with the GMC, despite being effectively controlled by it.

Whereas quality assurance of programmes in most higher education settings do not judge academic standards per se and most accreditation ensures minimum standards, the GMC approach is to ensure compliance across the board with the high expectations of provision set out in *Tomorrow's Doctors*, which includes academic standards and standards of competence as well as quality of the learning environment. The combination of continuous audit with evaluation of standards is thus relatively unusual.

The aim of the changes to medical education accreditation and quality assurance in the UK is to focus much more on continual improvement, reflection and encouraging real ownership of the quality improvement process. However, in the current political climate, with the regulator itself under intense scrutiny, anything but compliance might be seen as a risky strategy. This reflects the contradiction in the President's statement (BBC 2001) that policing the profession is both undesirable and essential. As enquiry and consultation pile on top of one another, the tenor of the times suggests that essentialism will overwhelm trust and dialogue. However, if medical education is to progress, to fully adopt a student-centred approach, new forms of pedagogy and continue to extend the portfolio of assessment techniques to better demonstrate competence of new doctors, which in itself will help to further protect the public, then the new approach will need to have an opportunity to flourish. The GMC prides itself on not prescribing delivery techniques and encourages student selection within the curriculum. If, in the aftermath of Shipman, the consultation supports a national examination, and effectively much closer control of the educational process, there may well be a retreat from innovation, the adoption of a compliance culture and the likelihood of the emergence of a controlling spiral similar to that in teacher education.

The emergence of the Teacher Training Agency (TTA), in the UK, as a regulator of teacher training followed high-profile concerns, mostly engendered through the reactionary press, that teacher education was encouraging innovation that 'endangered the education of our kids'. The TTA was developed to control teacher education and effectively bring in a national curriculum for teacher training, in line with the introduction of the national curriculum in schools. The approach adopted by the TTA became increasingly prescriptive. There was no question but that it required compliance, resulting in a lack of innovation and more importantly a lack of imagination, drive and innovation.

The Office of Standards in Education (OFSTED) undertakes periodic inspections of training provision alongside its ongoing inspection of the UK's schools.

> We do have to guarantee that students obtain the required standards, both in practical teaching and in academic assignments. To do this we have to provide our own assessment to external inspectors (OFSTED) and they then crawl all over us to demonstrate whether we are 'Good, with outstanding features' (Grade 1), 'Good' (Grade 2), 'Compliant, but needs substantial improvement' (Grade 3), 'Non-compliant with the Secretary of State's standards' (Grade 4). (Hoskyns 2000)

Indeed, teacher training in universities is about as far as one can get from the traditional model of academic freedom and autonomy (Harvey 2001).

The new proposals for the quality assurance of basic medical education offer a way of combining accreditation with quality improvement, which is relatively unique. There is a danger, though, that public accountability will be re-conceptualised as control rather than improvement and the new approach will founder on the rocks of national examinations and tighter prescription of curriculum content, leading to uninspired and conformist teaching and learning.

References

Accreditation Council for Graduate Medical Education (ACGME) (2005). http://www.acgme.org/acWebsite/home/home.asp. Accessed 12 June 2009.

British Broadcasting Corporation (BBC) (2001). New man in GMC hotseat. Tuesday, 6 November 2001, 13:16 GMT. http://news.bbc.co.uk/1/hi/health/1639097.stm. Accessed 12 June 2009.

British Broadcasting Corporation (BBC) (2005). Paediatrician's 'flawed evidence'. Tuesday, 21 June 2005, 16:40 GMT 17:40. http://news.bbc.co.uk/2/hi/health/4114688.stm. Accessed 12 June 2009.

Department of Health (DoH) (2007). Trust, assurance and safety: The regulation of health professionals. Cm 7013. Presented to Parliament by the Secretary of State for Health, 21 February, 2007. Norwich: HMSO. http://www.dh.gov.uk/en/Publicationsandstatistics/Publications/PublicationsPolicyAndGuidance/DH_065946. Accessed 8 March 2007.

General Medical Council (GMC) (1995). Good medical practice (3rd edition). May 2001. http://www.gmc-uk.org/about/council/papers/2001_05/12a_annexa.pdf. Accessed 12 June 2009.

General Medical Council (GMC) (1997). *The new doctor, 1997*. London: GMC.

General Medical Council (GMC) (1998). Report of the visit to the University of Leeds, 24–25 November. London: GMC.

General Medical Council (GMC) (1999). Implementing *Tomorrow's Doctors*: Report of the Education Committee's informal visits to UK medical schools between 1995 and 1998. London: GMC.

General Medical Council (GMC) (2000a) GMC moves to reform structure. Press release, 5 December. London: GMC.
General Medical Council (GMC) (2000b). Section 60 Order is laid before Parliament. Press release, 14 November 2002. London: GMC.
General Medical Council (GMC) (2002). Proposals for new quality assurance of basic medical education. http://www.gmc-uk.org/med_ed/default.htm. Accessed 12 June 2009.
General Medical Council (GMC) (2003). *Tomorrow's Doctors*. London: GMC.
General Medical Council (GMC) (2004a). Quality assurance of basic medical education evaluation report. Reporting document R001 v2.0. Evaluation of the QABME pilot programme 2004. Manchester: GMC.
General Medical Council (GMC) (2004b). Principles of Good Medical Education and Training Draft for Consultation, attached to a letter of 13 August 2004, from Professor Sir Graeme Catto. Manchester: GMC.
General Medical Council (GMC) (2004c). Licensing and revalidation: Formal guidance for doctors. Draft document, 29 September 2004. Manchester: GMC.
General Medical Council (GMC) (2004d). Annual review, 2003/2004. Manchester: GMC.
General Medical Council (GMC) (2004e). Annual report of the visitors to the University of Birmingham School of Medicine for 2003/2004. Manchester: GMC.
General Medical Council (GMC) (2004f). Report of the visitors to the University of Aberdeen Medical School for 2003/2004 (v 3.0). Manchester: GMC.
General Medical Council (GMC) (2004g). Annual report of the visitors to the University of Liverpool Medical School for 2003/2004 (v4.0). Manchester: GMC.
General Medical Council (GMC) (2004h). Evaluation of the QABME pilot programme 2004. Reporting document R 001 v 2.0. Manchester: GMC.
General Medical Council (GMC) (2005a). Quality assuring undergraduate medical education: An overview. Manchester: GMC.
General Medical Council (GMC) (2005b). *Developing medical regulation: A vision for the future*. Manchester: GMC.
General Medical Council (GMC) (2007). GMC medical education section of the website. http://www.gmc-uk.org/education/index.asp. Accessed 10 February 2009.
General Medical Council (GMC) (2008). Private UK based medical colleges. http://www.gmc-uk.org/doctors/registration_applications/join_the_register/uk_based_medical_colleges.asp. Accessed 10 February 2009.
General Medical Council (GMC) (undated). Quality assurance of basic medical education extranet. http://www.gmc-uk.org/qabme/index.htm. Accessed February 2005.
Harvey, L. (2001). The British experience of assessing competence. In C. A. Palumba, & T. Banta (Eds.), *Assessing student competence in accredited disciplines: Pioneering approaches to assessment in higher education*. Sterling, VA: Stylus.
Hoskyns, J. (2000). E-mail correspondence from Janet Hoskyns to Lee Harvey, 5 April 2000 12:59.
Liaison Committee on Medical Education (LCME) (2005). LCME. http://www.lcme.org/. Accessed 12 June 2009.
Quality Assurance Agency for Higher Education (QAA) (1998). University of Leeds, Medicine, Subject review report, November 1998, Q56/99. Gloucester: QAA.
Quality Assurance Agency for Higher Education (QAA) (2000). Subject overview report, Medicine. QO8/2000. http://www.qaa.ac.uk/reviews/reports/subjectlevel/qo8_00.pdf. Accessed 12 June 2009.
Teacher Training Agency (TTA) (1995). Interview between Mr. Phil Holden, Quality Team Leader, and Selena Mason, Centre for Research into Quality, 30 March 1995.

Chapter 14
Contracting for Quality Improvement and Financing in Public Universities of Catalonia, Spain

Josep M. Vilalta and Joaquim Brugué

Introduction

The university system in Catalonia and in all of Spain has, in the last few years, experienced spectacular numerical growth. As a result of the improvement in living standards, which are approaching those of Western Europe, access to public universities has become more widespread. At the same time, there has been a substantial increase in all elements of the universities' activities: from courses, to research groups, to support infrastructures.

For this reason, the requirements of university management have been radically modified in a short period of time. Today, it is more complex to manage the university system and each individual university, faculty, school, service or laboratory. This is the case not only because of the volume of resources or the size of the academic and student populations, but also because the university is converting itself into a key element in the new "knowledge society".

In only a few years the Catalan university system has undergone a major evolution. The number of universities has increased from 3 in 1986 to 12 in 2003 and the number of offered courses has increased from 35 to 160. The number of students has more than doubled in the last 20 years (from 105,706 in 1982 to 221,417 in 2002) and the floor space has almost tripled (from 580,983m^2 in 1985 to 1,636,544m^2 in 2001) (DURSI nd). The number of universities, the size of university floor space, and the number of courses offered are all examples of this rapid expansion in the university system. The sudden expansion has produced, on occasions, a disorganised growth, difficult to manage without the necessary resources.

In Table 14.1, the figures show the present state of the university system in Catalonia. The sector is important in its scope and even more important from a strategic point of view. Universities are the motors of social, economic, technological and cultural development, within a context of growing international competition.

J.M. Vilalta (✉)
Catalan Association of Public Universities (ACUP), Barcelona, Spain

D.D. Dill, M. Beerkens (eds.), *Public Policy for Academic Quality*, Higher Education Dynamics 30, DOI 10.1007/978-90-481-3754-1_14,

Table 14.1 Principal indicators of the Catalan public university system, 2001–2002

Students enrolled	First cycle studies	58,903
	First or second studies	101,965
	Second studies	13,911
	TOTAL	174,779
First-year students	First cycle studies	16,169
	First or second cycle studies	20,119
	Second cycle studies	6,186
	TOTAL	42,474
First cycle and second cycle graduates		23,063
Doctorate students (2000–2001)		8,825
Postgraduate students (2000–2001)		48,820
Doctoral theses read (2000–2001)		1,063
Teaching staff (number of people)		13,143
Teaching staff (full-time equivalent)		10,407
Administrative and service personnel		6,516
Budget (total liquidated funds 2001, million Euro)		962.7
Net investment (2001, million Euro)		57.4
University floor space (2001, m^2)		1,636,544

Source: DURSI. Note: Data includes only the public system, not UOC, and refers to official studies in constituent centres.

In general terms, there is no doubt that higher education is undergoing major changes and transformations in all Spain. These are mainly the result of new expectations and demands emerging in society: university education has gradually become mass education, and the university and its systems have had to respond to this new reality with innovative and occasionally imaginative new policies. The traditional teaching and research functions have become more complex and the various actors involved have highlighted new needs and adopted new strategies in keeping with their roles and the changing context in which they find themselves. All this has occurred in a setting of growing competition for public resources, increased demands for improved quality in public policy and management, and for more transparency with regards to the benefits obtained by the public.

These changes have also had an effect on the relations between the Government and the universities (Neave and Van Vught 1991). The Government has tended to become more regulatory (establishing the legal framework), more strategic (defining specific sector policy strategies) and more evaluative. The universities now tend to have more autonomy for developing their own institutional profile and projects. This has led to a tendency to establish objectives, analyse performance, and foster effectiveness and efficiency in the public universities within the context of an added-value inter-institutional network. In many countries evaluation of the university and the quality of its services has become a central function of public higher education policy (Neave 1998), and Catalonia is no exception.

Closely linked to this new situation and bound up with the concept of academic and administrative autonomy, we find the notions of accountability, transparency and the evaluation of the performance of public services.

All the above changes have been formally incorporated into the Catalan public university system since 1997. Programme-contract is one of the new policy instruments, which is designed to foster academic quality. The programme-contract has proved a useful instrument for the improvement of university quality through coordination between the Catalan Government's university and research policy and the Catalan universities' strategic planning. The contracts also try to promote transparency by making universities to publish their outcomes to the society. The Catalan universities have attempted to adapt themselves to this new framework and obtain both academic and management improvements.

The present chapter analyses what the programme-contracts represent for the Catalan public universities in the context of Spanish higher education system. It focuses on their function as tools for the new public administration (Paradeise et al. 2009) and as contractual initiatives between the government education authorities and the universities for the purpose of improving university activity and moving towards financial models which are increasingly based on objectives and results.

Content of the Policy Instrument

At the beginning of 1997, the *Generalitat* of Catalonia – represented by the Universities and Research Commission (CUR) of the Presidential Ministry – and the Technical University of Catalonia agreed to formalise the first university programme-contract. This represented a new university planning and coordination departure for Spain, linking additional funding to the attainment of mid-term objectives for improved quality. The programme-contract was then extended to the majority of Catalan public universities in a first phase which lasted until 2001. After 2002, it was a *fixed strategic* subvention feature in the funding distribution model for Catalan public universities.

At this time, the university system (and the public sector in general) was entering a period of public deficit control arising from the demands of European convergence. A process of reflection had also begun, in part motivated by the experiences of other countries and institutions, which pointed to the need to optimise the learning process and graduate employment, to improve teaching, academic performance and the general quality of university services, to use the new technologies to extend the scope of education and to innovate, to clarify and highlight the benefits of research for society as well as improving the quality and quantity of research and institutional management.

These new objectives required new instruments for observation and diagnosis, planning and decision-making. It was necessary to incorporate into daily university life the management criteria that are commonly found in other areas of society (Paradeise et al. 2009): *efficiency*, i.e. improved performance using existing assets; *transparency*, understood as accountability to society; *rationality*, establishing

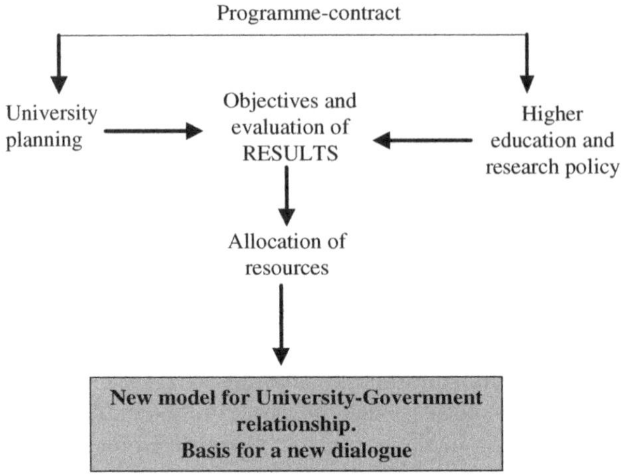

Fig. 14.1 The Programme-contract rationality in Catalonia

measurable objectives; and *quality*, understood as the measure of satisfaction achieved in responding to the needs of society by university teaching, research and services.

In response to these needs, the CUR introduced a set of multi-level instruments for cooperation between the Catalan universities and the government education policy bodies. These were founded on five basic pillars: the Pluriannual University Investment Plan (1995), the university teaching programmes, the beginnings of quality evaluation by the Agency for the Quality of the Catalan University System, AQU (1996), the Catalan Research Plan (1997) and the programme-contracts linking funding to achievement of objectives (1997).

A number of the Catalan universities had already started similar processes, as set out in their strategic or quality planning, in the form of new organisational structures designed to facilitate change-management, with phased implementation of specific sector plans and the development of innovative (at that time) management techniques and tools appropriate for the new public management.

The programme-contracts have become a strategic instrument for management and quality improvement in universities and the university system in general. They are the result of an institutional pact between the government and each public university. The main characteristics of the programme-contract are:

- It establishes specific objectives for improved quality in the services offered by the university to society, for more effective management and for an improved service to users (with pluriannual time scales enabling definition of mid-term policies and plans spanning more than a single academic year).
- It provides for evaluation of the extent to which the objectives are achieved by means of pre-established indicators, mainly quantitative in nature.

- It determines specific public funding according to the extent to which the objectives are achieved. (In the first phase of the programme-contracts until 2001, this funding was in addition to funding to cover the university's basic activities. Later, it was integrated into the university funding distribution model).
- It includes provision for annual revision of the objectives, in accordance with an evaluation of the results of the contract and the evolution of the government's higher education policy and the priorities of the universities themselves.

With varying degrees of effect, the programme-contracts have served for the development of different aims: budgetary purposes, improved quality and social transparency.

Implementation

The Government of Catalonia – first through the Commission for Universities and Research and subsequently through the Ministry of Universities, Research and the Information Society – signed programme-contracts with the following Catalan public universities for the 1997–2001 period:

- Universitat de Barcelona (UB), 1999–2002 (This period was later shortened by 1 year, and the new funding distribution model was applied from 2002 on);
- Universitat Autònoma de Barcelona (UAB), 1998–2001;
- Universitat Politècnica de Catalunya (UPC), 1997–2000 and a transition period for 2001;
- Universitat de Girona, 1997–1999 and another for 2000–2001;
- Universitat de Lleida, 1997–1999 and another for 2000–2001;
- Universitat Rovira i Virgili, 1997–1999 and another for 2000–2001.

Two programme-contracts were also signed with the Universitat Oberta de Catalunya (The Open University of Catalonia) for the 1998–2000 and 2001–2004 periods which, given the specific characteristics of the university, established public funding for each period on the basis of mixed parameters involving quantitative and qualitative evaluation based on pre-established objectives.

Two types of programme-contract can be distinguished in this initial period: *Type B* contracts or pre-contracts, and *Type A* contracts, or programme-contracts per se. *Type B* contracts had a shorter duration (3 years), and were established with the more recently created universities (University of Girona, University of Lleida and Universitat Rovira i Virgili). They were designed to contribute to reducing the structural deficiencies (human and material resources) linked to their process of creation, and to stimulate the establishment of overall quality policies. The increased public subvention for the purpose of structural consolidation and improvement was subject to the introduction of policies for quality and management.

Longer programme-contracts (*Type A*, 4 years in duration) were agreed with the three long-established universities (UB, UAB and UPC). To a large extent these contracts coincided with the governing periods of the rectoral staff. These contracts established specific objectives, annual evaluation of performance by a set of indicators, and additional funding linked to this evaluation. From the outset, these programme-contracts were defined as policy instruments designed to improve quality. Following the initial 3-year period, all the Catalan universities have had programme-contracts of this type.

The starting point for each contract was the individual university's Strategic Plan and the Government's specific university and research policy objectives. From these two pillars, a concrete number of objectives (between 14 and 19) were drawn up for each university. These were explicitly interrelated with the lines of action set out by the university's Strategic Plan and took into account the specific characteristics and needs of each university.

A number of basic lines of action were set out for each objective, along with a broad-based set of tools and instruments for their implementation and a set of between 50 and 70 indicators (mainly, but not exclusively quantitative). (See Table 14.2 for an example of objectives and Table 14.3 for indicators.) The few qualitative indicators employed corresponded to new actions. In later years, they were replaced by quantitative indicators. For each indicator, the starting point was specifically set out, namely: the situation immediately prior to commencement of the contract period and the values that were to be achieved for each of the contract years and by the end of the contract period. Annually, the programme-contract monitoring committee revised the required values in the light of the results achieved in earlier periods and the overall commitments made for the entire contract period.

Table 14.2 Objectives in a programme-contract, Technical University of Catalonia (1997–2000)

Programme-contract objectives	*Lines of action*	*Indicators*
1. To improve student flow by increasing the number of graduates and assuring that course content and teaching load are compatible with the requisites of educational quality.	5	4
2. To help graduates find work and evaluate the acceptance on the job market of their levels of qualification and preparation, and their ability to adapt to the needs of society.	4	4
3. To plan the study programmes on offer in terms of needs and demands of society	2	4
4. To reform the contents, norms and management of doctorate programmes in order to increase the number of doctors in technological fields, reassess the value of doctorates within the business world, and adapt training to the needs of the socio-economic environment.	6	3
5. To programme and promote quality continuing education adapted to the needs of society	4	3

Table 14.2 (continued)

Programme-contract objectives	*Lines of action*	*Indicators*
6. To consolidate quality R+D activity in the University's research teams, thereby ensuring that UPC as an institution achieves a reputation for excellence in the field of research and technology at the service of society	6	2
7. To expand R+D activity at UPC by increasing the number of academic staff and research teams working in research and technology transfer, and by promoting the degree of self-funding of R+D activities	7	3
8. To increase technology transfer to firms and other institutions by ensuring that R+D at UPC responds appropriately to social, industrial and technological needs and demand.	6	4
9. To increase UPC links with other institutions and strengthen its ties with society.	6	4
10. To provide graduates with the ability to carry out their professional activities with an awareness of the economic, social and cultural context of Europe. To consolidate and broaden European and international cooperation in research and technological development (RTD) as a guarantee for the future of scientific and technological standards at UPC	7	5
11. To develop, with society in mind, an integral model for environmental protection and sustainable development based on the potential present in UPC institutions.	5	3
12. To adapt the academic staff structure progressively to the established objectives in order to achieve the desired quality	3	3
13. To adapt the non-academic staff progressively to the strategic objectives by increasing professionalism and management efficiency and effectiveness	4	3
14. To develop and implement planning, assessment and resource assignment systems on the basis of quality control criteria in order to improve UPC's activities at the service of society	5	4
15. To evolve an active policy of obtaining resources which will provide new opportunities and collaborators willing to contribute to the funding UPC	4	3
TOTAL	**74**	**52**

To evaluate the achievement rates for each objective, each contract included its own weighting system. The different indicators within each objective were assigned different weights. The weighting reflected the strategic value of the different elements according to their impact on the improvement of quality. Strategic objectives in the areas of learning and teaching and research were given higher priority than the others.

Table 14.3 Indicators to measure the main objectives of the programme-contracts

First and second cycle courses:
- Percentage of students choosing courses as a first option relative to the number of places available
- PAU exam grade of a given percentage of students enrolled
- Student success rates in first and second cycle courses
- Student performance rates in first and second cycle courses
- Percentage of students graduating within pre-established time and/or within an additional year relative to initial cohort
- Percentage of total credits registered relative to total theoretical credits for each degree.
- Drop-out rate
- Percentage of students assigned a tutor
- Percentage of students/graduates completing work experience programmes in companies and institutions
- Credits registered for skills and universal competencies

Third cycle and postgraduate courses:
- Number of doctoral theses read
- Number of DEAs achieved
- Number of students registered on doctoral programmes/accredited doctoral programmes
- Number of accredited doctoral programmes
- Number of inter-university doctoral programmes
- Percentage of doctoral graduates/students with first degrees from other universities
- Total number of postgraduate class hours registered/number of students registered on postgraduate programmes

Research and technology–knowledge transfer:
- Resources obtained from competitive research funds
- Income from research agreements with companies and institutions
- Percentage of teaching and research staff participating in funded research projects
- Percentage of teaching and research staff participating in technology transfer projects
- Number of consolidated research groups
- Number of patents registered
- Number of predoctoral scholarships awarded
- Number of externally funded postdoctoral researchers

International dimension and cooperation:
- Student mobility: number or percentage of students on exchange visits to other universities/number or percentage of students from other universities registered in the Catalan university
- Percentage of graduates having studied abroad
- Teacher mobility: total months spent by teaching and research staff on exchange visits at other universities, and by teachers from other universities at the Catalan university
- Number of development cooperation and volunteer projects

Catalan and foreign language skills:
- Percentage of first and second cycle teaching in Catalan
- Number of books and manuals published in Catalan
- Number of first and second cycle subjects taught in English
- Number of doctoral and/or Master's subjects taught in English

Monitoring Committees comprised three representatives of the university and three of the government and they were appointed to monitor each programme-contract and determine the extent to which the objectives were achieved on an annual basis. These committees normally met once per semester.

Leaving aside the formal procedures for implementing the programme-contracts, the experience in Catalonia may be useful to identify some of the main difficulties for launching this new policy tool. In this sense, we would like to underline two main shortcomings: the difficulties for both designing clear and quantitative policy objectives and obtaining the commitment of the University actors (professors, students, staff personnel).

First, getting clear and quantitative policy objectives and indicators is both needed and difficult. It is needed because the programme-contract is designed over such objectives and indicators, and it is difficult because there is no immediate answer to the question about what are we expecting from a public university system (*policy objectives*). Neither is it easy to measure complex issues such as teaching quality, technological transfer or research excellence (*indicators*). In front of these difficulties, the programme-contract makes pressure and seems to force the political actors to assume a simplistic (and therefore ready for use) approach to the objectives of the universities. Sometimes, there is not enough analysis about such objectives. There is no answer for the crucial question about why we need a public university system. The system is sometimes assumed without such kinds of ontological questions, and the policy objective is simplified to instrumental and ready for-use terms: the objective is the economy and the efficiency of the system. Everybody understands such objectives and, moreover, they are easy to measure.

The second set of difficulties with the implementation phase is related to the lack of commitment among the majority of university actors. A contract is a commitment, and without the willingness of the partners to accept such a contract its force is limited. In our experience, it is often the case. Most of the academics, students and staff do not feel that it is "their contract" and, therefore, they are not committed with its terms. Even more, some of the internal actors look at the programme-contract like a sort of managerial extravagance. For them, it is often perceived as a loss of time and effort. Moreover, they have not been invited to discuss either the terms of the contract or their opinion about this managerial tool.

Confronted with these widespread attitudes, there is no doubt about the initial weakness of the process. It does not mean that the experience failed, but that it has been highly informative, not just for its own results, but also as a means of policy learning.

Impact

All the programme-contracts set out objectives and actions in four common strategic areas:

- Teaching, education and the learning process;
- Research and technology–knowledge transfer;
- University–society relations (the *third function* of the university);
- Internal university organisation and management.

A number of universities also included other strategic areas of their own.

In the area of *teaching/learning*, the objectives included improved adaptation of university education to the needs of society through sensitivity to social demands and the needs of the learning process, placing emphasis on methodological renovation, flexible teaching, improved teacher training and the quality of the teaching provided (Mora 2004). As we mentioned before, there are obvious difficulties to move from such generic objectives to particular indicators. The process of simplification that sometimes has been used to deal with such difficulty has not always been satisfactory. In this sense, some critics point out the distortion or danger of transforming the concept of "quality of teaching" into the indicators such as "academic success" (number of students passing the exams).

In *research and technology–knowledge transfer* the objectives concerned active collaboration of the universities in achieving high-quality, internationally competitive science, technology and innovation systems which would contribute to the progress of Catalonia. This would facilitate improvement and expansion in both the research itself and the impact of results. It was also meant to foster multidisciplinary approaches, integration in international networks and collaboration with companies, particularly companies which are technologically innovative. Here again the measurement difficulties have led, in some cases, to a certain degree of simplification. Moreover, the university culture – especially among academics – in Catalonia and overall Spain is quite individualistic and, therefore, it is complex to design incentives for transversal research approaches.

In the area of *university–society relations*, the challenges were: to improve communication with society, so as to enable society to communicate its needs to the university and for the university, and to highlight the contribution the university can make to society; to improve the attention given to the new students entering the university, to graduate employment, and the promotion of entrepreneurial spirit among graduates; improving university services to companies and institutions, and the adaptation of continuing education to the needs of society and the changing demands of the labour market. Other objectives sought to improve language skills in two areas: first, in fostering greater use of the Catalan language in university life and academic activities; second, in developing improved foreign language skills in response to increased international mobility in university life.

Given the difficulties of evaluating such objectives and also the traditional gap between the Spanish university and the society, most of the programme-contracts have been weak in aspects related to the university–society relations.

Finally, with regards to the *improvement of management*, the emphasis was placed on the need for effective and flexible organisations, oriented to constant improvement, with qualified, motivated and well-trained staff, and the need to develop management systems and tools facilitating improved quality, effectiveness and efficiency. In particular, it was important to continue to foster use of information and communication technologies in all spheres of university life so as to adapt to the new information society.

Despite the fact that the objectives were ambitious and that the difficulties were quite impressive, most of the commitments made in the various programme-contracts have been achieved, in many cases more than adequately. It is also the case, we have to recognise, that most of the commitments were designed to be satisfied – i.e. universities tend to sign contracts only when they are pretty sure about their capacity to satisfy or fulfil them.

In any case, if we refer to academic achievements, the improvements achieved included the following ten points:

- Improved quality in teaching, the learning process and teacher training, adapting university studies to the professional practice of graduates.
- Improved student performance, with a significant increase in numbers graduating and improved results in courses.
- Improved graduate employment rates due to the services provided to students by the university, the evaluation of labour market needs and, especially, due to equipping students with the flexibility required for continuous re-learning and to fostering contacts with the labour market.
- Improved range of doctoral courses and continuing education courses in response to needs for specialisation and updated skills. Also, an increase in the number of doctoral graduates in research and industry.
- Improvement in the activities, resources and results of research and technology–knowledge transfer.
- Increased "internationalisation" of the universities: More mobility programmes for students and teachers and exchange agreements with international research and teaching institutions and networks.
- Improved use of the Catalan language as the language of the universities and greater participation of the universities in their cultural setting. Improved foreign language skills among the university community.
- Improved harnessing of the professional skills of the teaching staff and the material resources in the university.
- Implementation of sustainable, integrated, environment-friendly, management systems, and development of professional health and safety policies.
- Strengthening and development of the use of information and communication technologies in all spheres of university life, thus harnessing the potential of the information society.

The gradual achievement of these objectives has been reflected annually by the programme-contract indicators. In all, the level of achievement has improved year after year, with annual percentages ranging between 85 and 95%. In some

Table 14.4 Main institutional outputs, Technical University of Catalonia (1997–2000)

	1997	2000
University graduates	3.418	5.126
University professors with Ph.D.	45.3%	49.7%
Number of students in continuing education courses	4.672	8.296
Students with experience at business (practicum)	3.540	6.156
Average time for students to start working	–	2 months
Incomes by R+D activities	4.288	5.430
Research impact	320	535
Percentage of research self-financing	45.0%	52.15%
Mass media impact at university activities	1.898	3.411
Incomes by companies agreements	1.509	2.345
International mobility at professors (in months)	82 months	210 months
Members at the University Association Friendship	1.304	2.627

universities, the last year saw achievement rates of 100%. (See Table 14.4 for an example.)

Such high level of achievement is debatable. On the one hand, it is obvious that most of the programme-contracts are very successful in terms of quantitative indicators. On the other hand, it is also possible to suspect that such high level of success could be spurious. Some critics could argue that the high level of success is, in fact, the proof of the irrelevance of the programme-contracts, contracts made – as we mentioned before – not for improving academic outputs but just to be satisfied and therefore getting some extra funding. Nevertheless, the real fact is that contract-programmes have led to a new university culture and policy orientation towards academic objectives, output-oriented, and to the interrelation between academic outputs and funding mechanisms.

Resources, Costs and Financial Benefits

From a budgetary point of view, the contracts represented additional funding of a total of more than €720 million for the universities over the 5-year period, 1997–2001. This represents an extra funding of around 5% of total higher education expenditures.

The experience gained over the first period up to the year 2001 showed the programme-contracts to be an effective instrument for improving quality. The instrument continues to function as part of the funding distribution model for all Catalan public universities, which has been in operation since 2002.

The general objective of the funding distribution model is to establish criteria for distribution of public funds among the public universities. It is based on the principle of the equality of all students in the public system and aims to be objective. Calculations are by a system of easily determinable common parameters.

The model includes five forms of subvention:

- *Fixed funding*, equal for all universities, covering the minimum structural expenses necessary for their operation.
- *Basic funding*, which provides resources for their ordinary academic activity and related operating expenses based upon common objective parameters.
- *Derived funding*, for expenses deriving from employment of teaching and research staff.
- *Strategic funding*, linked to Quality objectives in relation to the university strategy.
- *Competitive funding*, for certain measures established by the Ministry of Universities, Research and the Information Society (DURSI) and affecting all universities simultaneously.

The funding from the programme-contracts is included within the *strategic subvention* to the universities. Within the model, the programme-contracts assign funds to the universities for three purposes: first, in order to achieve objectives in the area of *quality*, second, for *specific purposes* in each individual university which cannot be included within a general model, and third, in order *to align the funding of each university to the requirements of the transition phase*, in order to guarantee convergence with the model's target funding framework. As already pointed out, in this second period the programme-contracts have been fully integrated into the instruments for public funding management, and the aim is to increase their relative weight in the funding provided to each university.

In the case of the first two purposes mentioned above, the DURSI establishes an annual maximum sum which is determined by variables of scale but also by strategic considerations in the case of each university. The final amount assigned is determined by the overall extent to which the programme-contract objectives have been accomplished as reflected by the indicators. This is expressed as a percentage which is then applied to the maximum funding permitted.

With regards to their structure, the new programme-contracts are much more homogenous. All are for a duration of 4 years (2002–2005) and all are type A. The number of objectives has been reduced significantly (to between six and ten) as has the number of indicators, (around 30). A similar weighting system has been maintained for objectives and indicators along with a system for evaluating the progressive accomplishment of objectives.

Regarding their contents, additional objectives have been introduced in order to adapt the system and the universities to the new European higher education and research space and to new legislation (Table 14.5). Objectives have been added as well to improve student intake, access and induction processes, to increase involvement by the universities in regional development through closer links with society, and to improve the use of Catalan and enhance third-language skills.

In this second phase the programme-contracts have been simplified, by identifying the main common strategic priorities of the university system and ignoring the bulk of ordinary activity. Similarly, a set of previously tested output/outcome indicators has been established and the work of the Monitoring Committee has been facilitated.

Table 14.5 Objectives of the new programme-contracts between the Ministry and each of the Catalan universities, period 2002–2005

Objective	*UAB*	*UB*	*UdG*	*UdL*	*UPC*	*UPF*	*URV*
Improvement of the quality of education provided and effectiveness and efficiency of the system	X	X	X	X	X	X	X
Improvement of measures to attract and enrol new students	X	X	X		X		X
Improvement of research and in attracting competitive external funding	X	X	X	X	X	X	X
Development of mechanisms for technology–knowledge transfer and for promotion of R&D&I	X	X	X	X	X	X	X
Promotion of high-quality third cycle education and employment for doctoral graduates	X	X	X	X	X	X	X
Strengthening active career guidance for graduates and fostering links with former students	X	X	X	X	X	X	
Consolidating the use of Catalan and improving the university's linguistic competence in third languages	X	X			X	X	X
Participation in regional development via closer links between universities and local society				X	X		X
Strengthening the international dimension of academic activities, university mobility programmes and development cooperation	X	X	X	X	X	X	X
Improvement of the quality of management processes and optimisation of human and material resources, with support of ICTs	X	X	X	X	X	X	X

In financial management and planning, the programme-contract aims to introduce new mechanisms for public funding of the university's day-to-day spending. Additional new resources were programmed which were linked to achievement of pre-established objectives. Underachievement led to no funding, and could lead to revision of the university's quality policy and its overall funding. Therefore, full and wide cooperation from the university community was needed in order to achieve the majority of the objectives established and thereby receive the additional funding.

Despite the fact that this additional funding was relatively low in comparison with the total funding resources provided to the universities, it did lead to a major effort to achieve rational use of resources and foster a results-driven culture of efficiency. From a situation in which the university's funding was the subject of annual discussions largely steered by an incremental principle, we have moved to negotiation and an institutional pact which sets out the improvements the university is to achieve over a given period in order to receive a given level of funding.

The programme-contract has also served to disseminate the challenge of improvement in the four strategic areas established (teaching and learning; research and technology–knowledge transfer; the university–society–region link; and management, organisation and resources). The first programme-contract fostered establishment of strategic quality policies, priority setting, associated action plans and evaluation.

The programme-contract also led to assimilation of the concept of each university being accountable to society. Being accountable in terms of improvements achieved is relatively easy, bearing in mind that the contracts explicitly set out the use to be made of the public funding provided and that results were measured by means of a simple system of objectives and indicators. Through the programme-contract, each university set out the pre-established objectives for which it received additional funding. The progress made was monitored by the Government. The Catalan Government was also able to attach priority to aspects which it considered important, while still respecting the principle of university autonomy.

Introduction of the programme-contracts also had a positive effect in some universities with regard to internal functioning and dissemination of institutional objectives, and permitted adaptation of activity at different levels of the organisational structure to the academic and service objectives established in conjunction with the *Generalitat*. Some universities have even used similar mechanisms internally for the purpose of resource allocation.

Evaluation and Expectations

The overall evaluation of programme-contracts in the Catalan university system is positive, although some difficulties and shortcomings have also been identified. Study of their application also enables us to draw conclusions that enrich the international debate on the role of performance contracts in the management of public universities (Chevaillier 1998; Höltta 1998; Jongbloed and Vossensteyn 2001). The following section highlights some of the most interesting considerations discovered in the application of contractual mechanisms in the Catalan university system and also the elements which have to be corrected or improved in order to achieve greater effectiveness and efficiency in university management and policy.

Among the positive effects are the *development of a policy of institutional co-responsibility and increased dialogue and cooperation between the government and the universities*. The programme-contracts have contributed to generating a climate of inter-institutional trust and cooperation which has led to a higher share of co-responsibility for universities in policy development, and multi-level cooperation based on negotiation, mutual understanding and increased government support. Nevertheless, some more efforts must be made to produce a real dialogue about the social role of a public university and on a public university system. Why do we need such a system? The programme-contracts must stem from the answer to that question and, now, it is not clear enough.

The contracts have also played a role in *promoting the autonomy, specific identity and strategy of each individual university*. They respect and promote university autonomy, which they link to the concept of accountability. At the same time they recognise the different contexts in which the universities operate. In addition to parametric criteria which are common to the entire system, the contracts are sufficiently flexible to incorporate the specific features, problems and projects of each individual university. It is also true that, comparing different universities, the substance of their programme-contracts is not very different. We can suspect, therefore, that the leading role assumed by politicians and experts has produced some standardisation in the process.

The contracts have proven to be an *effective tool for analysis, diagnosis and evaluation, in that they allow study of all aspects of the university's life*. Hitherto, standard practice was to focus on a given aspect of the university (teaching, research, mobility, teaching staff, resources and infrastructure, etc.) and it was rarely possible to carry out a comprehensive diagnosis of all aspects which would facilitate institutional decision-making and management. From another point of view, the participants in the definition, implementation and monitoring of the programme-contracts also consider them a valuable tool. This has reinforced the university authorities in their commitment to improvement and innovation. Such commitment, nevertheless, must be assumed for the whole university community. Some members of the academic community do not participate in such commitment and it is one of the main weaknesses of the policy.

As previously mentioned, the programme-contracts have contributed to the *modification and modernisation of the university funding system*. The funding mechanisms now take account of academic and other results, as opposed to the inputs which predominated in previous models. Now, the pluriannual programme-contracts enable long-term financial projections rather than single-year approaches. Since 2002 the Catalan system has had a model for the assignation of funding to public universities which utilises the programme-contract as one of its most important constituents. However, further consideration is required concerning what percentage of the total funding should be results-based.

Finally, there is less need to insist on the *importance of transparency and the communication to society of the actions and results of the universities*. The contracts have played an effective role in achieving transparency; however, it is evident that further improvements can be made in the area of university–society communication and public accountability.

To sum up, the programme-contracts can be improved in, among others, the following aspects:

The contracts should highlight more effectively the value of the higher education and public services provided by the universities. It is necessary to intensify and move forward the debate on the public value of university education, the challenges being faced, and the strategies which need to be developed.

The contracts are not directly linked with other higher education policies in the area of university quality evaluation. The initiatives taken and the outputs by the Catalan Agency for University Quality (AQU) and the Central Government (Ministry of Education and Science – National Agency for University Evaluation and Accreditation, ANECA) are not very much related with the experience of contracts-programmes.

We must also better harness the full potential of the contracts for innovation in university policy. They provide possibilities for the introduction of rarely employed mechanisms of benchmarking – between the institutions and within the general system. Clearly, this would bring the challenge of finding the balance between competition and cooperation among universities in a highly integrated system. Striking the correct balance will facilitate development of high-quality universities capable of serving society and being competitive at European and international level. To achieve this, the contracts will have to make increased provision for cooperation, synergy and implementation of large-scale, collaborative projects between the universities themselves and with other educational, economic and social institutions.

Work must also continue on improving various aspects of the indicators employed and overcoming problems with some of the indicators: problems such as the influence of environment and other random external variables and temporal factors. Work must continue to define common indicators which will be valid for the entire system.

Finally, one of the most frequently voiced criticisms has been the *low level of participation by the different university collectives in the definition of the objectives of the contracts and in working for their accomplishment.* In this regard, the situation and context varies widely from one university to another and each has taken the approach most suited to its own context. Once again, work must be done to establish the right balance between participation by university personnel and strategic and management leadership by the university's management. Several universities have taken the opportunity to improve their internal management, by developing internal contract mechanisms. These instruments have enabled increased objectivity in analysis of the academic activity of the various departments, faculties, schools and institutes and have fostered the development of new objectives agreed between the university management and each unit. This is a practice which should be extended to all elements of the Catalan university system.

References

Chevaillier, T. (1998). Moving away from central planning: Using contracts to steer higher education in France. *European Journal of Education*, 33(1), 65–76.

DURSI (nd). Ministry of Universities, Research and the Information Society, Government of Catalonia. www10.gencat.net/dursi. Accessed 4 February 2009.

Höltta, S. (1998). The funding of universities in Finland: Towards goal-oriented government steering. *European Journal of Education*, 33(1), 55–63.

Jongbloed, B., Vossensteyn, H. (2001). Keeping up performances: An international survey of performance-based funding in higher education. *Journal of Higher Education Policy and Management*, 23(2), 127–145.

Mora, J-G. (2004). A decade of quality assurance in Spanish universities. In S. Schwarz, D. F. Westerheijden (eds.), *Accreditation and evaluation in the European Higher Education Area*. Dordrecht: Kluwer.

Neave, G. (1998). The evaluative state reconsidered. *European Journal of Education*, 33(3), 265–284.

Neave, G., van Vught, F. (1991). *Prometheus bound: The changing relationship between government and higher education in Eastern Europe*. Oxford: Pergamon Press.

Paradeise, C., Reale, E., Bleiklie, I., Ferlie, E. (2009). *University governance: Western European comparative perspectives*. Dordrecht: Springer.

Chapter 15
The National Assessment of Courses in Brazil

Simon Schwartzman

In 1996, the Brazilian Ministry of Education introduced a National Assessment of Courses for Brazilian higher education. The exam – which became known as *Provão*, the big exam, or ENC – consisted of a national test applied to all the students graduating in each specific course program in the country. The results were published in a five-point scale, from A to E, according to their distribution in each field. In the first year, the test was applied to students graduating in Law, Administration, and Civil Engineering – the careers with the largest attendance. In 2003, the exam included 470,000 students graduating in 26 different fields in 6,500 course programs in the whole country.

The objective of the test was to provide information to the public on the quality of higher education courses, helping the students and their families to choose where to study, and to provide the Ministry of Education with information that could be used in the accreditation and reaccreditation of higher education institutions. Besides, the exam generated an intensive process of discussion and consultations among academics about the contents and standards of the different careers, which is supposed to have helped to improve the quality of Brazilian higher education throughout.

The exam was introduced without previous consultation, and was received with strong opposition from student associations, teachers' unions, and many higher education institutions. However, from the beginning, it received strong support in public opinion and in the press. The criticism ranged from specific objections to the way the tests were conceived and the results presented – a uniform test for the whole country, a national rank of outcomes without consideration of existing conditions and explicit standards – to broad objections to any kind of measurement of education outcomes. However, once in place, the results became widely used as references for students in their choice of institutions, and for the institutions themselves, particularly in the private sector, to publicize their results, or to try to improve them. Bad results, when persistent and associated with other indications of low quality, were supposed to lead to the closing down of the course programs by the education authorities, but, in practice, this has seldom happened.

S. Schwartzman (✉)
Insitute for the Study of Labor and Society (IETS), Rio de Janeiro, Brazil

D.D. Dill, M. Beerkens (eds.), *Public Policy for Academic Quality*, Higher Education Dynamics 30, DOI 10.1007/978-90-481-3754-1_15,

In 2002, the opposition Laborer's Party won the presidential elections, and the candidate's program for education announced the end of the National Exam. Once in power, the new Minister of Education established a Commission to examine the issue and to propose a new approach to higher education assessment. The Commission published its conclusions in September 2003, and, in December the Government announced its own proposal for higher education assessment, which changed the previous system very substantially. At the end, the new assessment system kept the national exam with a new denomination, the National Exam of Student Achievement (*Exame Nacional de Desempenho de Estudantes* – ENADE) and very substantial modifications which, in this author's view, has very serious methodological and conceptual flaws, and makes it much less significant than before. This chapter provides a detailed analysis of ENC, and also an overview of ENADE, implemented in the years 2004, 2005, and 2006 (Table 15.1). The last ENC took place in 2003.[1]

Table 15.1 Evolution of the National Assessments of Course Programs in Brazil

	Course programs	*Fields*	*Students examined (enrolled)*	*Type of exam*
1966	616	Administration, Law, Civil Engineering	59,343	ENC
1997	822	Administration, Law, Engineering (civil and chemical), Veterinary Medicine, Dentistry	94,296	ENC
1998	1,710	Added: Chemical and Electrical Engineering, Journalism, Literature, Mathematics	142,107	ENC
1999	2,151	Added: Economics, Mechanical Engineering, Medicine	173,641	ENC
2000	2,808	Added: Agronomy, Biology, Physics, Psychology, Chemistry	213,590	ENC
2001	3,701	Added: Pharmacy, Pedagogy	288,417	ENC
2002	5,031	Added: Architecture and Urbanism, Nursery, History, Accounting	386,095	ENC
2003	5,897	Phonoaudiology, Geography	471,659	ENC
2004	2,184	Agronomy, Physical Education, Nursery, Pharmacy, Physiotheraphy, Phonoaudiology, Medicine, Veterinary Medicine, Nutrition, Dentistry, Social Work, Occupational Therapy, Zootechnology	143,170	ENADE

[1] The official reports of ENC and ENADE are available for consultation at the site of the Institute for Education Research from the Brazilian Ministry of Education, INEP, at http://www.inep.gov.br.

Table 15.1 (continued)

	Course programs	*Fields*	*Students examined (enrolled)*	*Type of exam*
2005	5,511	Architecture and Urbanism, Biology, Social Sciences, Computer Sciences, Engineering, Philosophy, Physics, Geography, History, Literature, Mathematics, Pedagogy, Chemistry	277,476	ENADE

Context

Brazilian higher education developed late, and was based on the European, mostly French and Italian models. Until the early nineteenth century, Brazil was a colony of Portugal, and no higher education institutions existed – it was necessary to go to Coimbra in Portugal or perhaps France to get a degree. In 1808, the Portuguese King and his court moved to Brazil, fleeing from the invading Napoleonic troops, and Rio de Janeiro became, for several years, the capital of the Portuguese Empire, to become later an independent country. The first higher education institutions were established in those years – one military academy, later to become a school of engineering; two medical schools; and two law schools. They were all owned, financed, controlled, and supervised by the royal government. In the late nineteenth and early twentieth century, as the old Brazilian Empire was replaced by a decentralized Republic, other institutions were added. Some states – notably the state of São Paulo – started to create their own institutions, and private institutions began to appear. Until 1889, only 24 higher education faculties existed; between 1889 and 1918, 56 new, mostly private, faculties were established.[2] New fields, like pharmacy, dentistry, agriculture, and accounting, were introduced side by side with the old learned professions.

The first universities were established in the 1930s, and they were, mostly, a collection of old schools or faculties, with one important innovation, a new Faculty of Philosophy, Sciences, and Letters, which was to be, at the same time, the place for scientific and academic research, and for the preparation of secondary school teachers. The first University, the University of São Paulo, was established by the State government in 1934, and the *Universidade do Brasil*, now the Federal University of Rio de Janeiro, was established in 1939 by the National government. In the early 1940s, the Catholic Church created the first private university in Rio de Janeiro, and they all introduced course programs in the natural sciences, mathematics, history,

[2] For the early history of Brazilian higher education, see Azevedo (1971); Durham (2004); Schwartzman (1991); Teixeira (1969).

geography, social sciences, philosophy, and language and literature, which did not exist before.

Throughout the nineteenth century, holders of higher education degrees strived to assert their exclusive rights to practice their respective professions, and, after the 1930s, the principle that a university degree was tantamount to a professional license became firmly entrenched (see Coelho 1999). This created, at once, a problem of regulation, which was never fully solved. To deal with this, a new Ministry of Education was created together with a National Education Council formed by public personalities. The new ministry tried to establish a "model university" in the country's capital, based on a detailed description of the course contents of all disciplines, down to the assignation of textbooks and time tables, which all other institutions had to follow (Schwartzman et al. 2000). At the same time, the government created a complex system of professional councils, which, together with the business associations and the trade unions, were supposed to organize the country into a neat and coherent corporatist structure, integrating the professions, the entrepreneurs, the unions, and the education institutions (Malloy 1977; Schmitter 1974; Schwartzman 1988).

This tightly conceived system never worked in practice, and its limitations became all too obvious as higher education began to expand and new professions started to emerge after the Second World War. However, the basic assumptions established in the 1930s – that all higher education degrees should be equivalent to a professional certification, that all professions had to be regulated by law, controlled and supervised by a legally established professional council or association, and that it was the role of the Federal government to make sure that all course programs provided equivalent contents – remained and are still in place. The National Education Law of 1996 introduced more flexibility, and the legal requirement that each career should have a national "minimum curriculum" was replaced by more general "curriculum guidelines" (Ranieri 2000). In a sense, the National Assessment of Courses of the 1990s could be seen as a step backward in terms of centralization, although, in many cases, there was a genuine effort to limit the assessment to very central skills and competencies, allowing for local experimentation and variations.

The Policy Problem

In the 1940s and 1950s, the Federal government created a network of Federal Universities, established usually by the absorption, through legislation, of existing private and state-based institutions, based on political considerations, without any mechanisms of quality assurance. At the same time, new private institutions emerged, first as religious and community-based institutions, and later, predominantly, as profit-oriented endeavors. In 1968, there was an important university reform, introducing several innovations taken from the American context – graduate degrees, the credit system, departments, and institutes – with the assumption that all higher education should evolve toward a university model, based on academic research and a full-time academic profession. Simultaneously, however, the government responded to the growing demand for higher education by making it easier for private institutions to open up and offer degrees, without too much control and oversight. By the 1990s, higher education in Brazil had expanded very rapidly. The number of students doubled in 10 years, from 1.5 to more than 3 million, two thirds of them in private institutions. Some of these institutions tried to follow the 1968 model of university organization. Most of them, however, provided just one or a few course programs, particularly in business administration or law, without postgraduate education, and based on part-time lecturers, drawn from the professions or from retired or moonlighting academics from the public sector.

The pressures for and against opening up new institutions and controlling their quality comes from many sides (Schwartzman 1998). Brazil's higher education coverage, at about 11% of the 18–24 age cohort[3], is still very limited, and the social and economic benefits of higher education degrees and the entrance in the learned professions are very high, creating a growing demand for more places. In recent years, the provision of private higher education became a multibillion dollar business, employing about 200,000 people, among lecturers and administrative workers (Schwartzman and Schwartzman 2002). Side by side with small institutions, there are now very large private universities, with tens of thousands of students in many different locations, with considerable ability to lobby the government and Congress for freedom from control and regulation. Opposition to the expansion comes from the professional organizations, particularly in Medicine and Law, who are concerned about the watering down of their professional standards and job market privileges; and from the academics and students in public institutions, for similar reasons.

Quality assurance is not, however, a problem limited to the private sector. The Brazilian legislation grants full academic autonomy to universities, many of them public, which includes the right to create new course programs and to define the number of students admitted each year. The assumption is that universities are established according to strict academic standards, but in fact public universities can be created by Federal or state legislative acts. In principle, private institutions need to be accredited to get university status and be granted the same autonomy, but, in practice, accreditation has been granted case by case, without any systematic assessment. A new type of institution has been officially recognized in recent years, the "university centers", which are (mostly private) institutions dedicated solely to teaching, expected to be of good quality, which have almost the same autonomy as the

[3] This is the net rate. In 2005, according to the National Household Survey from the Brazilian Institute for Geography and Statistics (IBGE), there were 5.183 million students in higher education in Brazil, 6% of which were in advanced, graduate education. Of those, only 53% were in the expected age cohort of 18–24. The gross rate of enrollment, comparing all students irrespective of age with the corresponding cohort, was 21.2%, still a very low figure compared with other countries in the region.

universities.[4] Thus, the authority of the Ministry of Education is limited to the approval for the creation of new universities and university centers in the private sector, and to the minute oversight of non-university institutions, which have to apply for each new career they want to establish, and for the number of students they expect to admit.

The demands for a system of quality assurance, beyond the bureaucratic and ineffective procedures of the Ministry and the National Council of Education, on their different incarnations, has been clear since at least the Presidential paper on Higher Education of 1985 (Brasil Ministério da Educação 1985), and has led to several initiatives since then. They included a program to provide universities with resources to develop their self-evaluation (Ministério da Educacão and Secretaria de Educação Superior 1997) and the establishment of National Commissions of Specialists to define and revise the minimum core curricula of the different careers.[5] In the late 1995, under Minister of Education Paulo Renato de Souza, a comprehensive system of assessment of higher education was created. It included the development of a yearly census to provide quantitative information on the sector by region, state, fields of knowledge, and type of institution; qualitative assessments of each institution, large and small, looking at their installations, institutional development plans, research performance, and other indicators of quality; and assessment of individual course programs or careers, with two components. The first was an assessment of their resources in terms of academic personnel, infrastructure, and internal organization if they had clearly defined missions, self-assessment, and coherent pedagogical projects. This assessment was carried on by peers, who visited each course program to get the information and process them according to predefined template. The second was the National Assessment of Courses, an exam to which all the students had to present themselves before graduation. Postgraduate education (Masters and Doctoral programs), in the meantime, have been subject to a well-established assessment procedure which remained in place (Ministério da Educacão 2002).[6]

Implementation

The authority for the Ministry of Education to implement the assessment was established by federal law[7], which made it mandatory for students to complete the test if it is applied to their field in their last year of graduation, as a precondition to obtain their degrees. This was possible because higher education degrees in Brazil, to be

---

[4] By the end of 2003, the Brazilian government issued a Decree that forbade the creation of new university centers, and set a time limit for their transformation into universities or reversion to non-autonomous status (Brasil Presidência da República 2003).

[5] For a discussion of this program, see Amaral and Polidori (1999).

[6] In 2004, this system was replaced by a new one, the National System for Assessment of Higher Education (SINAES). For an official description, see http://www.inep.gov.br/superior/sinaes/.

[7] Federal law 9131/95.

legally valid, have to be registered with the Ministry of Education, usually through the office of a Federal university. However, there is no minimum pass grade for the students, since the goal is to assess the course program, not the student. In the first years, the National Student Union asked their members to boycott the exam, and, in some institutions, the students would just sit without answering the questions. This, however, led to a low ranking to their courses, which reflected badly among their colleagues who did participate, and this practice was abandoned almost completely in the following years.

The implementation for the National Assessment was carried on by an agency within the Ministry of Education, the National Institute for Education Research (INEP) following a very elaborate procedure.[8] First, an assessment committee was established for each field of knowledge. Members were chosen from lists prepared by professional associations, teaching and scientific associations, and by the Brazilian Council of Rectors and the Ministry of Education. They had to be also representative of Brazil's different regions, and different types of institutions – public and private, large and small. From these lists, the Ministry of Education would choose seven names in each area. Thus, for the year 2002, there were 24 such commissions, in Administration, Law, Civil Engineering, Chemical Engineering, Veterinary, Dentistry, Electric Engineering, Journalism, Language and Literature, Mathematics, Economics, Mechanical Engineering, Medicine, Agronomy, Biology, Physics, Psychology, Chemistry, Pharmacy, Pedagogy, Architecture, Accounting, Nursing, and History. They met in Brasilia, and their task was to define the general contents, scope, and goals of the assessment of their fields. For their work, the Ministry obtains all the course descriptions, pedagogical projects and teaching programs adopted by all institutions in the country, and organize this material in terms of their goals, objectives, basic bibliography, teaching procedures, and so on, identifying eventual differences in these orientations and goals. The Commissions work also with reports of the assessment of previous years, prepared by the Ministry and through seminars held with the participation of course coordinators and professors in each field. Based on this information, it is the task of the Commissions, each year, to revise and improve on the guidelines of the previous year in an interactive and continuous learning process.

Once ready, the guidelines prepared by the Committee are passed on to an external contractor, who has the responsibility of developing the tests, administering them, and tabulating its results. The choice of this external contract is made through open, competitive bids. In practice, two institutions working together, the Fundação Carlos Chagas in São Paulo and Fundação Cesgranrio in Rio de Janeiro, have won all these bids since 1995. They are experienced in administering large-scale assessments, having started with the entrance examinations for public institutions in São Paulo and Rio de Janeiro. They also recruit academics in the universities to develop the instruments, and persons in different institutions to deliver, control, and oversee the exams.

---

[8] See, for a detailed description, INEP (2002).

Before the exam, the institutions have to provide a list of all students likely to conclude their course programs in a given year. The exam takes place on the same day throughout the country, and is widely announced in the press. Observers from professional associations, teaching associations, and other entities are regularly invited to be present in the different locations where the exam takes place.

Just before the exam, the students receive from the Ministry of Education a magazine, explaining the purposes of the exam, the description of the procedures, and other materials. The core instrument is a written exam, which can be either of a multiple-choice test or open-ended questions, or both, according to the Committee's recommendations. The general orientation is to put emphasis on the mastery of key concepts, on the ability to think independently and to apply knowledge to new situations; rote learning and the accumulation of information for its own sake are discouraged. Another instrument is a survey questionnaire, in which the students are asked to provide socioeconomic information on themselves and their families, and their views and perceptions about their course programs. A third instrument is the student's assessment of the assessment – if they like the instrument, if they considered it too easy or too simple, inappropriate, etc.

The correction of the multiple-choice tests is done through the use of optical scan technology, and grades are provided after an assessment, by the Commission, of each item's discrimination, level of difficulty and reliability. For open-ended questions, a sample of the responses is used to develop an assessment protocol, which is then applied to the universe of respondents. The grades received by each student are established according to their relative place in the distribution of results for the whole country. According to the mean results of their students, each course program receives a grade – from A to E. The student's individual results are made available confidentially to each student, in a bulletin with information about his relative placement regarding his class, region, and country. The course's mean score, however, is made public.

There are several follow-ups, besides the establishment of the grades. Immediately after the exam, the correct answers to the questions are made public, so that the students can see what they did right or wrong, and the professionals in the field can assess the quality of the exam. Then, the aggregate results of the students' assessment of the assessment for each course are made available on the Internet to the course coordinators.

The next step is a series of national seminars, for each field of knowledge, to discuss the results of the last exam, with the cooperation of professional associations, course coordinators, and university. In these seminars, the results are presented, the Commissions share their views, complaints are aired, and the officers from INEP in charge of the whole process have an opportunity to hear the views of the academic community and express their perceptions of the whole process.

Meanwhile, the Ministry prepares a series of technical reports for each exam, and also a summary of the main statistics obtained with the socioeconomic questionnaire, which helps clarify the characteristics and attitudes of the students. There is a report with a synthesis of all the results, reports for each field of knowledge, and individual reports sent to the persons in charge of each course program.

Finally, some research institutions and independent researchers are asked to make more in depth analysis of the data, which may be disseminated by the Ministry, published as academic papers, or remain as technical reports of limited circulation.[9]

There is no estimation of the total cost of the operation. In 2002, the cost paid to the external contractor was about 36 million reais, or 12 million dollars (according to the prevailing exchange rate at the time). With these resources, they were required to prepare 24 different exams to be applied to 361,000 students graduating from 5,000 course programs in 627 municipalities. The per capita cost was, therefore, 100 reais or 33 dollars per student. There are many more course programs in the country, but these 24 accounted for about 90% of the students graduating in that year. There is no information about the internal costs for the Ministry of Education, which includes travel of the 168 members of the academic commissions to meetings in Brasilia, the time of the staff working in the preparation of the materials for the Commissions to work, the organization of seminars and other events, and contracts with external consultants for the analysis of the data. It is a sizeable effort, but not out of proportion, if one considers that the Ministry of Education spends about 5 billion reais – 1.6 billion dollars – a year in higher education alone.

Impact

The Brazilian legislation gives to the Minister of Education, with the support of the National Council of Education, the authority to accredit new higher education institutions and to renew their accreditation periodically. In practice, however, once a higher education institution is allowed to function, only in extreme cases will it lose its authorization or accreditation, and the process of periodical accreditation and reaccreditation of universities was never fully implemented. The government has intervened in a few private institutions in recent years, but never in a public university, and never because of a negative assessment of their academic quality (in some cases, attempts by the Ministry to close down bad quality courses and institutions were stopped by the judiciary or by appeals to the National Council of Education). The National Assessment of Courses was meant to be influential information for such decisions, but, since it refers to course programs, and not to whole institutions, it can be at most one element of information in a much broader assessment procedure, still to be implemented.

Because of this, the direct contribution of the National Assessment for the regulation of higher education has been minimal. Its indirect impact, however, is considered very important. One such impact was to encourage students to search for better-ranked course programs. According to a study done by the Ministry of Education, the number of new applicants for courses in Administration, Law, Civil Engineering, Chemical Engineering, and Dentistry, who received "D" and "E" in the

[9] The Center for Studies of Public Opinion of the University of Campinas was commissioned to prepare these reports, which were not confidential, but were not widely distributed. See Meneguello et al. (2002).

assessment, went from 35,000 to 18,000 between 1997 and 2001, while the demand for courses rated "A" increased by 6%. Another finding was that new courses in private institutions, established after the assessment was introduced, tend to be better than many old ones. Thus, both students and academic officers are taking the concepts into account, and changing their behavior, looking for better course programs, and trying to work according to higher standards (Ministério da Educação and Instituto Nacional de Estudos e Pesquisas Educacionais 2002).

The professional and academic associations in Administration carried out a detailed survey among course coordinators on the impact of the assessment in their institutions (Conselho Federal de Administração and Associação Nacional de Cursos de Graduação em Administração 2003). They asked whether the institutions introduced changes in their course programs in the last 3 years, and whether these changes were induced by the assessment or not. They found that about 65% of the course programs reported changes in the period, half of which attributed directly to the national assessment. In general, private institutions reacted more to the assessment than public ones, but the difference is not large – 38–30% of all the course programs in the sample. Not surprisingly, the most frequent innovation was to prepare the students to take the exam, followed by changes in pedagogical and teaching practices of different kinds (Table 15.2). Changes involving investments, infrastructure, and salary raises were much less frequent.

The few studies that exist on the socioeconomic characteristics of the students, with the information produced by the exams socioeconomic questionnaire, provide very interesting information, some of it unexpected (Meneguello et al. 2000). In general, achievement has to do much more with the characteristics of the institutions than with the characteristics of the students, and the correlations between socioeconomic status and achievement are not high. Part of the reason is that course programs in the private sector tend to be of lower quality than those in the public sector, but students in the private sector come from families with higher income than those in public institutions. The other reason is that, once the students are able to reach higher education, they have already overcome most of the disadvantages that would usually affect their academic performance. There are, however, important differences in careers choice: the parents of more than half of the students in journalism, law, engineering, and medicine have a higher education degree against less than 20% for those in teaching careers: mathematics, language, and pedagogy, where the percentage is under 10%. There are some differences among public and private institutions, but they are much less significant than those among careers. Finally, detailed regression analysis confirms that achievement depends, above all, on whether the student is in public or in private institutions, and on factors like age, knowledge of English, hours dedicated to study, work, and whether the student attended public or private secondary education (with best results for those coming from private schools).

One of the most important contributions of the National Assessment, not readily documented but very clear in the minds of those responsible for its implementation, was the opportunity it provided for course coordinators, academics, and professional associations to come together in a continuous process of discussion and negotiation about the quality standards of their respective fields. Beyond the efforts of many

Table 15.2 Main changes in administration of courses induced by the National Assessment of Courses

	Percentage of coordinators reporting the change
Using question items of the assessment in classroom	82.2
Changing teaching methods	68.3
Changing course contents	66.8
Assessing the students abilities to perform in the exam	61.4
Interdisciplinary work	55.5
Mock assessments	55.5
Improving the library	54.5
Upgrading the teaching staff	52.0
Working to improve the image of the institution	51.5
Improving the use of the library by the students	47.0
Strategic planning	41.1
Marketing	41.1
Hiring new staff	39.1
New multimedia resources	35.6
Internet access	34.7
Teacher training	33.7
Links with firms and business sectors	32.2
Trainee programs for students	32.2
Investments in computers	31.7
Interactions with the community	29.7
More working time for staff	25.7
Better equipment in classrooms	25.3
More fellowships for students	13.4
Higher salaries for academic staff	8.0

institutions to "learn the tricks" of the exam to get better grades, there are many stories of institutions looking for help to improve their courses, and others closing down because of lack of student demand.

Opposition and Criticism

From the onset, the National Assessment was received with strong opposition from the National Students Union (UNE) and some public universities. The Student Union asked the students to boycott the exam, and tried to disrupt its implementation. Both the Student Union and the Federal University of Rio de Janeiro went to Court trying to stop the assessment from taking place. The students argued, among other things, that the assessment would hurt the students from the institutions receiving lower ratings. The eventual shortcomings, however, were not the responsibility of the students, but of their institutions, or the government, which did not provide the institutions with the support they needed. The arguments coming from public universities were similar. If they did not perform well, it was because they were not getting the necessary support, and should not be punished for that. There were other criticisms, from general statements about the impossibility to measure and quantify quality, to a principled stand against establishing comparisons and competition among institutions and students, bringing a market mentality to the realm of culture and education.

These criticisms have to be placed in the Brazilian political context of the time. Both the National Student Union and the higher education teachers' association were in the opposition to the Fernando Henrique Cardoso government, and strong critics of whatever initiative came from the Ministry of Education, for good or bad reasons. Cardoso, a renowned sociologist and former professor at the University of São Paulo, had a history of strong opposition to the Brazilian military regime that lasted until 1985, and was elected President in 1994 after being able, as Brazil's economic minister, to bring the country's inflation under control. His mandate, which lasted until 2002, was characterized by very significant efforts to bring order to the economy and reduce the runaway expenses of the public sector. It was a period of economic stagnation, his government was accused of obeying the neoliberal orientations of the International Monetary Fund (IMF), and one of the strongholds of the opposition was the organized civil servant unions, including those working in public universities.[10]

Partisan reasons aside, several criticisms to the assessment are reasonable. The adoption of a single, unified exam in each field for all course programs in the country led all the institutions to adjust to the same mold, and may have thwarted their freedom to experiment and to diversify. By selecting a group of specialists to write up the exam, the Ministry made the particular biases of this group the national standard. This policy was coherent with the traditional view that all higher education course programs in a given field should provide the same contents and equivalent professional certifications. In areas with well-established academic and professional paradigms, this is not controversial; but this is the exception, rather than the rule, in a highly differentiated mass higher education system, with different types of students, institutions, and visions about what the contents of higher education should be.

The decision to make public the place of each course program in a five-point scale, based on the distribution of results, was related to a conscious decision not to establish clear references, or cutting points, in relation to which a given course program could be considered acceptable or not acceptable. So, in a field where all course programs are of very good quality, 12% of them or so would be ranked as "E", while in another, where all course programs are bad, 12% would receive an "A"[11]. In other words, all courses are ranked by uniform criteria, and the public is informed about their relative position in the rank, but not if they are of good quality or substandard. The reason for this was never spelled out very clearly, but it is not difficult to understand. The establishment of cutting points would be very

[10] See, on the period, Cardoso and Font (2001); Font (2003).

[11] Until 2000, the grades were distributed according to fixed percentages – 12, 18, 40, 18 and 12%, for A, B, C, D, E. Since 2001, the normal distribution was used, with courses above one standard from the mean receiving an A, and those one standard deviation below receiving an E.

controversial, and the official information that many, perhaps the majority, of the course programs in many fields are substandard – a very likely result – would create a crisis situation the Ministry could not possibly handle.

A third criticism is that the assessment may be measuring the cultural capital the students bring to the university, rather than the education value added to them by their courses. Prestigious institutions attracting very good students would have good results even if the courses were bad; hardworking and dedicated institutions accepting students with poor backgrounds would not be able to get higher marks, regardless of their effort. It would be possible to estimate the value added by the courses by taking into account the student's achievements on their entrance examinations to the university, or their achievements in a another national voluntary test, applied to students at the end of secondary school. A statistical analysis using information from the student's university entrance examinations in the state of Minas Gerais shows that, indeed, previous conditions affect the final outcome, but that, in general, this information would not change the final rankings in the national assessment, except in a few isolated cases (Soares et al. 2001).

A fourth criticism was that, by looking only at the student's results, without considering other variables related to the academic staff, installations, computer facilities, library resources, and so on, the National Assessment was not an complete assessment instrument, but at most a partial one. In fact, together with the Exam, the Ministry of Education developed another assessment procedure of these input variables with heavy weight given to the academic degrees of the faculty (the percentage holding doctoral and master degrees) and the percentage with full-time contracts, plus an assessment of their physical installations and their pedagogical project, if any. Initially, the Ministry ranked the course programs according to a combination of these instruments. The information on inputs is necessary and useful, but there are good reasons not to combine input and output effects in the same scale, since it is important to know, for instance, which inputs are more effective than others in producing the outcomes. Besides, most lecturers in public institutions are nominally full-time, while most in the private sector are not, and this introduced a bias in favor of public against private institutions.

The Commission established by the Ministry of Education in 2003 to propose a new national assessment system for higher education in Brazil presented a detailed criticism of NEC, and suggested a different path. Some of the criticism was technical, like the ones mentioned above, related to the lack of clear standards and the measurement of the education value added to the courses, and the lack of comparability of results through time. Others were more political and ideological, like the statement that the exams responded to "motivations coming from outside, rather than inside the institutions, leading to isolated distorted and wrong representation of the academic world", or that "its rationality was much more market-oriented ("mercadológica") and regulatory than academic and pedagogic." Other criticisms, finally, were related to the growing cost of the assessments. According to the report, the current costs are likely to grow, as higher education expands and new fields and disciplines were included in the assessment (Comissão Especial da Avaliação da Educação Superior 2003).

ENADE – The New Higher Education Assessment

With the change of government in early 2003, the original team responsible for the establishment and implementation of the National Assessment within the Ministry of Education was disbanded, and most of the institutional memory and experiences accumulated in recent years was lost. In 2003, the Ministry of Education still went ahead with the implementation of ENC, following the standing legislation, but without carrying on the usual procedures of analyzing the results with the participation of the academic committees. In December of 2003, at last, the government issued a "provisional act" (Brasil Presidencia da República 2003)[12] changing the legislation regulating the whole higher education assessment system, while the Ministry of Education issued another document spelling out how it intended to proceed. The provisional act created a new system for the assessment of higher education, based on two new National Commissions, one to provide guidelines and another to implement the new procedures. The members of both institutions are to be nominated by the government, the first among persons with recognized competencies and representatives of the "organized civil society" of students, teaching and administrative staff, and the second among civil servants from the Ministry of Education.

The new system is supposed to rank the "institutional quality" of higher education establishments on three levels, satisfactory, regular, and not satisfactory. The five-level ranking system disappears, and the new ranking will combine the results of four different assessments: institutional capabilities, teaching, knowledge production processes (presumably research), and social responsibility. In another document, the Ministry of Education spelled out the broad outlines of the new assessment it expects to undertake (Ministério da Educacão 2003). There will be an "Index of Development of Higher Education" (inspired in the Index of Human Development of the United Nations Development Program) which will combine the results of the four assessments. The National Assessment of Courses remains, with a different name, to assess the learning process. But, instead of a yearly universal assessment of all graduating students and course programs in specific fields, the assessments will be done now every 3 years through sampling procedures. And, instead of just one assessment, there will be two, one at the beginning, the other at the end of the course program.

Apparently, the new system was more comprehensive than the previous one, and more friendly to the institutions being evaluated. In practice, it was too ambitious, and impossible to implement. The former Minister of Education, Paulo Renato de Souza, in a press conference, indicated some potential problems, stating that, in practice, the government was shutting the assessment system down, and other observers also raised questions (Souza 2003). By making the participation in the

[12] In Brazil, it is possible for the Executive branch to create laws through provisional acts ("Medidas Provisórias") which are valid immediately, but can be changed or rejected by Congress within a short limit of time. This is supposed to be used only on extraordinary situations, but, in practice, it is used whenever the government wants to avoid the lengthy procedures of sending ordinary bills to be discussed in Congress.

assessment voluntary for the students, the Ministry would not be able to get them to participate; the proposed sampling procedures were not spelled out; it was not clear how the assessment of a sample of course programs could be combined with the assessment of institutions; it was not clear whether the assessments to be published would refer to course programs or to institutions as a whole, which seems to be the case; the new legislation bypasses the National Council of Education; and the new evaluation committees are likely to represent the existing unions of students, lecturers, and civil servants, rather than the country's existing academic and professional communities. Finally, by combining the results of the assessment of outputs with three other assessments, supposedly with the same weight, the new procedure was likely to obscure, for society, the main information it wants, the quality of the education provided in specific course programs, which may vary widely within the same institution. In spite of these criticisms, the new system was presented as an important improvement over the past.

The new assessment, now called the National Exam for the Assessment of Students – ENADE, was carried on in the years of 2004, 2005, and 2006. The 2005 results are available in the website of the Ministry of Education for each institution and as a very comprehensive technical report (SINAES – Sistema Nacional de Avaliação da Educação Superior 2006). The full availability of results was part of a new policy of the Brazilian Institute for Education Research, INEP, to provide detailed information on assessment results at all levels, to allow for comparisons and stimulate competition for quality, an important departure from the previous stand of the Ministry of Education against quantitative assessments and comparisons.[13] Another important function of this policy is to allow for a critical assessment, and perhaps improvement, of the assessment instruments in place. Looking at the data and the methodological explanations published in the official documents, it is possible to say that, so far at least, the new assessment for higher education did not overcome the flaws pointed out by observers when it was announced (Schwartzman 2005; Verhine et al. 2006).

The first flaw was the adoption of sampling, instead of universal coverage, for course programs with more than 20 graduating students. The justification was that it would be cheaper. However, the sample fraction of ENAD 2005 was about 50%. The additional cost of applying the test to all students in the same class in course program is probably not higher than the procedures of selecting one in two randomly; in any case, it is impossible to know, since no information is available on costs. One problem with the sampling is the possibility of selection bias, with only the best students showing up for the test. The Ministry of Education states that participation in ENADE is mandatory for the selected students, as a condition for receiving their degrees, but presents no information about the number of selected students who did not participate, and what happened with them. Two of the leading universities

[13] The information available both at the site of INEP and in microdata includes now, among others, data on basic education by school (Prova Brasil), the National Exam for Secondary Education (ENEM), the basic and higher education censuses, and ENADE.

in Brazil, the Universidade de São Paulo and the Universidade e Campinas, both state institutions, refused to participate in ENADE, arguing that they have their own assessment system, and it is doubtful that the Ministry of Education will sanction them or their students.

One criticism of ENC was that it did not measure the added value of education in the course programs, since it did not take into account the student levels of competence at the start. ENADE tried to overcome this limitation by applying the same tests for students both in the first and last year of their courses. It also developed a new test to measure the "general competence" of students, to be applied side by side with the specific assessments of each field. However, instead of using these tests to measure the value added, considering the differences, they decided to add them in one composite measure, to establish the final grade of the course program in a five-point scale. This final grade is a combination of the results of the graduating students in the specific test, weighting 60%; the results of the entering students in the same test, weighting 15%; and the combined results of entering and graduating students in the general test, weighting 25%. The rationale for this strange procedure is nowhere to be found, but the consequence is that it makes the final result more biased in favor of institutions that get the best students in the first place.[14] The consequence of this procedure was to raise the scores of the public institutions, which usually get the best students, even if they did not add much to their previous knowledge. A comparison between the results of ENADE 2004 and ENC 2003 for medical schools (Table 15.3) showed that their correlation was small ($r^2 = 0.16$), and that ENADE's results tend to be fairly homogeneous and much higher than those of ENC, which may please the institutions, but does not make the assessments more reliable.

Another problem is the validity of the tests. It is impossible to know what the test of "general capabilities" actually measures. In the legislation, and in the official document, it says that the test should measure "how far the student is being

Table 15.3 Results of ENC 2003 and ENAD 2004, 78 medical schools

Grade in ENC 2003	*Mean grade in ENADE 2004*	*Cases*	*Standard deviation*
1	3.45	11	0.93
2	4.13	8	0.64
3	4.19	37	0.57
4	4.69	13	0.63
5	4.33	9	0.50
Total ENADE	4.18	78	0.72
Total ENC	3.01	78	1.14

[14] Thus, in a hypothetical case, if the grades for entering and concluding students are 80 and 100, the added value would be 20, the combined value of these two components would be 80*.15 + 100*.6 = 72; if the grades were 50 and 100, the added value would be 50, but the combined value would be 60.7. So, in two courses with the same level of achievement for the graduating students, the one which added less to their previous background would have a higher final mark.

educated as a professional who behaves ethically, is competent and committed with the society in which he lives. The test should also measure the student's ability to analyze, synthesize, deduct, develop hypothesis, establish relations, make comparisons, detect contradictions and organize their ideas" (SINAES – Sistema Nacional de Avaliação da Educação Superior 2006, p. 12). However, the test is made up of just ten questions, of which seven are multiple-choice, and three require written answers. It is clearly impossible to measure all the expected dimensions with so few questions, and more so without very strong validation procedures, which did not exist. The validation of the tests for the specific areas did not exist either, except for the item's correlation and discrimination, a flaw that ENADE shares with the previous ENAC.

The 2005 report presents an extensive and laudable effort by INEP's statisticians to analyze the results in different ways, combining the test results with the answers to a large socioeconomic questionnaire the students also had to fill in. One of the most intriguing efforts was the development of a measure comparing the expected and actual achievement of the course programs, based on the general characteristics of the entering students. However, the meaning of the figures presented is impossible to ascertain. For instance, in a scale of 100 points (presumably before standardization), students in physics gained 1 point in the scale of general capabilities, while students in computer science gained 5.9 points; in the scale of specific capabilities, the gains were 7 and 10.4, respectively; but there is no way to know what these differences actually mean.

There are other conceptual and technical flaws with ENADE, but these are enough to draw the general picture. The Ministry of Education presents ENADE as just one component of a larger system of assessment, which should not be considered in isolation, and does not use nor recommend its use for policy purposes, or for the students to choose the places where they will study; but the results of the other components of the assessment are still to be published, and it is not clear how they will be used. (An obvious use of this assessment would be to select the institutions that receive students which benefit from "University for All", a government program to provide free higher education in private institutions for poor students). For all these reasons, the publication of ENADE results has been received with much less interest than what used to happen with the publication of the results of ENC.

The Future

The Brazilian National Course exam, in its two incarnations, is a unique and extraordinary experience, which has generated admiration and interest in higher education circles in different parts of the world, and received strong support in the Brazilian public opinion. Its future, however, is uncertain. On hindsight, it is possible to say that the main weakness of the original National Course Exam was its lack of proper institutionalization, and the absence of a clear sense of ownership of the exam within

Brazil's higher education and professional communities. The Exam started as a personal initiative of the Minister of Education, Paulo Renato de Souza (an economist who had been the rector of the University of Campinas and a high-ranking officer of the Inter American Development Bank) who had to start by convincing his own staff of its need, and implementation given to the most flexible and independent branch of the Ministry of Education, the National Institute for Education Research, INEP. In principle, other institutions could have taken this task – the Secretary for Higher Education within the Ministry, the National Council of Education, the National Conference of Rectors – and a new institution could have been created with this purpose, like the National Commission for the Assessment and Accreditation of Universities in Argentina.[15]

Had the Minister decided to work through one of these institutions, subject to all kinds of interest groups and administrative hurdles, or to create a new one, he might not have succeeded in moving so rapidly, and achieving so many significant results in such a short time. Acting on the power of his cabinet and thanks to his personal prestige, it was possible to move quickly through the complex legislative process to get the legal authorization, and to place the necessary human and financial resources in the hands of the able head of INEP, Maria Helena Guimarães Castro.

The price, however, was that no institution or segment of the academic community claimed ownership of the Assessment, except a small team within the Ministry of Education. Hundreds of academics were asked to participate in the Commissions and probably did a very important work, but they were there by the Minister's invitation. The statistical data generated by the exams remained under the Ministry's control; some qualified researchers and research centers were invited to analyze them, but they were not made publicly available to the academic community of education research specialists. In the effort to keep up with the complex procedures established for the Assessment, most of the energy of INEP's staff was dedicated to the preparation of technical documents and other materials for the Commissions, the students and the course program coordinators, with little left for the deeper reflection on the general important and significance of the Assessment. To conquer public opinion, the Minister had a competent public relation staff, which kept the press well informed of the achievements of the Assessment, and helped to win the battle of the public opinion against the organized opposition

Without clear ownership in society, established as just one activity within one sector of the Ministry of Education, the National Assessment did not have the strength to resist the impact of a change in administration. The new Evaluation Commission established by the government after 2002 to implement the new assessment system could have been a step in the right direction, if these Commissions could become truly independent and autonomous from vested interests. This did not happen, however, given their membership: representatives of the "organized society" – unions of students, lecturers, and university employees, known to have been the strongest opponents of the National Assessment of Courses as it existed, and

[15] Comisión Nacional de Evaluación y Acreditación Universitaria (CONEAU).

political appointees in the Ministry of Education. The new Commission did not have the strength to abolish the national exam of courses altogether, but was effective in reducing their strength as a reliable information for the public and an instrument for quality assurance of Brazilian higher education.

To become a stable and significant feature of Brazilian higher education, the Assessment would have to find a permanent institutional house, which can be neither the Ministry of Education nor the unions and corporations with vested interests against any kind of external assessment of their own work. Between these two extremes, a proper space will have to be found, if the experience of recent years is not to be lost.

Acknowledgments I am indebted to Gilda Portugal, Maria Helena Guimarães Castro, and, particularly, to Jocimar Archangelo, for the wealth of insight and information on the National Assessment of Courses. I am also grateful to Cláudio de Moura Castro, Jacques Schwartzman, and Maria Helena Magalhães Castro for detailed comments and criticisms on the first draft of this chapter.

References

Amaral, A., Polidori, M. (1999). Quality evaluation in Brazil: a competency based approach? *Higher Education Policy*, 12(2), 177–199.

Azevedo, F. de (1971). *Brazilian culture: An introduction to the study of culture in Brazil*. New York: Hafner.

Brasil Ministério da Educação (1985). Uma Nova Política para a Educação Superior Brasileira – Comissão Nacional para a Reformulação da Educação Superior – Relatório Final. Brasilia: Imprensa universitária da Universidade Federal do Ceará.

Brasil Presidencia da República (2003). Medida Provisória 147, de 15 de dezembro de 2003. Institui o Sistema Nacional de Avaliação e Progresso do Ensino Superior e dispõe sobre a avaliação do ensino superior. Pp. Seção 1, p. 2 in Diário Oficial da União. Brasília.

Brasil Presidência da República (2003). Dispõe sobre os centros universitários de que trata o art. 11 do Decreto no. 3.860, de 9 de julho de 2001, e dá outras providências. Pp. 50 in Diário Oficial da União. Brasília.

Cardoso, F. H., Font, M. A. (2001). *Charting a new course the politics of globalization and social transformation*. Lanham, MD: Rowman & Littlefield.

Coelho, E. C. (1999). *As Profissões imperiais: advocacia, medicina e engenharia no Rio de Janeiro, 1822–1930*. Rio de Janeiro: Editora Record.

Comissão Especial da Avaliação da Educação Superior (2003). Sistema Nacional de Avaliação da Educação Superior SINAES: Bases para uma nova proposta de avaliação da educação superior.

Conselho Federal de Administração and Associação Nacional de Cursos de Graduação em Administração (2003). Alterações, efeitos e influências do provão de administração na opinião dos coordenadores dos cursos de administração. Brasília: Conselho Federal de Administração.

Conselho Federal de Administração, Associação Nacional de Cursos de Graduação em Administração, and Ad Homines (2003). *Alterações, efeitos e influências do provão de administração na opinião dos coordenadores dos cursos de administração*. Brasília: Conselho Federal de Administração.

Durham, E. R. (2004). Higher education in Brazil – public and private. In C. Brock, S. Schwartzman, *The Challenges of Education in Brazil*. Oxford, UK: Triangle Journals.

Font, M. A. (2003). *Transforming Brazil: A reform era in perspective*. Lanham, MD: Rowman & Littlefield.

INEP (2002). Provão 2002 – Relatório Síntese – Resumo Técnico. Brasília: Instituto Nacional de Estudos e Pesquisas Educacionais.

Malloy, J. M. (1977). *Authoritarianism and corporatism in Latin America*. Pitt Latin American series. Pittsburgh: University of Pittsburgh Press.

Meneguello, R., Lourenço, F. A., Perez, J. R. R., Dentzien, P. A., Silva, A. M. A. C., Ferreira, C. S. da, Del Porto, F. B., Silva, F. A., Rezende Barreto, J. de, Galvão, J. H., Vilton Costa, J., Cencig, P. V., Gelape, R. S., Stocco, S., Silva Aranha, S. da, Vieira, V. L. C. (2002). *Relatório Final, Meta 4 – Análise dos dados do Exame Nacional de Cursos – Provão. In Projeto Estudos Socioeconômicos em Educação*. Campinas, SP: Universidade Estadual de Campinas, Centro de Estudos de Opinião Pública.

Ministério da Educacão (2002). Avaliação e informação como instrumento de política educacional. In Educação – Políticas e Resultados. Brasília: Ministério da Educação.

Ministério da Educacão (2003). Sistema Nacional de Avaliação e Progresso da Educação Superior: Uma nova sistemática de avaliação do Ensino Superior brasileiro: Assessoria de Comunicação Social.

Ministério da Educação and Instituto Nacional de Estudos e Pesquisas Educacionais (2002). Cai demanda por cursos com baixo desempenho no Provão. In Notícias do Provão, vol. November 5, 2003.

Ministério da Educacão and Secretaria de Educação Superior (1997). Programa de Avaliação Institucional – PAIUB.

Ranieri, N.B. (2000). Educação superior, direito e Estado na Lei de Diretrizes e Bases. Lei no. 9.394/96. São Paulo: EDUSP/FAPESP.

Schmitter, P. C. (1974). Still the century of corporatism? In F. B. Pike, T. Stritch (eds.), *The new corporatism: Social-political structures in the Iberian world* (pp. 85–131). South Bend, IN: University of Notre Dame Press.

Schwartzman, J., Schwartzman, S. (2002). O ensino superior privado como setor econômico. *Ensaio – Avaliação e Políticas Públicas em Educação* 10, 411–440.

Schwartzman, S. (1988). *Bases do autoritarismo brasileiro*. Rio de Janeiro: Editora Campus.

Schwartzman, S. (1991). *A space for science the development of the scientific community in Brazil*. University Park: Pennsylvania State University Press.

Schwartzman, S. (1998). *Higher education in Brazil: The stakeholders*. Washington, DC: The World Bank.

Schwartzman, S. (2005). O enigma do ENADE. IETS – Instituto de Estudos do Trabalho e Sociedade.

Schwartzman, S., Bomeny, H. M. B., Costa, V. M. R. (2000). *Tempos de Capanema*. São Paulo, Rio de Janeiro: Paz e Terra; Editora da Fundação Getúlio Vargas.

SINAES – Sistema Nacional de Avaliação da Educação Superior (2006). Resumo Técnico – ENADE 2005. Brasilia: Ministério da Educação.

Soares, J. F., Ribeiro, L. M., de Moura Castro, C. (2001). Valor agregado de instituições de ensino superior em Minas Gerais para os cursos de direito, administração e engenharia civil. *Dados – Revista de Ciências Sociais* 44, 363–396.

Souza, P. R. (2003). Observações sobre a Medida Provisória 147 Sobre o sistema de avaliação do ensino superior.

Teixeira, A. (1969). *Ensino superior no Brasil: análise e interpretação de sua evolução até 1969*. Rio de Janeiro: Editora da Fundação Getúlio Vargas.

Verhine, R. E., Dantas, L. M. V., Soares, J. F. (2006). Do Provão ao ENADE: uma análise comparativa dos exames nacionais utilizados no Ensino Superior Brasileiro. *Ensaio*, 14, 291–310.

Chapter 16
Reflections and Conclusions

David D. Dill and Maarja Beerkens

One important lesson to be drawn from the international financial tumult of the early part of the twenty-first century is the failure of existing government regulation and self-regulation to cope with innovations in economic transactions and the forces of globally competitive markets. Similarly, as our analyses suggest, the traditional institutions for assuring academic standards in universities have proven ineffective or inadequate to cope with the new demands of mass systems of higher education and the rapidly changing environment of global academic competition. As a consequence the leading nations are experimenting with new instruments for academic quality assurance and we have attempted to provide balanced assessments of the most interesting and innovative of these policies.

In this concluding chapter we reflect on the lessons to be learned from our studies of professional regulation, market regulation, and state regulation of academic quality. We also attempt to derive from these studies of individual instruments some general guidelines that may prove useful in designing national framework conditions for assuring academic standards in the university sector.

Lessons Learned About Professional Self-Regulation

As noted in our introductory chapter, we regard various forms of external professional self-regulation in higher education as instruments of public policy in that the opportunity for these institutions to play a role in publicly funded systems of higher education is a conscious choice by the state in lieu of alternative regulatory options. The conditional nature of academic self-regulation is illustrated by contrasting the external quality assurance function traditionally performed by state ministries of education in Europe, where as a consequence professional societies played a much more limited role, to the privilege of self-regulation of academic quality granted by the state in both the United Kingdom and the United States. It is also illustrated by

D.D. Dill (✉)
University of North Carolina, Chapel Hill, NC, USA

D.D. Dill, M. Beerkens (eds.), *Public Policy for Academic Quality*, Higher Education Dynamics 30, DOI 10.1007/978-90-481-3754-1_16,

the recent interventions by the national governments in both the United Kingdom and the United States, which have altered this historic grant of professional privilege in an effort to assure the integrity of academic standards.

While the practices currently employed in self-regulatory initiatives, such as UK external examining and innovative specialized accreditation models like TEAC, provide valuable guidance for the design of more effective quality assurance instruments, our policy analyses also revealed some of the limitations of this form of external quality assurance. First, without the authority of government self-regulatory institutions are often ineffective in defining and enforcing academic standards. For example, UK external examining existed for over 100 years, but not until the government intervened to strengthen the process was serious empirical research conducted to assess its effectiveness and formal guidelines published to help standardize practices in the university sector. Similarly, it was not until Congressional legislation in the 1950s and 1960s tied accreditation to federal student aid and ushered in an "era of quasi-regulation" (Orlans 1975) that the regional accrediting agencies implemented the practices now associated with US accreditation. These practices included limiting the duration of accreditation to a 10-year period, adopting seven general review standards, and requiring an institutional self-study, followed by an external peer review (Ewell 2008). Indeed, although the New England Association of Schools and Colleges (NEASC) had been the first accrediting agency to form in 1885, it had only accredited secondary schools until 1954. It was not until that year, two years after the approval of the 1952 GI Bill, that NEASC voted to formally accredit colleges and universities in its region and only "after sharp debate and over the determined opposition of a few leading universities" [i.e., Harvard and Yale] (Selden 1960, p. 37). Similarly, the TEAC accreditation, originally voluntary, has now become a part of the official licensing requirements in several states and seeking state endorsement is a conscious tactic of TEAC as it seeks to increase its impact.

Second, the external quality assurance processes conducted via voluntary self-regulation have often lacked a sufficient focus on academic standards. While from the outset US academic accreditation was advocated as a means of addressing academic standards (Selden 1960), unlike UK external examining it has never directly examined student academic work or assessed marking standards. Rather, the regional accrediting agencies have conducted comprehensive institutional reviews with an emphasis on resources and "managerial" needs and concerns (El-Khawas 1998). Visiting teams have consisted primarily of administrators, often with specialized backgrounds in areas such as libraries, information services, student personnel, and finance (Amaral 1998). It was only following criticisms and increasing pressure for reform from the states and federal government during the 1980s and 1990s that the US regional accreditation associations revised their standards and criteria to place greater emphasis on student assessment practices (Nettles et al. 1998). Nonetheless, a 1997 national survey of student assessment activities in US colleges and universities confirmed suspicions that regional accreditation reviews of institutional assessment efforts made little attempt to actually evaluate their educational impact (Peterson and Vaughan 2002). The new reviews had simply been added to existing reviews of the large number of institutional processes and attributes many of which lay outside the institution's core educational mission. The survey's authors concluded that while regional accreditation appeared to have stimulated the adoption of assessment activities by institutions, it had little influenced institutional support for or use of assessment to improve academic standards.

The issues of focus and scope are also challenges for voluntary professional or specialized accreditation processes. Because these agencies are closely associated with practicing professionals and address subjects that require national licensing exams, some professional accreditation agencies have in fact been more attentive to the design of academic programs and more responsive to the emerging public concern with student learning than the voluntary institutional accrediting agencies in the United States. But the specialized agencies exist only in fields with established professional associations; therefore their coverage of the academic subjects taken by the majority of university students is insufficient to successfully safeguard academic standards. The objectivity and independence of these accrediting agencies has also been questioned, because their close connections with private practitioners raise potential conflicts of interest (Orlans 1975). Furthermore, US university leaders have continually complained that dealing with multiple, uncoordinated, independent professional accreditation agencies is burdensome and time-consuming (Ewell 2008). Finally, the focus of specialized accreditation on particular programs provides inadequate incentives for universities to strengthen or develop institution-level processes for assuring academic standards in all programs, a limitation shared as we will note below with European-style subject assessments and academic accreditation.

A third weakness, as Lewis' analysis of external examining clearly reveals, is that the traditional forms of external self-regulation have been seriously compromised by the changes in academic structure associated with the development of mass higher education. For example, external examiners customarily reviewed the validity of individual student marks at the subject level as part of their assessment. But the worldwide adoption of modular teaching, continuous assessment, and credit-based systems – educational processes more similar to the academic structure in the United States – has altered the role that the examiners can play. With the decline of proctored examinations as the primary basis for awarding academic degrees, external examiners now must focus on evaluating the integrity of the overall subject marking system. Ironically, the reforms in external examining and other external quality assurance systems caused by these curricula changes parallel the earlier US experience with academic quality assurance (Ewell 2008). The introduction of the elective system in the United States during the 1880s, in which individual student choice of course modules largely replaced a required core curriculum, and the subsequent adoption of an academic accounting system based upon credit hours and the so-called "Carnegie Unit" were viewed by many as threats to the consistency and coherence of academic programs. These changes helped motivate the development of both regional and specialized accreditation in the US during this period.

Many countries are also now adopting degree frameworks that involve a hierarchically integrated system of first, second, and third level taught degrees (e.g., the Bologna Accords). Because higher level degrees are often associated with increased economic returns, the issue of fair and equitable access to advanced degrees will likely raise more concerns in the future about the equivalency of standards and equity of student assessments across subject fields in both first and second level degree programs. In this new context the limitations of subject-oriented external examining, which compares the equivalency of marking standards in the same field across universities, becomes clearer. Similarly, the rapid expansion of distance and IT-based academic degrees in the United States has underscored the limitations of the traditional capacity or "bricks and mortar" approach to academic quality assurance employed by the institutional accrediting agencies. The challenge of IT-based distance education to traditional higher education has further strengthened the case for an external quality assurance process that focuses on student learning and academic standards (Ewell 2008).

Another weakness of self-regulatory mechanisms is the financial challenges they confront in a mass system of higher education. Ewell (2008) makes the traditional argument that a self-regulatory approach to external academic quality assurance is more efficient for society, because it saves taxpayers the costs of a publicly funded process. Of course, much of the financial support for the existing voluntary instruments derives from the fees paid by publicly funded universities, which are essentially a public subsidy. Furthermore, as Ewell's (2008) overall assessment of the US accrediting system makes clear and as Lewis also notes in his analysis of external examining in the United Kingdom, voluntary mechanisms now lack the financial resources necessary to effectively train, supervise, and compensate the academics involved in external quality assurance processes. Finally, the user-pay system, which is used to support self-regulatory approaches, raises serious questions about the independence of external quality assurance agencies. This problem was highlighted in the collapse of the global financial system in 2008. A major contributor to this collapse was the failure of credit-rating agencies such as Moody's and Standard & Poor's to accurately assess the risks involved in new financial instruments and corporate actions. Because these agencies are directly supported by the issuers of the bonds they rated, they had every incentive to systematically disguise financial risk rather than expose it. As noted by critics of this practice, "[t]here should be a rule against issuers paying for ratings … *if public ratings are deemed essential, they should be publicly provided*" (Lewis and Einhorn 2009, emphasis added). In sum, for the external assurance of academic standards to be effective, it will likely need to be publicly supported or subsidized.

While acknowledging the limitations of self-regulatory instruments, the experience with external examining in the United Kingdom and with innovative specialized accreditation agencies such as TEAC in the United States provides valuable guidance for the design of more effective public policies on external quality assurance. As these instruments suggest a first priority is that any external quality assurance mechanism must be clearly focused on maintaining and improving the core academic processes by which universities assure academic standards. This will entail assessing the integrity of the institution's processes for approving and evaluating academic curricula, assessing and improving teaching and student learning, and monitoring the validity of student assessment. These are also the areas in which faculty members have distinctive expertise and where their time and effort in external peer reviews logically should be concentrated. In contrast, while there is a public interest in the quality of academic governance, institutional administration, strategic planning, financial affairs, and student services, there is little empirical evidence that these processes are as influential on academic standards as the core academic processes of curriculum design, teaching, and student assessment (Pascarella and Terenzini 2005). Developing an effective external quality assurance mechanism that focuses on the core academic processes should therefore logically precede in public importance and urgency the development of evaluations for the more administrative-oriented processes.

Second, the UK experience suggests that evaluations of the equity and integrity of student examinations and grading practices need to be an essential feature of external academic quality assurance. However, as noted above, these evaluations will likely prove more effective if they are "meta-evaluations" focused on the adequacy of institutionally based processes for assuring the validity of subject-level examinations and marking (Stensaker et al. 2008). The QAA Code of Practice cited in Lewis' analysis provides some useful initial criteria for this task.

Third, the learning-oriented accreditation process developed by TEAC offers guidance for assuring academic standards in professional fields traditionally vital to the public interest such as medicine (see Harvey's related analysis of the GMC in the United Kingdom), law, or teacher education, but also provides a potentially valuable model for the design of more effective external quality assurance assessments for institutions as well. As El-Khawas noted the TEAC reviewers have been consistently surprised to discover that faculty members often fail to apply to their academic courses and programs the same scholarly rigor they apply to their research. This problem was well summarized by Sir Eric Ashby over four decades ago:

> All over the country these groups of scholars, who would not make a decision about the shape of a leaf or the derivation of a word or the author of a manuscript without painstakingly assembling the evidence, make decisions about admission policy, size of universities, staff-student ratios, content of courses, and similar issues, based on dubious assumptions, scrappy data, and mere hunch. (Ashby 1963, p. 93)

An important contributor to the effectiveness of the TEAC approach, also reflected in the subject assessments in Denmark (see Stensaker) and the GMC accreditation process in the United Kingdom (see Harvey), is the application of widely accepted norms of scholarly inquiry in an evidence-based approach to quality assurance. Thus the TEAC self-study requires that program claims be supported by evidence and that the program's means of assessing students meet accepted standards of validity and reliability. Similarly, the TEAC review team emphasizes the systematic use of protocols and applies traditional scholarly rules of evidence such as the consistency and representativeness of the data and information provided.

By requiring those seeking accreditation to provide verifiable evidence on student learning and by grounding their evaluation of student assessment processes in scholarly criteria for evaluating evidence – e.g., the validity and reliability of the assessment method – TEAC has markedly strengthened the power of external quality assurance to assure academic standards and provided an important benchmark for the design of more effective quality assurance policies in the future.

While our review of professional self-regulation of academic quality suggests that such instruments alone are not a sufficient basis for a national quality assurance framework, these analyses do confirm that successfully assuring academic standards will continue to require the active participation of the academic profession(s). How the necessary balance among professional engagement, market forces, and state regulation can best be accomplished will be explored in the sections to follow.

Lessons Learned About Market Regulation

A necessary condition for an efficient market is that both consumers and producers have "perfect" information – rational choice requires that economic agents are well informed about both price *and* quality (Teixeira et al. 2004). Consequently, it is believed that if student consumers have sufficient information on the quality of university academic programs, they will make choices that will positively influence academic standards. This belief has motivated policymakers in many countries to seek to provide more effective information on academic quality to students. For example, a government White Paper on higher education in the United Kingdom (DfES 2003) argued that market competition could be an important driver of academic quality, if appropriate university information can be provided to help inform student choice. However, as we will emphasize below, in a competitive market information on academic quality not only may influence student choice, but also producer effectiveness. The provision of more valid information on educational quality also provides a needed incentive and input for academic staff to make genuine improvements in academic programs as a means of better competing in the market (Dill and Soo 2005).

In theory inadequate consumer information may provide incentives for commercial organizations to produce organizational report cards or rankings that will aid consumer choice (Gormley and Weimer 1999). From this perspective the proliferation of commercial university league tables around the world during the last 20 years is therefore a clear indicator of the growing role market competition is playing in higher education. Consumer expenditures on these university league tables can therefore be interpreted as an indirect measure of the inadequacy of existing information on academic quality. However, the accumulating evidence suggests that commercial guides and league tables do not effectively address information deficiencies in the higher education market in socially beneficial ways (Dill and Soo 2005). The cost and complexity of developing valid indicators of academic quality with relevance to student choice are significant and for-profit publications already enjoy substantial sales and influence among opinion leaders, higher achieving students, and even university personnel by focusing on readily available and/or highly subjective indicators of academic prestige. This focus on academic prestige, based primarily upon indicators of research reputation and quality of entering students, in fact distorts the assumed constructive link between information on academic quality and university efforts to improve academic standards. Given the influence of commercial university rankings, many universities have responded to market competition primarily by investing in academic prestige – emphasizing admissions marketing, "cream skimming" of high achieving student applicants, and increased expenditures on research – with limited attention to actually improving academic standards. As discussed, a primary motivation for the development of the four information instruments we have examined was the perceived "market failure" of commercial "league tables." The collective experience of these new information instruments provides additional guidance for the design of the national framework conditions necessary for effectively assuring academic standards.

The three instruments which provide institutional-level information to the market offer a valuable model for implementing socially useful academic rankings systems. As outlined in these policy analyses, the development and reporting of the CHE rankings in Germany, of the NSSE in the United States, and of the Australian student surveys provide a much more explicit and informative template for the design of valid academic quality information policies than do the best practices articulated in the Berlin *Principles on Ranking of Higher Education Institutions* (IHEP 2006). At the same time, while these instruments clearly provide more valid information on academic quality and information more relevant to the interests of prospective students than most of the commercial league tables, the analyses suggest, even among these carefully constructed instruments, some fundamental limitations to academic quality information as a guide to effective student choice. For example, the reliability of subject-level data in the CHE rankings and Australian surveys is debatable given the low and/or highly variable response rates among students surveyed in different fields and the discovered association between scores and institutional size. The experience with all three of the institution-level instruments also suggests the reported differences among subjects or institutions are modest and scores tend to be stable over time, thereby providing limited guidance to student decision making or encouragement for academic improvement.

Furthermore, while the public provision of relevant, valid, and reliable information on academic programs can be readily defended on the grounds of the consumer interest, the underlying belief that better informed student choice will help assure academic standards is likely overstated. Following Gormley and Weimer's (1999) model of effective report cards, information provision is likely to influence academic standards only if quality rankings utilize measures linked with societal valued outcomes, students use this information in their choice of subjects, and institutions respond to student choices by improving relevant academic programs. However, the reported impacts of these more valid instruments on academic quality are consistent with earlier international research on student choice, that is, quality rankings and ratings influence the educational decisions of a relatively small segment of the student population, primarily those of high ambition, achievement, and social class (Dill and Soo 2005). Most students report that academic quality rankings have little influence on their actual choice of a subject or an institution. The higher education choices of first degree students, particularly in mass systems, are in fact influenced by a wide variety of educational, social, and personal factors in addition to academic quality. Some students, for example, are attracted to particular universities because of consumption benefits that they value personally, but which may ultimately produce little benefit to society. These include the pleasures of living in attractive university surroundings, the appeal of university social life, and in the United States, the distractions of university athletics. The experience with rankings and student choice in Australia, Germany, and the United States in our analyses is generally consistent with these earlier observations, although the CHE analysis does offer some hope that more valid academic quality rankings may eventually create a closer link between student choice and academic improvement over time. Nonetheless, for the reasons noted, it is likely that the individual decisions of even better informed higher education applicants may have a limited impact on assuring academic standards.

Similarly, the *Measuring Up* report cards, designed to inform policymakers on the relative performance of state higher education systems in the United States, appear to have had little influence on higher education policy decisions related to academic quality. Although this initiative focuses on system-level performance, not on institutional performance, the problems with the instrument are comparable. Information provision alone is not sufficient to prompt significant change. Only if such information creates a positive pressure in the system and motivates universities to improve their performance will the instrument fulfill its mission. Similarly, the main challenge for the *Measuring Up* approach is the transition from collecting and providing information to creating effective incentives for state policy change. The issue of the reliability and validity of higher education indicators makes this challenge even more complicated.

More positively, the analyses of these information-oriented instruments offer some useful considerations for policymaking. First, as was the case with our instruments of professional self-regulation, the relative effectiveness of these instruments is clearly influenced by the role played by government. The valuable NSSE was developed with the support of a private foundation, but its voluntary nature means that there is no incentive for the leading US universities to participate in the survey. This deprives potential students of useful comparative information and also provides less motivation for the universities themselves to experiment with the survey for internal improvement. Only in those states where the governments have formally mandated use of the survey are all publicly supported universities participating. Similarly, while the development of the CHE rankings was supported by a private foundation, the stated interest and encouragement by the German Ministry of Science was undoubtedly essential to gaining the wide participation and involvement of the German universities. In Australia, the substantial costs of developing and implementing the CEQ and GDS surveys were directly subsidized and supported by the Australian national government, which thereby provided a strong incentive for the "voluntary" participation of all Australian universities. In the United States, while there has been little national effort to improve the validity of academic quality information at the first degree level, the development and publication of the highly regarded National Academy of Sciences research–doctoral rankings are subsidized by agencies of the federal government. As these examples suggest, unless government defines and/or subsidizes the development of more valid information on academic quality and encourages its use, it is unlikely to be produced by the commercial sector or to be actively applied by universities in internal improvement, a positive impact identified in both the Australian and German cases.

While the commercial sector may not have an incentive for the development of more valid measures of academic quality, if data is mandated and/or developed by the government, commercial publishers can sometimes make a contribution by publishing it. For example, the commercial publication *Die Zeit* is responsible for distribution and marketing of the CHE rankings, but the integrity of the information and the validity of its presentation are assured by the independence of the nonprofit foundation. A similar relationship is emerging in Canada between the Educational Policy Institute and the commercial *Globe and Mail. The Good University Guide* in Australia, a commercial publication, plays a similar role in disseminating the results of the CEQ and GDS surveys which have been compiled and analyzed by agencies of the Australian national government.

A second useful contribution of these instruments is the apparent emerging international consensus on indicators of societal valued outcomes (Gormley and Weimer 1999). While the public provision of valid information on the educational value added of academic programs has proven a daunting task, as the Australian surveys suggest, socially valued proxies of this value added such as indicators of program productivity and labor market outcomes are much more easily obtained and provided. Examples of these indicators include data on student retention and progression, graduate employment, and median salary level, as well as further study – all by subject field. These data may not only assist students in achieving a better fit between their preferences and the qualities of academic programs, but as recent research (Romer 2000) suggests, may also lead to a better fit between student choice and societal needs. As we have previously argued public subsidies for higher education degrees are provided in part because of the expected human capital that educated program graduates will provide to society, but incomplete information in the higher education market may lead to an additional market failure – a mismatch between societal needs and numbers of program graduates (Romer 2000). In the current environment potential students have inadequate information on the implications of program choices for their future lives, and universities face insufficient pressure to respond to changing market demands for different sets of skills. The mandatory provision of the data suggested, by university and subject field, would help students make more satisfying life choices and aid universities in improving the design and effectiveness of academic programs. In addition as we have suggested information on legitimate output and process indicators may have a value even if it fails to steer student choice. Universities ignore academic standards not only because of the lack of external incentives and intrinsic interest but because they also lack clear evidence of their weaknesses. The NSSE experience proves that information that is considered academically legitimate and valuable is useful to institutions for local planning and quality-enhancement purposes, and it is used by (some) institutions even if no sanctions are associated with the outcomes. Similarly, the CHE ranking has created a demand from universities for consultancy and guidance on how to improve certain aspects of their performance. It is unlikely that this interest is motivated entirely by competitive pressures since the association between student choice and performance is not very clear.

Finally, while the orientation and maturity of new first degree-level students in mass systems of higher education may ultimately limit the influence of better consumer information on improving academic programs, the international market for research doctoral students is more perfectly competitive and suggests academic program rankings at this level could make a beneficial contribution to assuring academic standards (Dill 2009). Doctoral applicants are an older, more educationally experienced set of consumers, who are pursuing advanced degrees primarily for vocational reasons. Furthermore, US universities and increasingly universities in other countries use high paying fellowships to compete aggressively for the very best international doctoral students. Doctoral applicants therefore are less likely to be swayed by consumption benefits, social factors, geographical considerations, and institutional reputation in their choice of academic programs and more likely to be influenced by valid information on doctoral program quality. In addition, academic staff are more psychologically invested in the quality of their doctoral programs than in their first level degree programs, in part because doctoral graduates are more visible products of the individual mentor and subject field – particularly to academic colleagues at other universities – than are largely invisible first degree recipients. Consequently, in this more perfectly competitive market, it is not surprising that the well-designed National Research Council quality rankings subsidized by the federal government have motivated demonstrable improvements in US doctoral programs (Dill 2009). Given the acknowledged positive influence of research–doctoral graduates on economic growth in the developed countries (Aghion 2006), government support for such rankings appears to be a particularly well-justified component of a national academic quality assurance policy.

In sum, as in other fields such as food or automobile safety where there is a strong public interest in valid, reliable, and socially useful information provision, government policy can influence the quality of information provided the public by subsidizing the production of more valid indicators, by stipulating appropriate formats for information provision, and by holding information producers accountable for the accuracy of their information. The development and provision of socially beneficial information on academic quality therefore is best understood as a public good, which must be regulated and subsidized by government.

Lessons Learned About State Regulation

If both professional self-regulation and regulation by the market are insufficient safeguards for assuring academic standards in the new environment of globally competitive mass higher education, what can be learned from experiments with new forms of state regulation? In assessing these new regulatory instruments, it is important to emphasize that similar to professional self-regulation or market regulation, poorly designed government regulations may also fail. For example, the design of the initial process of subject assessments implemented by the Higher Education Funding Council in the United Kingdom was highly influenced by former school inspectors and, therefore, came to reflect their professional interests in particular approaches to instruction rather than the public interest in assuring academic standards (Brown 2004). Consequently, the assessments included direct inspections of teaching and encouraged adoption of specific teaching practices rather than the improvement of student learning.

The analyses of state regulations suggest that efforts to publicly define the learning outcomes of academic programs make a modest contribution to assuring academic standards. The rapid expansion of new academic programs that accompanied massification and the growing autonomy of universities in the newly deregulated context of higher education motivated the development of instruments such as national degree frameworks, the so-called Dublin Descriptors, and the UK Subject Benchmarks Program. While some policymakers clearly hoped and some academic staff clearly feared that these guidelines could become an effective regulatory device for assuring the fitness of purpose of academic degrees, the impacts of these instruments appear much more limited. The complexity and increasing specialization of academic knowledge as well as the rapid development of new interdisciplinary fields of study in the university sector compromise national attempts to prescribe academic content. Reaching agreement on the core content of a field proved difficult in the United Kingdom due to substantial variations discovered within an academic subject – a problem that characterizes particularly nonparadigmatic disciplines. As a result, the frameworks and subject benchmarks proved more broad and general, more formative and developmental, than regulatory. The most significant contribution of qualifications frameworks is to encourage a focus on student learning outcomes rather than course content in national debates about academic standards, while the subject benchmarks assisted some universities in planning new courses of study.

A major new development over the last several decades is the introduction of state-sponsored subject assessments and subject accreditations. In contrast to the specialized accreditations in the United States, these evaluations are comprehensive, covering the majority of subject fields in which university students are enrolled. While they employ peer review, they are carried out under the auspices of the government and therefore are designed to better reflect the public interest in assuring academic standards rather than the special interests of professional associations. As will be noted below, the experience with these new instruments provides valuable guidance in the design of the framework conditions necessary to assure academic standards, but the instruments themselves have significant weaknesses as long-term policies. The subject assessments developed in Denmark, the Netherlands, and the United Kingdom made important contributions to the improvement of teaching as well as the structure and content of academic curricula in these traditional university systems with restricted numbers of universities, disciplines, and fields. Subject assessments, however, are a less effective instrument in expanding systems with new fields of study. Comprehensive subject accreditations, as implemented in Germany and a number of the other nations influenced by the Bologna reforms, better address the development of new fields and degrees, but similar to subject assessments these peer review instruments are extremely labor intensive, costly in terms of academic time and effort, and consequently unsustainable over time. More critically, by their focus on the subject level, both of these instruments continue the tradition of centralized state control of academic subjects and do not build the capacity of the overall university to design new programs as well as improve the academic quality of all fields of study. By focusing on subject fields, these instruments of external quality assurance provide few incentives for the universities themselves to develop the internal, collective processes for assuring academic standards in all the subjects they offer, processes that will be essential in the new competitive world of deregulated higher education featuring more autonomous institutions. Moreover, when the overall university is not invested in subject assessments and does not incorporate the results in its internal quality assurance procedures, the impact of such an expensive exercise may remain quite limited. As shown by the Danish case, the external follow-up actions cannot be easily implemented and the results of a subject assessment may therefore be poorly enforced. These weaknesses are reflected in the shift in emphasis in both Denmark and the United Kingdom from subject assessments to an institution-oriented academic audit process as well as in the current proposals in Germany for the adoption of an institution or process-oriented form of accreditation.

A crucial problem revealed by these new state policy instruments is the challenge of developing national and/or institution-wide indicators of academic outcomes as a primary means of assuring academic standards. The most ambitious attempt to develop valid measures of the outcomes of academic programs is the national examinations policy adopted in Brazil, and the inability to continue this innovative policy suggests the fundamental limitations of this approach. The ability of the Brazilian government to unilaterally implement these exams stemmed from a unique legal context in which all academic degree programs are considered equivalent to professional certification and, therefore, may be regulated by law. Consequently, all graduates could be required to take national examinations in their subjects as a condition for receiving their degrees. As noted in the analysis, this unique structure might have eventually encouraged the development of more valid academic value-added measures for each program utilizing state-mandated entrance exams. However, in countries with more mature and complex university systems, it is quite unlikely that such a policy is educationally effective, politically feasible, or financially sustainable, as eventually proved true in Brazil as well. Such instruments are also vulnerable to dysfunctional effects. For example, universities in Brazil began to focus on training students for the test, which is not necessarily the same as contributing to their academic and professional development. Furthermore, this type of state policy, if fully implemented, could retard or undercut the fundamental contribution universities make to society through the development of innovative new fields of study and research. For example, the adoption of a single, unified exam in each field for all course programs in Brazil led all the institutions to adjust to the same mold and discouraged experimentation and diversification.

The analysis of performance contracts in Catalonia lends further weight to this point. While the improvement of academic quality was an important component of the performance contracts implemented in Catalonia, the analysis suggests the quantitative indicators of quality available were too generic and the understandable contractual emphasis on measures of student retention and graduation created potential incentives for reducing or simplifying academic standards. For this reason performance-based funding or contracts, which are usually based upon available input, process, and output measures have consistently proven to be an inadequate instrument for assuring academic standards (Jongbloed and Vossensteyn 2001). As supported by the evidence from Catalonia, performance contracts seem a conceptually attractive model – they respect institutional autonomy, recognize the diversity of institutions, and link funding allocations to actual outcomes rather than inputs and processes. Unfortunately, performance contracts rarely deliver all they promise. Compared to "light touch" instruments such as academic audits, performance contracts are more strongly enforced by government, yet their effect on processes within universities seems more limited. In Catalonia, it has been difficult to commit the academics and staff in universities to the agreed targets and therefore the contracts have not brought the expected changes in core academic processes. This experience illustrates that in the case of quality assurance initiatives not only the strength of enforcement matters but also the legitimacy of the initiative in the eyes of academic staff. Therefore, performance contracts alone are not sufficient to ensure academic standards, and they need to be supplemented by government mandates on quality information provision and external quality assessments. It is worth noting that in Spain, as in other countries experimenting with university performance contracts such as Denmark and Finland, these contracts are reinforced by subject assessment, accreditation, or academic audit processes (Mora 2004).

The significant challenge of developing more valid and useful measures of academic value-added has led to experiments with standardized tests of general knowledge and skill such as the Graduate Skills Assessment (GSA) in Australia and the Collegiate Learning Assessment (CLA) in the United States. However, as Williams notes in his discussion of the Graduate Standards Program in the United Kingdom an attempt to define such general skills or abilities in a systematic empirical study of subject fields proved unsuccessful. Furthermore, even if such indicators could be developed, there is little reason to believe that these generic indicators would prove influential on student choice of academic programs or relevant and useful to academic staff in assuring and improving academic standards within those programs. Finally, even the well-regarded CEQ and NSSE measures developed respectively in Australia and the United States have limited utility as institutional indicators of academic quality. The validity of the CEQ has been questioned, since it may be more closely associated with particular pedagogical approaches than student learning and may thereby mislead institutions in their efforts to improve academic standards. The NSSE, while apparently a more valid instrument, does not discriminate effectively at the institutional level. Therefore, it may ultimately prove more useful in systematic surveys within universities designed to identify academic quality issues and best practices at the subject level.

These limitations lead us to conclude that the continued search for universal, valid measures of academic value-added, particularly in the university sector, is similar to the hunt for the Holy Grail, that is, it is an elusive goal, the pursuit of which is as likely to distort or diminish academic standards as to assure them. Instead, these collected analyses of state regulations suggest the major focus of an effective academic quality assurance policy should be in providing incentives and support for the development of valid measures of learning outcomes *at the subject level within universities*. As Pascarella and Terenzeni concluded in their exhaustive review of the available empirical research on teaching and learning in higher education:

> Assessment of department-specific learning outcomes can be a useful vehicle for change. *Assessment plans and activities developed and approved by faculty* can provide an empirical foundation of systematic and ongoing rethinking, redesigning, and restructuring programs and curricula. For faculty members, trained to be skeptical about claims, evidence is the gold standard in the academy, and they are unlikely to adopt new ways of thinking or behaving without first being convinced that the new pedagogies and organizational structures are better than the old. In addition, *the findings of assessment studies specific to faculty members' academic units will generate more interest and action than general or institution-wide evidence*. (Pascarella and Terenzeni 2005, p. 648) (emphasis added)

Overall, this analysis suggests that an academic audit, or process-oriented, approach to assuring academic standards is superior to other external quality assurance instruments in several respects. First, it reinforces the culture of continuous quality monitoring in universities and emphasizes institutional responsibility for assuring academic standards. Second, it recognizes the diversity in the system and respects universities' autonomy. Third, it is a "light touch" mechanism, yet the experience from Hong Kong shows that only limited external enforcement is needed to make universities respond to audit results. The academic audit system in Hong Kong faces one criticism though – that it fails to provide information on actual quality in institutions and does not show quality differences in the system. Thus it fails to provide information that would help students and their parents in choosing an academic program or college. While this desire may indeed be a wish for a magic bullet that simply does not exist, as noted by Massy, it is a criticism that is difficult to ignore. Disregarding the public demand for comparative quality information encourages superficial presentation of data by private providers that can negatively affect the sector. As noted in the analyses of Denmark and Hong Kong, the private media are eager to publicize information on universities' relative quality, making use of sources that were not originally intended for such use. This suggests that there is a strong interest for comparative information and if more valid data is not provided in an accessible form by public agencies, it will be provided in some manner by private players.

Despite the problems and weaknesses with the state regulatory instruments, the analyses provide valuable guidelines for the design of external quality assurance processes. First, these analyses make clear that developing a stronger culture of quality for teaching and student learning and creating conditions for the continual assurance and improvement of academic standards within universities will require actively engaging both the collegial leadership of an institution and academic staff in departments and programs. The positive impacts of the studied subject benchmarks, assessments, and accreditations as well as the academic audits were most clearly visible in the increased discussions about academic quality as well as changes in curricula organization, student assessment, and modes of instruction that took place within academic programs. It is at the subject level that academic standards are demonstrably assured and improved. At the same time, as noted, an effective external quality assurance process must create conditions in which the collective university assumes ongoing responsibility for maintaining academic standards and implements rigorous and effective collegial processes for assuring and improving academic quality in all the institution's academic programs. For this to occur, the university's core academic processes for assuring academic quality must be externally evaluated by competent peer reviewers and these evaluations must include an assessment of the impact of these processes at the subject or program level. As Massy suggests, this will require academic audits which include a review of a representative sample of academic programs – what are sometimes termed "audit trails."

A second design principal that can be deduced from these instruments is the core academic processes that must be evaluated. Based upon the focus of the state-mandated subject assessment, subject accreditation, and academic audit processes as well as the innovative self-regulatory processes we have analyzed, the core university processes that must be assessed include: the design and approval of new course modules and programs of study; procedures for reviewing academic programs; procedures governing grading and marking standards; procedures influencing the evaluation of teaching; procedures affecting student assessments; and other relevant components of the university's overall processes for assuring and improving academic standards. Particular attention must be paid as well to the university's processes for identifying and sharing best practice in assuring academic standards (Dill 1999). Some existing academic audit and institutional accreditation processes cast a much wider net than that suggested here and they may thereby compromise the rigor and impact of the evaluations. It is clear from the focus and influence of regulatory policies on academic quality that have emerged over the last decades that the integrity of the mentioned academic processes is a primary public concern. Also, these core processes are equally important for assuring academic standards in all types of higher education institutions, be they public or private, profit or nonprofit, full-time or part-time, residential or distance learning.

A third design consideration is the organization of these external institutional reviews. The most effective and legitimate instruments in the views of academic staff possess methodologies similar to those adopted by TEAC in the United States, the British Medical Council in the United Kingdom, and the subject assessments in Denmark, that is, peer reviewers are trained, supported during the review process by professional staff, and employ systematic, standardized procedures, and protocols. The reviews are rigorous and evidence-based, assessing the efficiency, validity, and reliability of the quality assurance processes. In this sense a feasible goal for these external reviews would be to ensure that the universities' internal processes for assuring academic standards are at least as rigorous, valid, and evidence-based as their existing institutional processes for assuring and improving the quality of academic research.

A final design component is assuring the efficiency of academic quality regulation. A known contributor to ineffective regulation is "regulatory capture" (Laffont and Tirole 1991) in which those whose interests are affected by the relevant regulation gain influence over the regulatory agency and promote their private interests over those of the public. The documented limitations of existing state and self-regulatory approaches to external academic quality assurance have motivated many countries to establish new national agencies for academic quality assurance similar to the Australian Universities Quality Agency, the Danish Evaluation Institute, the German Accreditation Council, and the Quality Assurance Agency (UK) discussed in our policy analyses. In the design of these agencies, serious efforts have been made to ensure that they reflect the larger public interest by structuring them, insofar as possible, to be independent of both the universities and the government. But because these agencies are granted public authority for designing, conducting, and/or overseeing required external quality assurance processes, they are essentially government regulators, which as our policy analyses indicate incur substantial direct costs to the public as well as indirect costs to the university sector. The efficiency of this regulation, as in other public sectors, therefore generates significant public debate. Briefly put, the question is posed "who guards the guardians" (Blackmur 2008) and the typical policy response to this question is to require public evaluation of the agencies themselves as a means of protecting the public interest in efficient regulation. However, the adopted process for actually evaluating national academic quality assurance agencies provides evidence of the problem of regulatory capture. The design and conduct of the evaluations is often controlled by the agencies themselves in cooperation with associations of agency professionals and/or selected representatives of those regulated (Szanto 2005). As Blackmur (2008) argues in his critical analysis of the external review of the Australian Universities Quality Agency, this type of evaluation lacks independence, fails to employ a suitably relevant and robust method of validation, and ignores the critical issue of value for money. Similar to the problems identified in our analyses of voluntary external examining in the United Kingdom and college and university accrediting in the United States, this type of self-regulative process also may limit the development of the "science" of external quality assurance, which as we have discovered in our analyses of the processes of academic audit, subject assessment, and accreditation already exhibit substantial variation in objectivity and rigor. Therefore, the public interest in efficient regulation of the university sector is likely to be better served if newly designed academic quality assurance agencies are themselves publicly evaluated by established and respected national evaluation or audit agencies similar to the Government Accountability Office in the United States, such as the Australian National Audit Office, the German Federal Audit Office, and the United Kingdom National Audit Office. Similarly, the legitimacy of the quasi-regulatory activities of regional agencies such as the European Association for Quality Assurance in Higher Education (ENQA) or international associations such as the International Network for Quality Assurance Agencies in Higher Education (INQAAHE) would be better established if they were publicly evaluated by respected regional or international organizations such as the European Court of Auditors, the OECD, or the World Bank.

Before applying the insights gained from our analyses of the new instruments of state regulation, professional self-regulation, and market regulation to the design of national framework conditions for assuring and improving academic standards, we need to acknowledge some of the recent international experiments in this field.

Academic Quality and International Pressures

This volume presented a number of innovative instruments that aim to assure and improve academic standards at the national level. Higher education, however, functions increasingly in an environment that crosses national borders. International mobility of students, globalization of the labor market, and cross-border provision of academic degrees have posed a new set of risks and challenges that cannot be addressed by national quality policies only. As a result, international cooperation between quality assurance agencies has intensified and several international and regional associations have emerged such as the previously mentioned INQUAAHE, ENQA, and the Nordic Quality Assurance Network in Higher Education (NOQA). Many countries have bilateral agreements to recognize each other's accreditation agencies and/or accreditation decisions. Most recently, the European Ministers of Education established an international roster of legitimate quality assurance agencies – *The European Quality Assurance Register* – that is expected to take international cooperation and recognition in quality assurance to the next level. These are all examples of how national quality assurance systems attempt to address new international challenges. Most recently, however, new innovative instruments have appeared that bypass the layer of national quality assurance and directly target the international higher education landscape. Several of the instruments discussed earlier in this volume have an analogue that crosses national borders.

The Tuning Project is a Europe-wide initiative that tries to specify expected competencies of main subject fields, and it is thus a clear analogue to the subject benchmarks in the United Kingdom. Since 2000, the Tuning Project has brought together academics and stakeholders from different countries in order to develop "reference points" for subject fields at each educational level, i.e., to specify relevant learning outcomes and competencies, both subject specific and generic. The main aim of the project is to improve quality in the system, by offering a tool that facilitates curriculum development and evaluation (Tuning project 2005). While the project is very careful in avoiding any suggestion of standardization of programs, the rationale of the initiative is greater comparability of degrees, which is a necessity in an international higher education area. While massification of higher education created a new challenge for national governments to assure the equivalence of higher education degrees in the entire system (as we saw in the case of the United Kingdom, Brazil, and Germany), internationalization in higher education extends the same challenge to the international community. The comparability of degrees not only among different types of universities but also among different countries has become a problem that needs to be addressed.

The challenges of the Tuning Project appear to be similar to those of subject benchmarks. The exercise must be able to specify learning outcomes that are relevant regardless of the country context but which at the same time are not too general. An evaluation of the Tuning Project noted that the statements of competencies that universities had developed often remained either broad and vague or stated the obvious (CoRe 2007). The evaluation also found that sometimes there was no clear link between expected competencies and related course work, indicating that reference points alone are not sufficient to guarantee academic standards. In the second phase of the Tuning Project, one of the objectives is to strengthen the link between Tuning outcomes and quality assurance and assessment. There are some examples how the "reference points" can obtain a stronger role than a mere guidance and advice to universities. In some fields disciplinary associations have approved Tuning "reference points" as a foundation of "Eurobachelor" degrees. Universities can use the "Eurobachelor" label if the program follows the agreed structure of competencies. The impact of the Tuning Project on academic standards is thus not yet clear. It is certainly an interesting exercise because it targets directly academic practices in universities, skipping the level of national quality assurance. The proponents of Tuning see it as an important mechanism to ensure transparency and accountability in a higher education system (Adelman 2009). Similar to subject benchmarks, "it has stimulated a process of reflection, development and innovation in higher education programmes" (Tuning Project 2005, p. 10). Whether the effort could have a sustainable impact that inspires universities for a continuous improvement still needs to be seen.

Several interesting initiatives have also emerged as a response to international university rankings. Rankings such as the Shanghai Jiao Tong *Academic Ranking of World Universities* and the *Times Higher* ranking of *Top 200 World Universities*, which started in 2003 and 2004, respectively, have caused quite a stir among policymakers and have received much interest among potential students. It is widely acknowledged that the rankings are of questionable validity, but the massive interest in the ranking results indicates a lack of needed information on academic quality. The international community has thus started to look for a more valid instrument to fill the gap and to neutralize the effect of dubious rankings. The most ambitious initiative in these lines is OECD's AHELO project (*Assessing Higher Education Learning Outcomes).* It is an ambitious attempt to measure actual learning outcomes of students in universities, and thereby provide comparative information on institutional quality (see OECD 2009). Its main goal is to provide a more in-depth picture on academic quality in world universities and balance the skewed picture created by commercial rankings. The initiative is currently in the piloting phase to test the feasibility of such an endeavor in two fields – engineering and economics. The main challenge of the initiative is of course to develop measures that are valid for all cultures and to make learning outcomes internationally comparable. In the long run, the project aims not only to collect information on final learning outcomes but also to develop effective and efficient national measures of the value added by academic programs. Above we expressed skepticism about broad-scale attempts to measure the value added by universities. It has yet to be seen whether the AHELO group will be successful in developing accurate measures and can help offset the negative effects of highly questionable university rankings. Nevertheless, the project results are likely to feed interesting discussions for years to come on what institutional and systemic characteristics contribute to quality in higher education.

Recently, two additional interesting initiatives in Europe address the problems caused by questionable world rankings. In 2008, the European Commission announced its plans to design and test the feasibility of a world ranking (RAPID 2008). This initiative is expected to produce a multidimensional university ranking that includes not only European universities but also universities in America and Asia. Another highly visible initiative by the European Commission is an attempt to classify European universities – again an initiative that hopes to balance the distortive effect of unidimensional commercial rankings and maintain the diversity of European universities (see van Vught 2009).

The challenges the EU faces in the higher education sector due to internationalization are thus not unique. Comparability and transparency of degrees in a highly diverse system or consumer information on quality differences in universities are issues that characterize developments in many countries. The international dimension only adds another scale to the problem. The experiences from national policies outlined in this volume can therefore also usefully inform international experiments.

Toward a National Framework for Assuring Academic Standards

While the recent work of the OECD, UNESCO, the World Bank, and the other mentioned global and regional organizations suggest that international agencies and compacts may eventually prove influential on matters of academic quality, for the near as well as foreseeable future we anticipate that national policies will remain the primary force in assuring academic standards. Summing up the lessons learned from our Public Policy for Academic Quality Project, we believe the substantial changes in systems of higher education over the last decades require a major redesign of the framework conditions for assuring academic standards at the national level. The effects of massification on higher education, particularly the growing international market competition among universities, the commercial provision of quality information and the resulting academic arms race for research reputation and prestige, the rapid development of new academic subjects and fields, and the associated deregulation of academic program development have radically altered the balance among professional, market, and state forces by which academic standards were traditionally maintained. In this new environment, the public interest will best be served by a new institutional framework for assuring academic quality. While our analyses have focused on individual innovative instruments of academic quality assurance that have emerged during this period of rapid change, in this concluding section we draw upon these assessments to provide some guidance to the design of a national framework that is likely to prove effective in the new environment.

The analyses of these new instruments clearly confirm our strong belief that while new forms of government regulation and a better informed student market can play important roles, the public interest in the quality of higher education will best be achieved by designing an institutional framework that encourages the development of strong, effective, collegial mechanisms of academic quality assurance within all institutions of higher education. As the Conference of Ministers responsible for higher education in the EU noted in a 2003 Communiqué:

> ... consistent with the principle of institutional autonomy, the primary responsibility for quality assurance in higher education lies with each institution itself and this provides the basis for real accountability of the academic system within the national quality framework.

Given this belief, the evidence from our collected policy analyses suggests the need for a policy of "enforced self-regulation" (Ayres and Braithwaite 1992). In this approach external academic quality assurance processes are required and financially subsidized by the state, but designed and implemented with the active engagement of the academic community. Valid information on academic quality for both students and academic staff is also a necessary condition, the definition and provision of which should be guided by state policy. The primary challenge for each nation is therefore to design and implement a rigorous, evidence-based method of external quality assurance, which is focused on enhancing and improving the core university processes for assuring academic standards. The ultimate goal of such external quality assurance should be for universities themselves to become genuine "learning organizations" (Dill 1999), in which the institution's assurance of academic standards demonstrably involves: evidence-based decision making utilizing accepted canons of scholarly inquiry; peer accountability for the quality of academic programs and the rigor of relevant unit-level decision-making; and systematic identification and dissemination of best practice in improving teaching and student learning. We therefore recommend the following critical components of a national framework:

- A national degree framework outlining in broad descriptors the expected learning outcomes for the major academic degrees at different levels.
- National support for the provision of valid information to guide student choice, particularly rankings of research doctoral programs, and a mandate for the provision of data on student retention, student progression, and graduate outcomes by subject field by all institutions of higher education.

- A publicly subsidized national agency for assuring academic standards, independent of both the government and the academic profession, to publicly evaluate as needed institutions providing academic degrees and/or other agencies responsible for external academic quality assurance. This agency should itself be subject to regular public evaluation by an appropriate national government audit or evaluation agency.
- Articulation by the national agency of the criteria and standards for a rigorous academic audit process to serve as the basis for the external quality assurance of all institutions of higher education that receive public funds. The audit process should be focused on the core processes that universities employ to set, monitor, and assure their academic standards. These core processes would include at a minimum: processes for designing, approving, evaluating and improving academic programs; processes for evaluating and improving teaching and learning; processes for assuring the integrity of grading and marking standards; and processes governing student assessment. This academic audit process should be clearly evidence-based, evaluating the validity and reliability of academic quality assurance mechanisms and information at the institutional level and assessing the effectiveness of these mechanisms with an appropriate sample of academic programs.
- Establishment of specialized quality assurance agencies, independent of both the government and the professions, to accredit academic fields of critical public interest, e.g., medicine, veterinary medicine, and teacher education.

In closing as we have emphasized the public has entrusted the academic profession with its future human capital. Universities have been awarded substantial public support and professional autonomy with the expectation that they will in turn provide efficiently and effectively academic programs in which students genuinely learn the knowledge, skills, and values essential to society. Designing and implementing framework conditions that will strengthen the core academic processes within universities by which academic standards are best maintained and improved is the surest means of protecting the public interest in higher education in the coming years.

References

Adelman, C. (2009). The Bologna process for U.S. eyes: Re-learning higher education in the age of convergence. www.ihep.org/assets/files/EYESFINAL.pdf. Accessed 12 June 2009.

Aghion, P. M. (2006). A primer on innovation and growth. Bruegel Policy Brief 2006/06. http://www.bruegel.org/uploads/tx_btbbreugel/pbf_061006_innovation.pdf. Accessed 12 June 2009.

Amaral, A. (1998). The US accreditation system and the CRE's quality audits: A comparative study. *Quality Assurance in Education*, 6(4), 184–196.

Ashby, E. (1963). Decision making in the academic world. In P. Halmos (Ed.), *Sociological studies in British university education* (pp. 93–100). Keele, UK: University of Keele.

Ayres, I., Braithwaite, J. (1992). *Responsive regulation: Transcending the deregulation debate.* . Oxford, UK: Oxford University Press.

Blackmur, D. (2008). Quis custodiet ipsos custodes? The review of the Australian Universities Quality Agency. *Quality in Higher Education*, 14(3), 249–264.

Brown, R. (2004). *Quality assurance in Higher Education: The UK experience since 1992.* London, UK: Routledge Farmer.

Conference of Ministers Responsible for Higher Education (2003). Realising the European Higher Education Area. Berlin, Germany, 19 September 2003. http://www.bologna-berlin2003.de/pdf/Communique1.pdf . Accessed 12 June 2009.

CoRe (2007). Competences in Education and cross-border recognition: Evaluation of the usefulness of learning outcomes and competences for international recognition. Final Report. http://www.dashe.nl/publications/documents/CoRe20Final20Report.pdf. Accessed 12 June 2009.

Department for Education and Skills (DFES) (2003). *The future of higher education.* London, UK: HMSO

Dill, D. D. (1999). Academic accountability and university adaptation: The architecture of an academic learning organization. *Higher Education*, 38(2), 127–154.

Dill, D. D. (2009). Convergence and diversity: The role and influence of university rankings. In B. M. Kehm, B. Stensaker (Eds.), *University rankings, diversity, and the new landscape of higher education.* Rotterdam, The Netherlands: Sense Publishers.

Dill, D. D., Soo, M. (2005). Academic quality, league tables, and public policy: A cross-national analysis of university ranking systems. *Higher Education*, 49(4), 495–533.

El-Khawas, E. (1998). Accreditation's role in quality assurance in the United States. *Higher Education Management*, 10(3), 43–56.

Ewell, P. T. (2008). *U.S. accreditation and the future of quality assurance.* A tenth anniversary report from the Council for Higher Education Accreditation. Washington, DC: CHEA.

Gormley, W. T. Jr., Weimer, D. L. (1999). *Organizational report cards.* Cambridge, MA, Harvard University Press.

Institute for Higher Education Policy (IHEP) (2006). Berlin Principles on Ranking of Higher Education Institutions. Washington, DC. http://www.ihep.org/assets/files/publications/a-f/BerlinPrinciplesRanking.pdf. Accessed 12 June 2009.

Jongbloed, B., Vossensteyn, H. (2001). Keeping up performances: An international survey of performance-based funding in higher education. *Journal of Higher Education Policy and Management*, 23(2), 127–145.

Laffont, J. J., Tirole, J. (1991). The politics of government decision making. A theory of regulatory capture. *Quarterly Journal of Economics*, 106(4), 1089–1127.

Lewis, M., Einhorn, D. (2009). How to repair a broken financial world. *New York Times*, January 4, 10.

Mora, J-G. (2004). A decade of quality assurance in Spanish universities. In S. Schwarz, D. F. Westerheijden (Eds.), *Accreditation and evaluation in the European Higher Education Area.* Dordrecht, The Netherlands: Kluwer.

Nettles, M., Cole, J., Sharp, S. (1998). The landscape: Tracking a subtle storm: Assessment policies in higher education. *Change*, 30(2), 47–51.

Orlans, H. (1975). *Private accreditation and public eligibility.* Lexington, MA: Lexington Books.

OECD (2009). The Assessment of Higher Education Learning Outcomes (AHELO). http://www.oecd.org/dataoecd/3/13/42803845.pdf. Accessed 12 June 2009.

Pascarella, E. T., Terenzeni, P. T. (2005). *How college affects students, vol. 2. A third decade of research.* San Francisco, CA: Jossey-Bass.

Peterson, M. W., Vaughan, D. S. (2002). Promoting academic improvement: Organizational and administrative dynamics that support student assessment. In Banta, T. W., Associates (Eds.), *Building a scholarship of assessment* (pp. 26–48). San Francisco: Jossey-Bass.

RAPID (2008). Ranking Europe's universities. IP/08/1942, 11 December 2008. http://europa.eu/rapid/pressReleasesAction.do?reference=IP/08/1942&format=HTML&aged=0&language=EN&guiLanguage=en. Accessed 12 June 2009.

Romer, P. M. (2000) Should the government subsidize supply or demand in the market for scientists and engineers? In A. B. Jaffe, J. Lerner , S. Stern (Eds.), *Innovation policy and the economy*, (Vol. 1, pp. 221–252). Chicago, IL: University of Chicago Press.

Selden, W. K. (1960). *Accreditation: A struggle over standards in higher education*. New York: Harper & Brothers.

Stensaker, B., Brandt, E., Solum, N. H. (2008). Changing systems of external examination. *Quality Assurance in Education*, 16(3), 211–223.

Szanto, T. (2005). Evaluations of the third kind: External evaluations of external quality assurance agencies. *Quality in Higher Education*, 11(3), 183–194.

Teixeira, P., Jongbloed, B., Dill, D., Amaral, A. (Eds.) (2004). *Markets in higher education: Rhetoric or reality?* Dordrecht, The Netherlands: Kluwer.

van Vught, F. A. (2009). *Mapping the higher education landscape: Towards a European classification of higher education*. Dordrecht, the Netherlands: Springer.

Author Index

D.D. Dill, M. Beerkens (eds.), *Public Policy for Academic Quality*, Higher Education Dynamics 30, DOI 10.1007/978-90-481-3754-1,

Subject Index

CPSIA information can be obtained at www.ICGtesting.com
Printed in the USA
LVOW01s1321061013

355632LV00002B/80/P

9 789400 732223